Mara,

FIELD GUIDE TO THE PLANTS OF

NORTHERN BOTSWANA

INCLUDING THE **OKAVANGO DELTA**

This field guide is dedicated to our parents, Basil and Dorothy, Jack and Diana, who encouraged us in the outdoor life and an appreciation of the natural world from a very early age.

FIELD GUIDE TO THE PLANTS OF
NORTHERN BOTSWANA
INCLUDING THE OKAVANGO DELTA

USEFUL IN COUNTRIES AND GEOGRAPHICAL AREAS ADJACENT TO NORTHERN BOTSWANA IN THE ZAMBESI BASIN

PLANTS AND BELIEF SYSTEMS

When enjoying looking at plants in Botswana and reading about their local names, uses and beliefs, the information gathered in this book should be treated with great respect. Nature and especially plants and their uses are an integral part of the belief systems of people living in Botswana. Many medicinal uses reported here will not be effective without contemplation of the wishes of the gods and ancestors, and without meditation as to the result that is sought. Batswana rarely use hallucinogenic plants: their belief systems and meditative powers are considered the most effective means of achieving desired results.

Alison and Roger Heath
Edited by David Goyder

Kew Publishing
Royal Botanic Gardens, Kew

PLANTS PEOPLE
POSSIBILITIES

© The Board of Trustees of the Royal Botanic Gardens, Kew 2009
Text and photographs © Alison and Roger Heath

The authors have asserted their rights to be identified as the authors of this work in accordance with the Copyright, Designs and Patents Act 1988.

All rights reserved. No part of this publication may be reproduced, stored in a retrieval system, or transmitted, in any form, or by any means, electronic, mechanical, photocopying, recording or otherwise, without written permission of the publisher unless in accordance with the provisions of the Copyright Designs and Patents Act 1988.

Great care has been taken to maintain the accuracy of the information contained in this work. However, the publisher, the editor and the authors cannot be held responsible for any consequences arising from use of the information contained herein.

First published in 2009 by
Royal Botanic Gardens, Kew
Richmond, Surrey, TW9 3AB, UK

www.kew.org

ISBN 978 1 84246 183 9

British Library Cataloguing in Publication Data
A catalogue record for this book is available from the British Library.

Production editor: Sharon Whitehead
Typesetting and page layout: Margaret Newman
Design by Publishing, Design & Photography,
Royal Botanic Gardens, Kew

Printed and bound in the United Kingdom by Henry Ling Ltd

For information or to purchase all Kew titles please visit
www.kewbooks.com or email publishing@kew.org

Kew's mission is to inspire and deliver science-based plant conservation worldwide, enhancing the quality of life.

The paper used in this book contains wood from well-managed forests, certified in accordance with the strict environmental, social and economic standards of the Forest Stewardship Council (FSC).

Contents

Foreword vi

Preface 1

Acknowledgements 2

How to use this book 4

Area covered and habitat overview 7
   Where is Selinda? 7
   Sources of water 7
   Selinda's habitats – overview 10
   Description of habitats 11

Plants of Northern Botswana and the Okavango Delta
   Blue or blue-purple flowers 21
   Green or greenish flowers 62
   Orange, brown or red-brown flowers 111
   Pink, red or red-purple flowers 147
   Yellow or cream flowers 198
   White or whitish flowers 323
   Grasses 425
      Introduction to the grasses 426

Simple spike 428	Open panicle 465
Spike of spikes 430	Subcontracted panicle 491
Digitate spike 447	Contracted panicle 502
False raceme 457	Spike-like panicle 507
Plumose panicle 464	

   Sedges and sedge-like plants 516
      Introduction to sedges and sedge-like plants 518

Capitate 520	Branched pseudolateral 554
Umbel-like clusters 527	Panicle 555
Umbel-like with single spikelets . 549	Terminal elongate 557
Unbranched pseudolateral 552	

   Ferns and aquatic plants with no visible flowers 558

Contributors 562

Bibliography 564

Glossary of terms 567

Synopsis of plants in this book 572

Common names index 578

Scientific names index 588

Foreword

Under the feet of the great lion and the trampling herds of buffalo there is an often hidden side to Botswana, sometimes over shadowed by our intense interest in these iconic species. But down there is the lifeblood of everything, without which the leopards and elephants would disappear altogether and this beautiful Botswana would shrivel up into a dust bowl. It is of course, the plant kingdom.

Grasses, small plants, trees, sedges, reeds . . . we see them as the canvas that life is painted on, and we are right.

For many years Botswana has encouraged knowledge and more recently we encouraged Roger and Alison Heath to give up their cosy life in the United Kingdom and venture out into the wild with their unique knowledge and enthusiasm. At first they started in the Selinda Reserve in the north, collecting hundreds of plants, drying and cataloguing, and developing a computer program to make it easy for mere mortals like me to navigate easily if we need to identify a plant we come across. It is not my specialty, but I keep coming across people who are intensely interested in understanding the intricacies of the plant life here, and to my delight, I am told that the Okavango has a rich diversity of plants.

Of course why anyone would spend so much of their day with their noses buried in the grass, searching for a plant no larger than a coin sometimes is a mystery, at least until one realizes that this diversity is much more valuable than any coin, or currency, because it is the currency of the planet in its own system, one that keeps every part of it ticking over healthily.

This book will be a testament to their hard work as well as our appreciation as a nation.

His Excellency Lieutenant General Seretse Khama Ian Khama PH, FOM, DCO, DSM
President of the Republic of Botswana

Preface

Neither author of this book was formally trained as a botanist, though we both have qualifications in other disciplines and, as a hobby, botany has always been part of our life. We were fortunate enough to be able to take a relatively youthful early retirement and to be able to indulge our love of the bush by doing voluntary conservation work.

We were located on the Selinda Concession in the Linyanti and started looking at a project concerning elephants. It soon became obvious that observations of the impact of these mammals on the bush were being hampered by a lack of modern handbooks to aid the identification of the affected plants. The Royal Botanic Gardens, Kew in the UK edits and publishes the Flora Zambesiaca, but this is a huge multi-volume work with few illustrations and is really most useful for specialist botanists in herbaria. Although begun in the mid-1950s, this Flora is not yet complete. The alternative is to employ a library of books, many of which were written for other areas of Africa. We discovered that we were not the only researchers with this problem, and thus decided to produce the needed field guide.

The plants that appear in this volume are those identified during systematic botanising in the area. We feel that we have spotted most of them, but doubtless we have missed some. We want our book to be useful to researchers, tourist camp staff, staff of government departments, guides, rangers, tourists and anybody else wanting to know and enjoy what they see. There are sure to be omissions and errors, so we welcome comments that will allow us to up-date the electronic checklist and future editions. Contact us by e-mail at plantsandpeopleafrica@yahoo.co.uk.

We have set out to make this book a practical tool by using high-end digital photography to illustrate a finely focussed text. At the same time, we have aimed to make the book useful to those who do not have English as their first language or who are not professional botanists. To avoid ponderous essays, however, technical words have been allowed to creep in, and so we have drawn up a glossary.

The book has been written from original field research involving the collection, photography, recording, drying and pressing of plant material. The detailed data arising out of this activity is available to professional researchers and keen botanists as a searchable checklist, which includes images of the dried specimens. Dried specimens of most plants have been supplied to:-

1. The National Museum of Botswana — National Herbarium (GAB);
2. The Herbarium at the National College of Agriculture, Botswana (BACH);
3. The Herbarium at the Harry Oppenheimer Okavango Research Centre, Maun, Botswana (PSUB), part of the University of Botswana;
4. The Royal Botanic Gardens, Kew, UK (K);
5. Seed collections of a number of plants have been supplied to the Millennium Seed Bank Project in Botswana.

Acknowledgements

We could not have produced this field guide without the unstinting encouragement and support of a whole group of organisations and individuals. We therefore owe an enormous vote of thanks to all concerned.

We would like to start by thanking the Government of Botswana and Linyanti Explorations for allowing us access to the wonderful plant life of the Linyanti Area. We also thank the Office of the President, the Ministry of Labour and Home Affairs, and the Ministry of Environment for permission to research in the Linyanti.

We started our life in Botswana, like many people, as tourists, flying in and out from camp to camp. Soon, we wanted more and were introduced to MAP Ives, an almost legendary character. We shall never forget our mekoro safaris into the Okavango Delta, staying in our own camp and spending the days walking in the bush, learning as we went. Thus our education in matters of the wild started, and with it a love of the bush. MAP then completed the process by suggesting, after several years of safaris, that we should think about doing our own project in the bush to put something back in return for what it had given us. Thus, he planted the seed from which this project grew.

At the end of each stay with MAP, we always spent a period on the Selinda Concession, CHA/NG16, owned by Linyanti Explorations, initially at the Old Selinda Camp and then at its successor, Zibalianja Camp, for a few days of pampering after our bush existence. Linyanti Explorations are exemplary in the strong conservation ethic that underlies their approach to running the concession. Following early retirement, we approached Andre Martens, then a director of Linyanti Explorations, to ask if there was a conservation project that we could undertake for them. Our discussions resulted in a proposal to establish an inventory of the plants on the Selinda Concession together with a herbarium, for use as a management tool.

We thank Andre and Angela Morgan, then Conservation Officer, for shepherding us and for providing all sorts of advice, a base camp and much of the locally produced specialist field equipment. Directors of Linyanti Explorations, Brian and Jan Graham, Andre and Marianne Martens and Grant Nel fed and housed us and smoothed the way. We must highlight the enthusiastic support that we received from the whole staff at Selinda. Our contributors chapter enlarges on this topic. Linyanti Explorations, and especially Andre, could not have been more enthusiastic, motivating and helpful.

Part way through the production of the book, the ownership of the Selinda Concession changed hands. Dereck and Beverly Joubert, the new owners, have a charisma and clear identity of purpose that shows through in their breathtaking wildlife films. They have continued to give us every encouragement and we thank them for making our camp at Selinda into our Botswana headquarters.

The project started with the idea of making an inventory of the plants on the concession as a management tool. It soon became obvious, however, that there was a major problem in identifying plants other than well-known species using current publications. We therefore turned to the National Museum of Botswana (BNM) and, through them, the Royal Botanic Gardens, Kew, UK (Kew) who agreed to provide identifications of the plants in return for pressed dried samples. Our thanks go to the staff of both institutions for this core help. BNM has also smoothed the way for us in Botswana, helping with the official paperwork, enabling us to be there to do the research and sending plant samples to Kew. Kew has undertaken the task of editing and publishing this field guide. These two institutions were a joy to work with and we cannot thank them enough. We learned so much!

In talking of BNM, we thank the former Directors, Tickey Pule and Soso Lebekwe-Mweendo, and the current Director Mr G. Phorano for solving many of our problems and keeping us on the straight and narrow. We also thank the previous Keeper of Natural History, Bruce Hargreaves, for his huge knowledge of local plants, the current Keeper of Natural History, Nonofo Mosesane, and Daniel Mafokate.

At Kew, we were considerably encouraged and tutored by our mentors Gerald Pope, Kaj Vollesen, David Goyder, Tom Cope and David Simpson. They are botanists with vast knowledge and expertise who earned our huge admiration and grateful thanks by dealing patiently with two people starting a complex venture from scratch. The staff at Kew have long experience of working with volunteers; thanks to their encouragement, what began as bringing a few plants for identification developed into the production of this field guide.

The Harry Oppenheimer Okavango Research Centre (HOORC) in Maun very kindly allowed us to work in their library and to use their herbarium facilities. We thank the Director, Sue Ringrose and the past Director Professor Lars Ramsberg, Librarian, Monica Morrison, Research Fellow, Caspar Bonyongo, Herbarium Officer, Joseph Madome and other HOORC staff.

Our thanks must also go to Kew and BNM for persuading us to produce both this field guide and the 'Plant Finder' electronic database and checklist of the Plants of Northern Botswana. Their suggestions expanded the horizons of the project and heightened our sense of achievement. Our thanks go to John Harris and Gina Fullerlove of Kew Publishing who have guided us through our first book and left us with the intention to do more. Our thanks also to the Kew production team of Sharon Whitehead, Margaret Newman and Lloyd Kirton. Again, special thanks to David Goyder, Tom Cope and David Simpson of Kew Herbarium for their scientific editing skills and unfailing support.

We thank Heidi Allmendinger who accomodated us at Kubu Lodge whenever we were in Kansane. Kubu is balm for the soul after many hot dry weeks in the bush. We also thank Wilderness Safaris and Kwando Safaris for permission to go onto their concessions bordering Selinda, and Orient Express for accommodating us at their camps.

We made Maun our jumping off point, and there we received a high level of practical help from a lot of 'can do' people. Peter and Pauline Perlstein and Ronnie and Hilary Crous, who accommodated us in their family homes, and Brian Bridges of Ngami Toyota rank next to MAP Ives and Andre Martens in their central position in our lives in Northern Botswana. We also thank Alison Brown, Christiana Stolhofer, Paul Scheller, Pam Shelton, Clint Gielink, and Anna-Katherine and Shane Seaman.

Ngami Toyota have been essential to the safe and smooth running of each collecting expedition. They rebuilt the suspension of our Landcruiser to extreme bush standards. They store the vehicle and make sure that it is ready for each Selinda session, repairing the damage produced on the previous expedition. We thank Brian Bridges and Mark Muller, directors of Ngami Toyota, for their great friendship. Shane Scott, Oaitse Mosiane and Pat Hagen, staff members, come in for special mention and thanks. Shane made sure that we were equipped to be safe in the bush and gave us instruction on driving in sand and mud, and Oaitse is a lioness when it comes to official paperwork.

During the lifetime of this project, we started to gather seed for the Millennium Seed Bank (MSB) initiative and this has given us further material for this book. We thank Paul Smith, Director of MSB, who has given us encouragement and support by making time in his busy schedule to talk through problems and for the supply of equipment. We are grateful to Birgitta Farrington, MSB Botswana Coordinator, for her help on the ground. This activity brought us into contact with the Botswana Department of Agriculture, specifically with Tlhaloganyo Ofentse, who kindly allowed us to use his unpublished work on Cucurbitaceae. Ofentse sadly died during the final stages of production of this book. Work with the MSB also introduced us to the College of Agriculture and Pearl Lebatha.

We thank Ingrid and Coen van Graan who, at their home in Windhoek, supervised and encouraged us as we first equipped ourselves for life in the African bush; Eva and Ørnulf Lauritzen, June Wakefield and Norman Lazarus for their warm friendship and valuable advice in all matters scientific; and our nephew Nick Hoath who supervised our forays into the world of computer software.

We have used only one photograph that was not not taken by us and that is of the flowers of *Orbea huillensis*, which somehow were never open when we were around. This photograph was taken by Joanne Stone, a member of Linyanti Explorations staff. We thank her for the generous donation of this picture.

We would like to give a special vote of thanks to our very dear friend and mentor Tickey Pule. Since her move from the National Museum, Tickey has continued to take a personal interest in our wellbeing, guiding us through the various official mazes that we have encountered. But more than that, she has made sure that we understand and appreciate Botswana, not just as passing researchers, but as members of her family, living at her home when we are in Gaborone. It is this gift that makes writing this book especially rewarding and we thank her and her family unreservedly.

How to use this book

Geographical scope of this book

We have set out to provide information to help identify some 530 flowering herbs, trees, shrubs, ferns, grasses, sedges and sedge-like plants in northern Botswana. The majority will also be found in nearby countries of the Zambezi Basin: Angola, Mozambique, Namibia, Zambia and Zimbabwe. Many of the plants also grow in other neighbouring geographical areas, including the rest of Botswana and as far south as South Africa.

Equipment

To really enjoy the wonderful world of plants and to be able to see all the features in the descriptions, a hand lens ×20 magnification is a great help.

What you will find in this book

In the main, we have classified the entries according to the Preliminary Checklist of the Plants of Botswana by Moffat P. Setshogo.

This field guide is arranged in four sections: (1) 'Plants other than grasses and sedges', (2) 'Grasses', (3) 'Sedges and sedge-like plants' and (4) 'Ferns and aquatic plants with no visible flowers'.

The first section is ordered by flower colour and then by family and subfamily. Remember back to when you first saw the flower and what colour you thought it was in normal daylight. Early morning and late evening light can have a pronounced blue or orange colour bias. When it comes to deciding between say mauve and purple or blue-green and green-blue, individual observers may have differing conceptions of colour, so if you do not find what you are after in one colour section, then look in the other. Also you should look about you to see if there are other plants with flowers that are the same shape but differently coloured.

Pink or blue flowers can have white variants and we have illustrated some. So, if you are looking at a white flower and cannot find it in the white flowers section, then look in pink/red/mauve and then blue.

Within each colour, plants are organised by family, then subfamily, genus, species, sub-species and variety.

The two sections 'Grasses' and 'Sedges and sedge-like plants' are each ordered by inflorescence structure in completely new arrangements created by Tom Cope and David Simpson at Kew, then by family, genus and species.

Each plant is illustrated with photographs and a brief focused text describing the main features that aid identification and additional information. Each text can include the following sections:

1. **Plant names**
   The scientific name for the plant is given first, then common names in the languages used locally followed by the name in English. The plant families are as shown in the Botswana checklist.

   With regard to the Setswana names, Professor Desmond Cole's 'Setswana – Animals and Plants' and Dr Moffat Setshogo's 'Common Names of Some Flowering Plants of Botswana' have been invaluable, as has the late Tlhaloganyo Ofentse's as yet unpublished work on Cucurbitaceae, which he kindly allowed us to quote.

   Great care must be taken in using local names for plants because the same name may be used for different plants in different districts of Botswana. For example, *Cucumis metuliferus* is known as 'mokapana' in the Sengwato district, whereas *Acanthosicyos naudianus* is known as 'mokapana' in Setswana, the dialect of Tawana spoken by the people of northern Botswana. Likewise 'bogoma' is a generic name for plants that attach themselves to animals and humans as a means of distributing their seed; at least six plants are given this name.

2. **Derivations**
   Derivations are given for as many of the scientific and vernacular plant names as possible. The sources for these will be found in the bibliography, but we have also consulted various websites, including Calflora.net and plantzafrica.com.

3. **Key identifying features**
   A brief description of the plant is subdivided into the parts of the plant, as indicated by bold type. Complementing this are photographs that generally include a close-up of the flower or, for a grass, the inflorescence; a photograph of the whole plant in its environment; and other photographs showing the structure and distinguishing features of the plant.

   Descriptions of plants can sometimes be misleading as plants do not always conform to a specific shape or flower colour, being influenced by a range of external and internal factors. We outline a few potential difficulties below to put the user on guard:

   3.1 Height – plants are influenced by growing conditions, one of the most influential being water supply. As an example, in a year of poor rains *Crotalaria platysepala* might only reach 30 cm, whereas in a good year it may exceed 1.4 m in the same location. Remember that either an excess of minerals in the soil or nutrient-deficient soil could also change plant growth, even in years of good rains.

   3.2 Flowers – blue-, pink- and maroon-flowered plants often have white variants and can exhibit colour variations between the extremes. *Striga gesnerioides* and *Jacquemontia tamnifolia* are good examples with the former appearing in most shades from maroon through pink to white and the latter in white or sky blue.

   3.3 Leaves – First carefully inspect the plant to determine the shape of mature leaves, which are probably to be found halfway up the plant. Leaf shape varies within many species according to the age of the plant, growing conditions and where the leaf is on the plant. The photographs of dried samples below show the variation that can occur in the leaves of *Hibiscus sidiformis*.

4. **Habitat**
   We quote the habitat in which we have found the plant. Do not rule out others because time of year, rainfall, soil type and so on can affect where plants succeed. Plant-specific factors, such as shade for shade-loving plants, will bias the distribution of some plants towards locations with a strong content of that factor. We have worked in 26 different habitats in the north of Botswana and one on the Chobe. These are described in the section 'Area covered and habitat overview'. Similar environments occurring in other countries can also host the plants found in these habitats.

5. **Flowering period**
   We have quoted flowering periods relative to the main rains, temperature and light levels and not by the calendar. Late rains will postpone flowering and an early finish to the rains brings an early cessation of flowering. Extensive rains also postpone flowering as plants have grown for a longer period, and are thus larger, when they flower as the rain eases. Conversely, poor but timely rains will produce early flowering on small plants.

6. **Uses and beliefs**
   Please do not try these suggested treatments by yourself as many of the plants may be poisonous if incorrectly prepared.

   We have recorded the uses and beliefs concerning the plants where we can confirm them as belonging to Botswana. We have not set out to record information from surrounding countries.

   We have checked our information directly with local people and have found frequently that the parents of the current generation are the final repository of many customs concerning plants. Their children are ceasing to use them as they migrate to live in towns where medical care is more widely available, leading to a loss of traditional knowledge and medicine.

7. References
We provide references for further reading concerning the plants and highlight the local people who have provided information in the 'Uses and beliefs' section. Local people are identified in the 'Contributors' chapter towards the end of this book and publications are listed in the 'Bibliography'.

8. Abbreviations

c.	approximately, about
cm	centimetre(s)
dia.	diameter
inc.	including or inclusive of
m	metre(s)
mm	millimetre(s)
p.	page
pp.	pages
pt	part
subsp.	subspecies
syn.	synonym
var.	variety
Vol.	volume of a book

Area covered and habitat overview

Where is Selinda?

The Selinda Herbarium, upon which this field guide is based, largely contains plants collected in a part of Botswana known officially as CHA/NG16 and to others as 'The Selinda Wildlife Management Area' or 'The Selinda'. The Selinda forms part of the Okavango Delta wetlands, a significant Ramsar site. It is situated in northern Botswana, bordering Namibia, where the eastern end of the Caprivi Strip widens into a south-pointing triangle. Selinda is at the western side of the southern apex of this triangle, with its north-eastern boundary running along the Botswanan border, which is influenced by the course of the Kwando River. It then stretches south-west down the Selinda Spillway towards the Okavango Delta. Both the Selinda Spillway and the Savute Channel branch off from the Kwando River system on the Selinda Reserve.

Sources of water

The Selinda, part of the Okavango Delta System, is situated in the Kalahari Desert. This delta is a true inland delta with no outlet to the sea. Over 90% of the water arriving down the rivers or from rainfall is lost through being taken up into plants and by evaporation from the soil rather than by flowing into the sea. The rest is lost to underground seepage and so on. Surrounding the delta are the arid conditions of the desert.

The Kwando rises in the highlands of eastern Angola, flows across the Caprivi and along the border between Botswana and Namibia. It changes its name twice, to the Linyanti and then to the Chobe, before flowing into the Zambezi east of Kasane at Kazungula, just at the point where four countries, Botswana, Namibia, Zambia and Zimbabwe, meet. The Kwando River is joined to the central Okavango Delta river system by the Selinda Spillway which, depending on the relative levels of the two systems, can flow south-west from the Kwando to the Delta or vice versa; or, as in recent history, not at all, being dry. In 2009, the Selinda Spillway flowed from the Kwando right into the Okavango Delta for the first time since 1985. Also, historically, the Kwando flowed eastwards from the Zibalianja Lagoon into the Savute Marsh via the Savute Channel.

Thus, flora in the Selinda benefits from two sources of water, derived from flooding from the Angolan highlands and from a cycle of annual rains. The level of the Kwando and the extent of its flood depend on the level of the rains in the highlands of eastern Angola during the period from December to March. These rains usually arrive at the Selinda as flooding from the Kwando in June to August.

The whole of the Okavango Delta lies on a plateau at a height of about 950 m and is characterised by having very little change in height from one end to the other. Indeed, over a length of 250 km the main delta falls just 65 m. The delta owes its existence partly to this fact and to two geological fault lines that are an extension of the Rift Valley, one of which, the Gomare Fault, is followed by the Kwando as it leaves the Selinda. Thus small geological movements can influence the extent of flooding. Indeed, the end of 2005 almost certainly saw such a movement as the Kwando started to flood up the Selinda Spillway and the Savute Channel almost the whole year round, having been in a retreating phase for the previous 20 years.

The rains at Selinda, like the rest of southern Africa, arrive in mid to late summer, generally falling from late September to March but sometimes not starting until late November or continuing through to sporadic falls in April. Their timing and extent can be very variable and this affects the species of plants that can be seen, their size, frequency and timing of flowering. Total annual rainfall varies from around 344 mm to over 800 mm and there can be successions of years of good or bad rains. Overall, there seems to be a slight short-term upward trend buoyed up by two recent good seasons, 2003/4 and 2005/6 (as indicated by the blue line on the graph on p. 9).

The rains are split into two periods, the 'early rains' in November, December and early January and the 'main rains' from late January onwards. Either or both rains can fail or be good, and in some years, there is a drought between them that can be up to four weeks long.

The flatness of the land means that during the rains not all the water can get away to rivers. Although a lot soaks into the ground, much collects in shallow depressions where the ground is not so permeable to form small lakes and ponds called pans, which are dotted all over the landscape. The overwhelming majority are usually seasonal, drying out before the next rains even in good years.

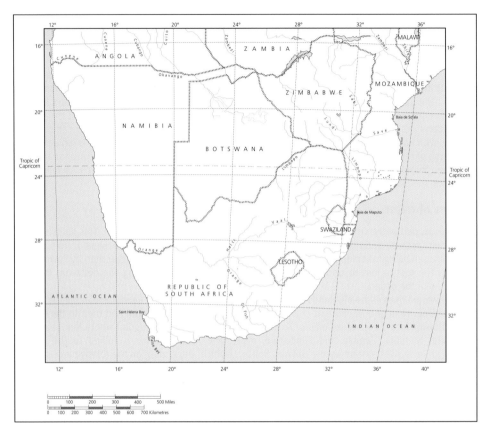
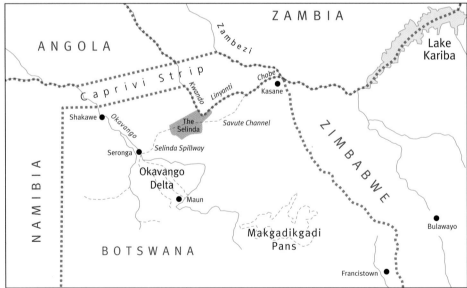

Map of Northern Botswana showing the location of The Selinda CHANG16 (shaded area)

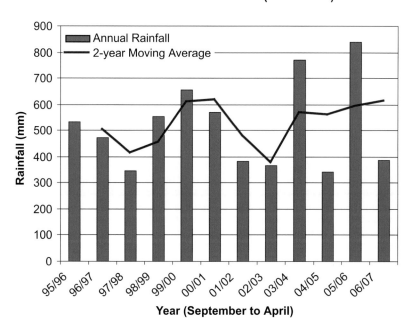

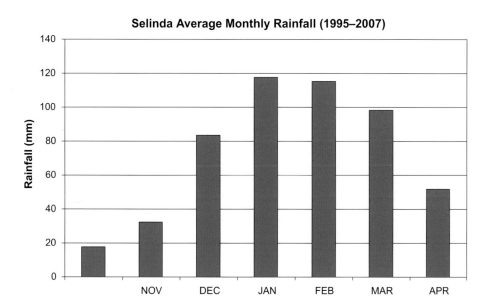

Source: Linyanti Explorations

Selinda's habitats — overview

The shape, situation and orientation of The Selinda gives rise to a wide variety of habitats for the flora and fauna ranging from flowing river water to near desert. As described above, The Selinda stretches between two river systems situated in the Kalahari Desert. Near a river there are flood-influenced marshy wet conditions, but upon moving away from the river the desert conditions have more of an influence and rainfall is more important.

A broad picture of the available habitats can be obtained by beginning at the Kwando and moving south-west from it.

The banks of the Kwando River are lined with marshes. In recent times, the Kwando has flooded only the first kilometre of the Selinda Spillway to Selinda Camp and has established a marshy riverbank. Since late 2005, however, the Kwando has almost continually flooded into this watercourse pushing a permanently filled waterway nearly 12 kilometres from the river itself and creating large marshy areas, notably from the Selinda Spillway near Selinda Camp south-eastwards towards the Zibalianja Lediba or Lagoon. During the same period, the river has been flooding into the Zibalianja Lediba towards Selinda camp and the Zibalianja Lediba has been starting to fill the Savute Channel.

Moving away from the river there is a mixture of regularly flooded and permanently dry flood plains. Dotted around the plains are islands each with either scrub or woodland vegetation, depending on their size and proximity to water.

These islands are not areas of higher land created by erosion or geological land movement but are created by a more subtle mechanism. An island can begin from a small group of flood-tolerant plants which includes some trees or bushes, sometimes on a termite mound. Over 90% of the water arriving in the area is lost into the atmosphere via transpiration by plants. Minerals that arrived dissolved in the water are left behind to accumulate in the soil beneath plants. These minerals build up below the surface and eventually produce ground heave, pushing the soil higher than the flood level. As an island becomes established, so the trees in its centre pull in more water from the surrounding land and cause a concentration of minerals at the centre which eventually reaches a level that becomes toxic to plants. At the edges of the island, however, the inward water flow dilutes the chemicals present. A mature island will have a balding centre with vegetation growing on the periphery, much like a monk's tonsure. Accumulation of salts from the innermost plants causes the island to grow further and, as it does so, the vegetation keeps to the new outer edges. Geological movement or drought can cause areas of floodplain to dry out but the islands are still referred to as islands even if they are in a dry floodplain no longer subject to annual flooding.

At the edge of the plains is a strip of riverine woodland where *Acacia nigrescens*, *Kigelia africana* and *Garcinia livingstonei* grow. Behind this strip, drier conditions prevail with *Colophospermum mopane* forests relieved by sandy areas growing *Philenoptera violacea*, *Philenoptera nelsii* and *Boscia albitrunca* and by areas with a high level of clay where seasonal pans are created. Much of the vegetation is severely pruned, or 'farmed', by grazing and browsing animals and thus shade from the sun can be in limited supply. This landscape is dotted with occasional coppices where the soil may be more fertile or where trees which have escaped damage to grow into fair-sized specimens. In years of low rainfall, pruning in the proximity of rivers becomes more severe due to the concentration of elephant and other animals journeying to the river waters when the backwoods pans are empty.

In the dry winter season, the middle of the Selinda away from the flooded part of the Spillway provides a tough environment for plant growth. Even in the rains, many plants make use of the shade afforded by bushes or grow right at the foot of the farmed *Colophospermum mopane*, hence obtaining shade for at least part of the day.

The Selinda Spillway, once away from the Kwando, leaves the wide marshy flood plain and cuts a channel with shallowly sloped banks. These banks are exposed to a greater or lesser extent dependent on the flood from the Kwando. Historical flow patterns have led to the deposition of sand banks in the channel and on its edges.

As the water floods down from the Kwando in the north, so seasonal flood waters from the Okavango's Nqoga River flow up the Spillway from the south-west. It has been unusual in recent years for the two to meet. In 2006 there was a straight line gap between the two of 19 km and the two actually joined for the first time in more than 20 years in 2009. If the main rains are heavy then the northern Spillway advance is temporarily augmented due to runoff of rain water. After the rains finish, the waters at the northern end can withdraw until the inflow from the Kwando regains the balance with loss by seepage, evaporation and plant transpiration.

Description of habitats

This section is devoted to describing the general features of each of the 27 habitat types used in this book.

The plants given in the summaries below are typical of the environment concerned. The mix of plants in a specific habitat will be influenced by the stage of maturity which it has reached and will continue to change until the climax community evolves.

1. Acacia scrub: Open sandy grassy areas on the floodplain and sand ridges in the *Colophospermum mopane* woodland dotted with trees and bushes of the *Acacia* family, notably *Acacia hebeclada*, *Acacia erioloba* and *Acacia luederitzii*. Left to their own devices, the acacias will merge to form impenetrable masses.

2. Bush, riverine: Mixed riverside bush interspersed with *Acacia nigrescens* bordering on the Chobe River at Kasane. *Pteleopsis myrtifolia*, *Grewia flavescens* and *Commiphora merkeri* are typical.

3. Edge of Spillway: The space between the marshy area along the water's edge and the forest edge often much poached by animals. Generally, this has a gentle slope and the interface between it and the water margin habitat can move with changes in seasonal water levels.

4. Floodplain: The Selinda reserve does not have a regime of seasonal flooding as in the Delta, and what is referred to as 'floodplain' here are plains which have historically been flooded but currently are often dry. The soil is sandy clay with grassland dotted with sand-filled depressions and sand ridges.

After a few years of dry conditions, pioneer plants such as *Pechuel-loeschea leubnitziae*, *Vernonia glabra* and *Laggera decurrens* start to colonise the area. In time, *Cymbopogon excavatus* can grow over large areas. Opportunist species like the acacias, *Philenoptera violacea* and even *Colophospermum mopane* will start to colonise.

5. Floodplain margin: The edges of the flood plain are typified by slightly rising ground and the establishment of a belt of shrubs such as *Combretum hereroense*, browsed (farmed) *Philenoptera violacea* and *Croton megalobotrys*.

Sometimes this belt is not well defined and there may be a gradual transition to forest or island vegetation. A feature of the marginal geology may be in the form of a sandy tongue, which provides a shrubby route for animals to penetrate the floodplain and then goes on to the water beyond.

6. Flowing waterway: The Selinda Spillway and Zibalianja Lediba (lagoon) are supplied from the Kwando River by channels of strongly flowing water, which can be restricted by the growth of vegetation to a metre or so in width. These are continually worked by hippopotamus, which contribute to keeping the channels open. *Cyperus papyrus*, which is a major factor in the blocking of channels, and *Vigna luteola* frequent these waterways.

7. Island: A raised area of land formed by the process described above. Plant environments vary greatly with the age of the island. Young islands offer much shade and denser vegetation including *Berchemia discolor*, *Philenoptera violacea* and *Hyphaene petersiana*. As the years pass, so the central vegetation is dominated by chemical-resistant hardier grasses, leaving trees and shrubs round the edge.

8. Island margin: A gently sloping area typified by a fringe of shrubs and bushes, such as *Croton megalobotrys*, *Ziziphus mucronata* and *Diospyros lycioides*, giving way to the floodplain vegetation. An environment sheltered from the sun by the trees at the island margin and supplied by rainwater run-off from the island encouraging a relatively luxuriant undergrowth.

9. Pan, seasonal: A pan forms in a depression with an impermeable lining in which rainwater run-off gathers. These can be quite small and short-lived but some grow large enough, depending on rainfall, to attract hippopotamus. They become the focus for animals during and after the wet season. Seasonal water plants such as *Neptunia oleracea* and *Marsilea vera* appear in the longer-lived pans. At the water's edge, there is a belt, well worked by the feet of animals, where *Sesbania rostrata* is to be found. As the pan dries up, plants such as *Heliotropium supinum* and *Sphaeranthus peduncularis* take advantage of the moisture remaining in the soil before it finally dries in the sun.

10. Pan, seasonal margin: When the pans are at their maximum levels, water intrudes into the surrounding vegetation enabling the growth of plants not seen in poorer rainfall years. In these conditions, *Commelina subulata* and *Kalanchoe lanceolata* can be found in *Colophospermum mopane* woodland. If the pan only partially fills, then the surroundings are drier and more akin to the drying pan described above.

11. River bank: The first area of solid soil after the wetland bordering the river. *Phragmites australis*, *Setaria sphacelata* and *Pennisetum macrourum* are typical.

12. Riverine forest: This term is used to describe the thick forest containing evergreen trees that grows on slightly raised land adjacent to rivers and large areas of water such as Zibalianja Lediba. The land has to be high enough to drain and not be wetland. Trees such as *Acacia nigrescens*, *Garcinia livingstonei* and *Diospyros mespiliformis* grow with an understorey of *Combretum mossambicense*, *Loeseneriella africana* and *Diospyros lycioides*. In dry land areas, this border gives way to *Colophospermum mopane* woodland not far away from the water.

13. Riverine forest margin: At the interface where the land starts to slope down to the water and the trees end, the light and moisture enable the growth of a thick resilient belt of bushy shrubs such as *Diospyros lycioides* and *Croton megalobotrys*. In their shelter, young trees have a chance to grow without being grazed. *Croton megalobotrys* is unpleasant to eat and is thus left alone except in the driest of years when food is really scarce. This offers deep shade for many varieties of herbs.

14. Riverine forest margin, land side: Away from the water, the forest trees give way first to *Combretum mossambicense* and interspersed with open grassy areas then to *Colophospermum mopane*.

15. Sand ridge: Areas of deep sand with vegetation typified by *Philenoptera nelsii*, *Terminalia sericea* and *Clerodendron uncinatum*. Driving can be difficult during dry periods.

16. Scrub mixed: An area of mixed bushes frequently occurs at the interface between two environments. Such areas can contain a wide range of shrubby plants, such as *Combretum imberbe*, *Combretum hereroense*, *Philenoptera violacea* and *Pechuel-loeschea leubnitziae*.

17. Shallow lagoon: As water flows out of the river along the main channels, it fills up large open areas to form shallow lagoons. There, sediment is deposited as the water spreads out and the speed of flow drops. *Nymphaea lotus*, *Nymphoides forbesiana* and *Potamogeton thunbergii* grow in profusion amongst rich sedge beds.

18. Spillway: The formation of the Spillway and its function have been described above. As the Spillway leaves the Kwando it passes through flowing waterways and open shallow lagoons.

Soon, it begins to assume the appearance of a meandering shallow river which, at its north-eastern end is lined with riverine forest dominated by *Acacia nigrescens*. Towards the south-western end of the Selinda, the banks of the Spillway become lined with forest dominated by *Combretum imberbe*. In both cases, the forest border gives way to forest dominated by *Colophospermum mopane* upon moving away from the spillway.

Between the northern and southern fronts of water the Spillway is a dry land environment. *Pechuel-loeschea leubnitziae* and even *Colophospermum mopane* grow in the Spillway bed.

The two advancing water fronts have different characteristics. The southern goes through a complete cycle each year, advancing rapidly many kilometres each year fuelled by the flooding Okavango River as it flows out of Angola. The level of the floods determines how far along the Spillway the water advances. As the Okavango floods pass, so the waters retreat.

By contrast, at the northern end, the advance is not punctuated by substantial withdrawals and takes place more slowly. Ahead of the water, ground seepage begins to kill off the dry-loving *Pechuel-loeschea leubnitziae* and *Geigeria schinzii*, which are replaced by *Laggera decurrens* and *Conyza stricta* growing in profusion. They, in their turn, are killed by the arrival of surface water and water plants such as *Nymphaea lotus* and *Potamogeton thunbergii* take over.

19. Termite mound: The mound itself, usually pointed, and an area immediately around the mound limited to the zone where a slope has been formed from material eroded by the action of rain and animals.

20. Termite mound, decayed: Generally a gentle mound formed by the remains of a long-dead termite mound. An area of hard grey soil supporting an arid environment, which may even be virtually bare of vegetation.

21. Treeline: A term used to describe any area where open grassland meets the trees, such as on the edge of floodplains after the floodplain margin. Viewed from out on the plain, it looks like a line of mixed trees. Lacking the moisture present in riverine forest margin, there is no dense bushy marginal vegetation but a thinner growth affording views into the trees and including *Acacia nigrescens*, *Acacia erioloba* and *Philenoptera violacea*.

22. Water margin: The area immediately back from the edge of standing or flowing water, which generally comprises a spectrum ranging from floating reed bed to river bank and includes the plants seen within *Cyperus papyrus* beds. A riotous profusion of plants including sedges, *Vigna luteola*, *Ludwigia stolonifera* and *Cyclosorus interruptus* may grow. Exit channels are formed by elephant, hippopotamus and crocodiles.

23. Wetland: An area inundated by subsurface or very shallow water to give a marshy environment. There may be a central channel, or the wetland may be formed by overflow from a spillway or river system into an adjoining basin. Bird and insect life abounds and red lechwe find their shelter there.

Frequently, the wetlands are surrounded by substantial reed beds which may even appear to separate them from the channel and contain stranded dryland bushes which act as nesting sites for water birds. The plants include *Ludwigia stolonifera* (on the edges), *Nymphaea lotus* and *Potamogeton thunbergii*.

24. Wetland, dried out: In the event that water losses by evaporation and plant transpiration exceed the inflow, then plants such as *Ludwigia stolonifera* and *Heliotropium ovalifolium* are left as the water retreats, and eventually more land-based plants start to grow.

25. Woodland, mixed: Generally the type of woodland existing beyond the treeline if the habitat is not riverine. It tends not to have such luxuriant growth. Trees such as *Acacia nigrescens* and *Philenoptera violacea* grow with *Combretum mossambicense*, *Croton megalobotrys*, occasional mature *Terminalia sericea* and *Combretum herreroense* as an understorey. *Urochloa trichopus* and *Chloris virgata* typically provide cover for open ground.

26. Woodland, mopane: Also known as cathedral mopane woodland because it includes full sized mature *Colophospermum mopane* trees. Growing in amongst the trees, plants such as *Barleria mackenii*, *Duosperma crenatum*, *Blainvillea acmella* and *Blepharis integrifolia* may be found.

27. 'Farmed' mopane: Areas in which *Colophospermum mopane* trees have been constantly pruned back by feeding elephants and other browsers. The result is what looks like a well-tended plantation of pruned *C. mopane*, about 2.5 m tall but much shorter in unfavourable growing conditions. Locally this is known as '*gumane*'.

Blue

blue or blue-purple flowers

Acanthaceae (Acanthus or spinyflower family)

Barleria mackenii Hook.f.

Common name: Setswana gagara.

Derivation: Named after Rev. J. Barrelier M.D. of Paris and M.J. McKen, plant collector and herbarium curator.

Identification: Compact erect **woody perennial** c. 50 cm tall. **Leaves** dark green and slightly leathery, opposite and decussate, oval with a sharp point, c. 7 × 3.5 cm, tapering to form a short leaf stalk, margins smooth. **Flowers** blue, with 5 lobes, the 2 uppermost smaller than the rest, fused to form a tube, a deeper blue/purple throat emerging from a pair of light green leafy bracts, c. 3.5 cm dia. **Fruit** light green, within a double calyx, exploding to release the seeds.

Habitat: Found occasionally growing locally in light shade on islands and in the tree-line along the floodplain, also in mopane woodland.

Flowering: Middle to end of the main rains.

Uses and beliefs:
- The steamed wood is used to treat conjunctivitis.

Ellery p. 157, Hargreaves p. 62.

Acanthaceae (Acanthus or spinyflower family)

Blepharis integrifolia E.May. & Drège var. *integrifolia*

Common name: Afrikaans rankklits.

Derivation: *Blepharon*, eyelid [Greek], referring to the bracts; *integri*, entire, whole [Latin], *folia*, leaf [Latin], having entire leaves.

Identification: Small branching **woody herb**, c. 17 cm tall, with a **rhizomatous rooting** system. **Leaves** sessile (stalkless), blue-grey, linear and slightly recurved with smooth margins, c. 2.5 × 0.3 cm, borne in whorls up the stems; leaves covered in fine **hairs** whereas stems are covered in coarser erect hairs. **Flowers** stalkless, borne at leaf axils, brilliant blue, very asymmetrical; corolla tube split, 3 main lobes form the lower lip, the other 2 very small and green; bracts covered in spiny bristles.

Habitat: Found occasionally locally in damp clayey areas of mopane woodland.

Flowering: Throughout the rains.

B Van Wyk Flowers p.242, WFNSA p.394.

Acanthaceae (Acanthus or spinyflower family)

Justicia bracteata (Hochst.) Zarb
(syn. *Monechma debile* (Forssk.) Nees)

Derivation: *Justicia*, in honour of the Scottish gardener, James Justice; *bracteata*, referring to the showy bracts from which the flowers emerge.

Identification: Erect **herb** up to c. 30 cm tall but more in rainy years. The plant is covered with **short soft hairs** overall. **Stem** buttressed forming a square section. **Leaves** opposite, decussate, narrowly oval, net veining, c. 11.5 × 3 cm, narrowing towards base to form short leaf stalk. **Flowers** mauve, in attractive pairs of clusters in the leaf axils, individual flowers emerge from leafy bracts outlined with silver hairs, c. 0.6 × 1 cm.

Habitat: Found occasionally, more commonly in damper years, growing in partial shade in the shelter of shrubs in the mopane woodland and in scrub on sand ridges.

Flowering: Middle to end of the main rains.

Acanthaceae (Acanthus or spinyflower family)

Justicia divaricata (Nees) T.Anderson
(syn. *Monechma divaricatum* (Nees) C.B.Clarke)

Common name: Setswana phuduhudu.

Derivations: *Justicia*, for James Justice, a Scottish gardener; *divaricata*, spreading apart at a wide angle [Latin]. *Phuduhudu*, steenbok.

Identification: A woody **herbaceous plant** with many branched stems, up to c. 80 cm tall. **Leaves** insignificant, strap-like, opposite and decussate, margins slightly serrated, fine hairs on both surfaces, folding at the mid-vein, slightly recurved along their length, c. 5 × 0.5 cm. **Flowers** borne singly in leaf axils, corolla tube with a lower petal enlarged to form 3 lobes with a darker mauve 'herring-bone' marking in the throat.

Habitat: Locally common in drier years, growing on the floodplain and sparsely shaded islands.

Flowering: From the middle of the main rains into the early part of the dry cool period.

GM.

Acanthaceae (Acanthus or spinyflower family)

Justicia heterocarpa T.Anderson subsp. *dinteri* (S.Moore) Hedren

Derivation: *Justicia*, for James Justice, a Scottish gardener; *hetero*, different, varying [Greek], *carpus*, seed [Greek], producing different colours of seed; *dinteri* for Prof. Kurt Dinter (1868–1945), a German botanist who collected extensively in Namibia.

Identification: Annual **herbaceous plant** up to c. 90 cm tall when supported in other vegetation. **Stems** ribbed, suffused maroon above axils. Stems and leaves covered in short hairs. **Leaves** opposite, decussate, oval, margins entire, net veining, c. 10 cm (inc. c. 2.5 cm leaf stem) × 4 cm. **Flowers** sessile, borne in groups in leaf axils but usually only flowering singly, lavender to pink with strong 'herring-bone' veining in throat and on the lower lobe, upper lobes much reduced having similar stronger colouring. In dry weather, emits a **strong odour** if crushed.

Habitat: Found occasionally locally growing in light shade on islands and in the tree-line along the floodplain.

Flowering: Middle to end of the main rains.

Acanthaceae (Acanthus or spinyflower family)

Ruellia prostrata Poir.

Derivation: *Ruellia* for Jean Ruel, herbalist to François I of France; *prostratus*, prostrate [Latin].

Identification: A shade-loving **woody perennial**, c. 90 cm tall, short **hairs** over whole plant. **Stems** square. **Leaves** opposite and decussate, ovate with slightly serrated margins and net veining. **Flowers** with 5 pale lavender-blue lobes forming a short tube, c. 2 cm dia., short-lived, opening in the early morning. **Fruit** fiddle-shaped.

Habitat: Locally an uncommon plant growing in deep shade on islands and in the riverine forest.

Flowering: Early morning during the main rains.

Blundell p.395

Acanthaceae (Acanthus or spinyflower family)

Ruelliopsis setosa (Nees) C.B.Clarke

Derivation: *Ruell* for Jean Ruel, herbalist to François I of France; *opsis*, like, resembles [Greek], hence like *Ruellia*; *setosus*, full of bristles, bristly [Latin], referring to its hairy appearance.

Identification: Perennial creeping herb with a stoloniferous **rooting** system. **Stems** and **leaves** dark green, tinged with dark pink, covered with coarse **hairs**. **Leaves** opposite, linear, c. 13.5 × 0.2 cm, with smooth margins, tending to fold at the mid-vein. **Flowers** lavender blue, 5 petals fused to form a cone that has a darker blue centre, c. 2.5 cm dia., stalkless, borne at the leaf axils. Leafy **bracts** below the flowers.

Habitat: Found occasionally in sandy soils on the margin of *Combretum* woodland.

Flowering: During the cool dry period.

Asteraceae (daisy family)

Epaltes alata (Sond.) Steetz

Derivation: *Alata*, winged [Latin], referring to the stems of this species.

Identification: Grey-green, branching, erect, **annual herb**, up to c. 40 cm tall. **Stems** winged. **Leaves** linear, alternate and sessile, with slightly dentate margins and velvety hairs, c. 2.0 × 0.3 cm. Composite **flowers** borne in clusters on long stems from leaf axils, purple, c. 0.5 cm dia., florets discoid. Anthers whitish. The plant emits a strongly aromatic **scent** when crushed.

Habitat: Common and widespread in full sun on the floodplain and by seasonal pans.

Flowering: Throughout both the early and the main rains.

Asteraceae (daisy family)

Erlangea misera (Oliv. & Hiern) S.Moore

Derivation: *Erlangea* named in honour of the Bavarian University of Erlangen; *misera*, wretched or miserable [Latin], possibly referring to the small size of the flowers.

Identification: Erect branching herb, up to c. 75 cm tall (more usually c. 50 cm). **Stems** often tinged dark pink. **Leaves** alternate, narrowly oval, margins serrated, sessile or clasping the stems, c. 12.5 × 5 cm. **Flowers** composite, disc florets purple, no ray florets, c. 1 cm dia.

Habitat: Occasionally plentiful in limited areas of deep sand, especially on sandy ridges through the mopane woodland, but also on sandy banks at floodplain margins.

Flowering: Middle to end of the main rains.

FZ Vol.6 pt1

Asteraceae (daisy family)

Ethulia conyzoides L.f.

Common names: English blue weed, Carter's curse.

Derivation: *Ethulia*, possibly from *aithon*, meaning fiery or sparkling [Greek]; *conyzoides*, resembling the genus *Conyza*.

Identification: Slightly aromatic, erect, annual, **aquatic herb**, c. 50 cm tall. **Leaves** spiral up the dark red/brown stems, dark green, elliptical, margins toothed, folding at the mid-vein, curving downwards along their lengths. **Flowers** composite, pale lilac-grey, with only disc florets, in clusters at the apex of the plant.

Habitat: Locally uncommon in dried-out areas on the edges of waterways and lagoons. [Introduced.]

Flowering: Almost throughout the year.

Uses and beliefs:
• Used internally to treat roundworm, parasites and opthalmia.
• Used externally to relieve skin irritations.

FZ Vol.6 pt1, Ellery p.161.

Asteraceae (daisy family)

Laggera crispata (Vahl) Hepper & J.R.I.Wood

Derivation: *Laggera* for the Swiss physician and botanist Dr Franz Josef Lagger (1802–70); *crispa*, finely waved [Latin], possibly referring to the leaf margins.

Identification: Single-stemmed **perennial herb**, up to c. 2 m tall. The plant is **sticky** overall. **Leaves** oval, alternate, spiralling up the stem, c. 15 × 7.5 cm, with dentate margins that are decurrent, forming 'wings' the length of the stem. **Inflorescences** nodding composite heads of purple disc florets, borne in panicles from the upper leaf axils.

Habitat: Locally rare found growing in the shelter of acacia scrub in the floodplain.

Flowering: During the cool dry period.

Uses and beliefs:
- Seemingly palatable to insects but not to animals.

Asteraceae (daisy family)

Nicolasia pedunculata S.Moore

Derivation: *Nicolasia* for Dr Nicholas Edward Brown (1849–1934), of the Royal Botanic Gardens, Kew, plant taxonomist and authority on succulents; *pedunculata*, with a well-developed flower stalk or peduncle [Latin].

Identification: Small, rather lax, multi-branched, aromatic, **annual herb**, c. 20 cm tall. **Stems** pink. Stems and **leaves** covered in sparse matted **hairs**. **Leaves** are alternate, linear and stalkless, c. 5 × 0.7 cm, leaf margins wavy. **Flowers** purple, composite with no ray florets, c. 0.5 cm dia.; stigmas emerge well beyond the florets.

Habitat: Locally uncommon in damp well-grazed grassland near seasonal pans.

Flowering: During the cool dry period.

Asteraceae (daisy family)

Pechuel-loeschea leubnitziae (Kuntze) O.Hoffm.

Common names: Setswana mokodi, omotonanyana, motlalemetsi, mokojane; **Hmbukushu** dikori; **Kalanga** thitha; **Subiya** dixombombo; **Shiyeyi** mokode; **Afrikaans** bitterossie; **English** wild sage, stinkbush, sweat bush.

Derivations: *Pechuel-Loeschea*, named for the German naturalist and geographer, Eduard Pechuël-Loesche (1840–1913); *leubnitziae* for the maiden name of Pechuel-Loesche's wife. *Mokodi*, the generic name for aromatic members of the daisy family in Setswana. *Ômotonanyana*, the male one, referring to the fact that this is the largest of the wild sages. *Motlalemetsi*, the plant comes with the water.

Identification: Erect shrubby **perennial herb**, up to c. 180 cm tall. **Bark** pale brown or yellowish. **Leaves** grey-green, alternate, linear with smooth, decurrent margins, covered with velvety hairs on both surfaces, c. 7 × 1.5 cm. **Flowers** narrow composite heads of purple disc florets, borne in clusters at the leaf axils. The whole plant has a pungent camphor-like **smell** even without crushing.

Habitat: Widespread pioneer colonising floodplains as they dry out and gradually moving onto islands.

Flowering: During the cool dry period.

Uses and beliefs:
- Hmbukushu burn this plant on a metal sheet with hot coals in chicken sheds, effectively eradicating fleas for weeks.
- Batswana use the boiled roots as an antiseptic mouthwash for children with oral sores and as a hot poultice for toothache; the poultice is replaced when it cools until the toothache is gone.
- Wayeyi use a decoction from boiled roots to treat coughs; the roots are also used to improve the metabolic system and in, the form of a strong infusion, as an abortifacient.
- San use the root to treat *diphate*, a sexual disease that is passed to men from women when they have unprotected sex during menstruation or in the first days after a miscarriage. They smear the roots with animal fat, burn them and inhale the smoke. Wayeyi also mix the roots of *Pechuel-loeschea leubnitziae*, *Gloriosa superba* and *Ziziphus mucronata* to make an effective treatment for gonorrhoea.
- In the Okavango, the roots are chewed and the saliva swallowed to cure stomach pains, whereas a decoction of the whole plant is used to treat peptic ulcers.
- Wayeyi believe that if someone is sick, evil spirits can be kept away by the strong aromatic smell of this plant burning.
- The crushed leaves rubbed on the body make an excellent insect repellent. San use the same method to hide their scent from animals when hunting.
- There is also a parasite that grows on this plant (a kind of mistletoe) that brings good luck. Either a powder of the dried plants is mixed with vaseline and applied to the body or an infusion of this powder is used as a body wash.
- Hmbukushu use the stems to make shelters and fishing baskets, also for toilet paper. Kalanga use the stems to make brooms for sweeping.

Ellery p.128, Flowers Roodt p.49. BB, MM, RM, PN, KS, SS, BT, IM, MN.

Asteraceae (daisy family)

Pseudoconyza viscosa (Mill.) D'Arcy

Derivation: *Pseudo*, false [Greek], false conyza; *viscosa*, sticky, clammy [Latin], alluding to the stickiness of the plant.

Identification: Small erect **annual herb**, c. 30 cm tall, covered in short velvety glandular hairs. **Leaves** stalkless, obovate with deeply toothed margins in the upper section and deeply incised below, oily and sticky when touched. **Flowers** narrow, composite heads of purple disc florets. A strong eucalyptus perfume.

Habitat: An introduced plant spreading locally in dry lightly shaded areas on islands and in mixed woodland.

Flowering: Late in the main rains and into the cool dry period.

Germishuizen p.182.

Asteraceae (daisy family)

Vernonia anthelmintica (L.) Willd.

Common name: Shiyeyi xaa.

Derivation: *Vernonia* for the English botanist, William Vernon, who collected in Maryland in 1698; *anthelmintica*, acting against and expelling intestinal worms.

Identification: Erect **composite herb**, c. 80 cm tall, sparingly branched above. **Stems** ribbed and hairless. **Leaves** alternate, narrowly oval, narrowing to form a leaf stem, with sharply serrated margins, slightly hairy, c. 11 × 3.5 cm. **Flowers** purple, mainly disc florets and occasional ray florets, c. 1.5 cm dia. **Seeds** have a hairy pappus.

Habitat: Locally found rarely in deep shade in the riverine forest.

Flowering: Late in the main rains.

Uses and beliefs:
- Plants with the species name *anthelmintica* usually have an anthelmintic or worming action.

FZ Vol.6 pt1

Asteraceae (daisy family)

Vernonia glabra (Steetz) Vatke var. *laxa* (Steetz) Brenan

Common names: Setswana phiho, philo, phefo; **Shiyeyi** shôi?; **English** cornflower vernonia.

Derivations: *Vernonia* for the English botanist, William Vernon, who collected in Maryland in 1698; *glabra*, without hairs, glabrous [Latin]. ?*Shôi*, 'the parrot', probably referring to the bright flowers.

Identification: Vigorously growing, erect, **perennial herb**, up to c. 140 cm tall. **Leaves** blue-green, oval, alternate, margins have widely spaced teeth. Composite **flowers** brilliant cornflower-blue, borne in clusters at the apex of the stems; later in the season as the upper flowers fade, new clusters grow from the leaf axils. **Seeds** small and attached to a group of hairs, wind distributed.

Habitat: A common and widespread pioneer plant growing in the drying-out areas of the floodplain in full sun.

Flowering: Late in the main rains and into the cool dry period.

Uses and beliefs:
- The roots are used as an abortifactant.
- Ash from the leaves is rubbed onto burns as a disinfectant and healing agent.
- Wayeyi boil the roots and leaves, cover themselves with blankets and sit in the steam as a general tonic.
- In Khwai, a boiled infusion of the roots is used to treat gonorrhoea.
- In the Delta, the stalks of the plants are used to weave fish traps.

FZ Vol.6 pt1, Flowers Roodt p.57. BB.

Asteraceae (daisy family)

Vernonia poskeana Vatke & Hildebr. subsp. ***botswanica*** G.V.Pope

Common names: Setswana morethothomi, magoleng, njunjuripa.

Derivation: *Vernonia* for the English botanist, William Vernon, who collected in Maryland in 1698; *poskeana*, possibly derived from the name of Gustav Adolf Poscharsky (1832–1914), head gardener in Dresden; *botswanica*, from Botswana.

Identification: Erect much-branched **annual** or **short-lived perennial herb**, up to c. 80 cm tall, very thinly covered with **fine hairs**. **Leaves** silver green, stalkless, linear, margins entire, spiralling up the stem. **Flowers** composite, deep purple with only loosely held ray florets, no disc florets.

Habitat: Locally widespread in sandy areas of the floodplain.

Flowering: During the main rains.

B Van Wyk Flowers p.246, Germishuizen p.180, Turton p.117.

Campanulaceae (bluebell family)

Lobelia erinus L.

Common names: English wild lobelia, edging lobelia, garden lobelia.

Derivation: *Lobelia* for the Flemish physician and botanist, Matthias de l'Obel, (1538–1616); *erinus*, an ancient plant name from Dioscorides [Greek].

Identification: Delicate, erect, branching **annual herb**, up to c. 40 cm tall. **Leaves** linear with slightly dentate margins, c. 2.5 × 0.2 cm. **Flowers** pale blue with 5 lobes fused to form a tube with a white throat, upper 2 lobes are like small horns, the lower 3 enlarged with a darker blue median stripe.

Habitat: Locally uncommon in damp areas of the floodplain near water.

Flowering: During the main rains.

FZ Vol.7 pt1, B Van Wyk Flowers p.258, Germishuizen p.420.

Capparaceae (caper family)

Cleome hirta (Klotzsch) Oliv.

Common names: Setswana malomaarotho, malomaagwerothwe, wasiwa; **Subiya** mokazanmolotho; **Ovambenderu** omutjaitjai; **English** pretty lady.

Derivations: *Cleome* from Cleoma, a name used in the Dark Ages for a strong-tasting plant growing in a damp place; *hirta*, hairy [Latin]. *Malomaarotho* means 'the maternal uncle of Rothwe'. *Rothwe* is the name for *C. angustifolia* subsp. *gynandra* and *C. kalachariensis*. *Mokazanmolotho*, a beautiful girl. *Omutjaitjai*, if you take the pods and shake them you get the sound of the name.

Identification: Erect **annual herb** branching low down, usually c. 120 cm tall but up to c. 2 m in years of heavy rain. **Stems** and **leaves** covered in glandular **hairs**, aromatic, slightly sticky. **Leaves** alternate, palmate with 5–9 lobes which are linear with smooth margins; main leaves tend to fall as leaf-like bracts appear and spiral up the flower stems. **Flowers** in terminal racemes, c. 2 × 1.3 cm, with 4 erect petals; petals pink, with a yellow nectar guide, surrounded with a dark blue line on the central upper petals. There are rare white variants. **Fruit** cylindrical, slightly curved pods, oval in cross-section, ribbed, up to c. 14 cm long.

Habitat: Common and widespread in full sun on the floodplain and in the woodland margin.

Flowering: From the middle of the main rains into the cool dry period.

Uses and beliefs:
• In the area of Lake Ngami, the dry stems are frequently used to fence around the houses.

FZ Vol.1 pt1, Blundell p.26, Roodt p.61, Turton p.99, WFNSA p.144. BB, BN.

Capparaceae (caper family)

Cleome monophylla L.

Common names: Setswana chinyevana; **Afrikaans** rusperbossie; **English** single-leaved cleome, spindlepod, spider flower.

Derivation: *Cleome* from Cleoma, a name used in the Dark Ages for a strong-tasting plant growing in a damp place; *mono*, single, one [Greek], *phyllon*, leaf [Greek], one-leafed, alluding to the leafy bracts of the inflorescence.

Identification: Erect, branching, **annual herb**, up to c. 130 cm tall. Both the **stems** and the **leaves** covered in fine short glandular **hairs** that make the plant slightly sticky. **Leaves** cordate, up to c. 11 cm long, alternate, with smooth margins. **Flowers** borne in long terminal racemes, asymmetrical, c. 1 cm dia., with 4 erect petals, petals open pink and turn blue on aging. Some flowers have a yellow nectar guide defined by a dark blue line on the 2 upper petals. Below each flower there is a leaf-like, stalkless, **bract** that is similar in shape to the leaves. These are interspersed with occasional long hairs. **Fruit** straight cylindrical pods, narrowing to a point, c. 10 cm long.

Habitat: An annual weed in many areas but locally rare, only occurring in years of heavy rainfall in damp areas of mopane woodland.

Flowering: During the main rains.

Uses and beliefs:
- The leaves are edible and the seeds can be used for making mustard.
- The caterpillars of *Belenois aurota* feed on this family of plants.

FZ Vol.1 pt1, B Van Wyk New p.198, Blundell p.26, Botweeds p.34, Germishuizen p.35, Turton p.95.

Capparaceae (caper family)

Cleome rubella Burch.

Common names: Setswana malomaarothe; **Afrikaans** mooinooiientjie, wilde-bos-ganna; **English** pretty lady.

Derivations: *Cleome* from Cleoma, a name used in the Dark Ages for a strong-tasting plant growing in a damp place; *rubella*, pale red, becoming red [Latin]. *Malomaarotho* means 'the maternal uncle of Rotho'. *Rotho* is the name for *C. angustifolia* subsp. *gynandra* and *C. kalachariensis*.

Identification: A much-branching **annual herb**, growing up to c. 50 cm tall. **Stems** and **leaves** are glandular and sticky. **Leaves** palmately compound with fine needle-like linear leaflets. **Flowers** pink to mauve with 4 upright petals with basal claws; stamens curved. **Fruit** are long narrow tubes that are strongly **aromatic** when crushed. Easily confused with *Cleome hirta* but much smaller overall and lacking the yellow markings on the petals.

Habitat: Uncommon locally growing in full sun on the floodplain.

Flowering: During the main rains.

FZ Vol.1 pt1, B Van Wyk Flowers p.198, Germishuizen p.76, WFNSA p.144.

Cappaaceae (Spiderwort family)

Commelina benghalensis L.
(This family is currently under revision and consequently these scientific names may change.)

Common names: Setswana mafavuke; **Afrikaans** blouselblommetjie; **English** Benghal commelina, wandering jew.

Derivation: *Commelina* for the Dutch botanists Johan Commelijn (1629–92) and his nephew Caspar (1667–1731); *benghalensis*, of Benghal.

Identification: A rather lax, slender, branching **annual**, up to c. 75 cm tall, depending on support. Perennial **rooting system**. Broken **stems** produce a **watery sap**. Stem, flowers and spathes covered in hairs. **Leaves** velvety, c. 5.5 × 2.5 cm; **leaf sheath** striped with red and has a rim of long hairs. **Flowers** brilliant blue, c. 2 × 1.5 cm, borne at leaf axils, protruding from folded hairy **spathes**, 3 sepals, 2 petals with long claws and 1 petal that is deeply divided into a false stigma. Flowers open only in the morning.

Habitat: Locally uncommon in damp areas of the floodplain, especially the bottom of dry channels.

Flowering: During the main rains.

B Van Wyk Flowers p.248, Botweeds p.114, WFNSA p.32.

Commeliniaceae (Spiderwort family)

Commelina diffusa Burm.f.
(This family is currently under revision and consequently these scientific names may change.)

Common name: Setswana kgopo.

Derivation: *Commelina* for the Dutch botanists Johan Commelijn (1629–92) and his nephew Caspar (1667–1731); *diffusa*, widely spreading [Latin], alluding to the arrangement of the leaves.

Identification: Scrambling marginal **herb** more than c. 1 m long, sheltering in reeds and papyrus. **Stems** hairless. **Leaves** hairy, c. 7 × 1.8 cm; leaf sheaths around the stem have a hairy rim. Up to 3 mid-blue **flowers** emerge from each leafy bract, c. 1 × 2.5 cm, each flower has 3 overlapping petals. Flowers only open in the morning.

Habitat: Locally common growing in the margin of fast-flowing channels.

Flowering: During the main rains.

Ellery p.157

Commeliniaceae (Spiderwort family)

Commelina forsskalii Vahl

(This family is currently under revision and consequently these scientific names may change.)

Derivation: *Commelina* for the Dutch botanists Johan Commelijn (1629–92) and his nephew Caspar (1667–1731); *forsskalii* for Per Forskål (1737–63), a Swedish botanist.

Identification: A low-growing **annual** from a perennial **rooting system**, forming mats c. 1 m across. **Roots** grow from the leaf nodes when they touch the ground. **Leaves** c. 3 × 1.3 cm, smooth margins, **leaf sheath** with a rim of long hairs on its upper edge. **Flowers** mid-blue, c. 1.8 cm long, borne in pairs from **leafy bracts** in which water is present, 3 sepals, 2 petals with long claws and 1 that is deeply divided into 2 false stigmas. Flowers open only in the morning.

Habitat: Common and widespread in full sun in sandy areas of the floodplain and mopane woodland.

Flowering: During the main rains.

Commeliniaceae (spiderwort family)

Commelina macrospatha Gilg & Ledermann ex Mildbr.

(This family is currently under revision and consequently these scientific names may change.)

Derivation: *Commelina* for the Dutch botanists Johan Commelijn (1629–92) and his nephew Caspar (1667–1731); *macro*, long or big; *spatha*, spathe or blade [Latin], long-spathed.

Identification: Lax herb of the water margin with a network of smooth **hairless stems**, c. 1 m long, creeping through the reeds. **Leaves** hairy, lanceolate, with wavy margins, c. 9 × 1 cm; leaf sheath has a hairy rim. **Flowers** mid-blue, c. 5.5 cm long, 1–2 emerging from a long spathe; 2 upper petals on short claws, the third petal is smaller and curved. Flowers open only in the morning.

Habitat: Locally common growing in the margin of fast-flowing channels.

Flowering: During the main rains.

Commeliniaceae (spiderwort family)

Commelina petersii Hassk.

(This family is currently under revision and consequently these scientific names may change.)

Derivation: *Commelina* for the Dutch botanists Johan Commelijn (1629–92) and his nephew Caspar (1667–1731); *petersii* in honour of the 19th century German botanist Prof. W. Peters.

Identification: An erect **annual herb**, c. 75 cm tall, from a **perennial rooting stock**. **Leaves** c. 4 × 14 cm long. **Flowers** sparse, pale blue, each having 2 petals with long claws; flowers emerge from **spathes**, c. 1 × 0.5 cm, which are borne at the leaf axils and at the apex of the stems. Flowers open only in the morning.

Habitat: Found occasionally locally on lightly shaded islands in the floodplain.

Flowering: During the main rains.

Commeliniaceae (spiderwort family)

Cyanotis foecunda DC. ex Hassk.

Common name: English blue powder puff.

Derivation: *Kyanos*, blue [Greek], *otion*, a little ear [Greek], referring to the colour and form of the petals; *foecunda*, fruitful, fertile [Latin].

Identification: A rather lax straggling **annual herb**, c. 30 cm tall. **Leaves** alternate, sessile, narrowly ovate, c. 5 × 1 cm, covered in fine hairs, leaf margins entire and decurrent forming a stem sheath. **Flowers** in the many leaf axils of the plant, intense purple-blue, c. 1 cm dia., 3 petals, 6 stamens with yellow anthers and a furry blue outer sheath.

Habitat: Widespread in years of high rainfall in sandy areas of woodland, especially mopane.

Flowering: During the main rains.

Blundell p.414.

Convolvulaceae (morning glory family)

Evolvulus alsinoides (L.) L.

Common names: Setswana molemowanonyane, masupegane, mosupogane; **English** blue haze, wild evolvulus.

Derivations: *Evolvo*, unravel or unwind [Latin], referring to the fact that the plant does not twine; *alsinioides*, resembling the genus *Alsine* (*Caryophyllaceae*). *Molemowanonyane*, medicine for birds.

Identification: Small branching **semi-perennial herb**, c. 25 cm tall. **Stems** and **leaves** usually covered in long silky white **hairs**. **Leaves** alternate, sessile, narrowly oval or linear, c. 2 × 0.8 cm. **Flower stalk** c. 1.5 cm long, jointed c. 3 mm below the flower where there are 2 minute bracts. **Flowers** borne at the leaf axils, tubular with 5 lobes, sky blue with white throat and stamens, c. 7 mm dia.

Habitat: Widespread on sandy areas of the floodplain in full sun.

Flowering: From the middle of the main rains into the cool dry period.

Uses and beliefs:
• Batswana mix the dried leaves with vaseline and use the resulting jelly as a treatment for the fontanelles (Setswana, *phogwana*) of babies.
• It is also used as incense and hung over shop doors to attract customers.

FZ Vol.8 pt1, Botweeds p.40, B van Wyk New p.250, Hargreaves p.57, Pooley p.468, Turton p.81. BB.

Convolvulaceae (morning glory family)

Ipomoea nil (L.) Roth

Derivation: *Ips*, worm [Greek], *homoios*, like [Greek], referring to the trailing habit.

Identification: An annual **herbaceous climber**, up to c. 1.5 m tall, with twining coarsely hairy **stems**. **Leaves** palmate, trilobed, with a heart-shaped base, covered with rather **flattened hairs** on both surfaces. **Flowers** large, pale blue with a white corolla tube; sepals linear and rather uneven, c. 2 cm, very hairy towards the base; 2 bracts immediately below the calyx, these bracts are also hairy where they join the plant; flowers open overnight and are already closing at 10.00 hrs.

Habitat: Locally rare, found only in years of heavy rainfall, in deep shade in the riverine forest.

Flowering: Late in the main rains.

FZ Vol.8 pt1.

Convolvulaceae (morning glory family)

Jacquemontia tamnifolia (L.) Griseb.

Derivation: *Jacquemontia* for Victor Jacquemont (1801–32), a French naturalist and traveller; *tamnus*, the Latin name for a family of climbing plants; *folius*, leaved [Latin], referring to the shape of the leaves which are similar to those of genus *Tamnus*.

Identification: A scrambling, branching, twining, **annual herb** found in grass. **Stems** up to c. 70 cm long, no **tendrils** or **latex** present, **hairs** usually white. **Leaves** on long stems, alternate, heart-shaped with wavy margins. **Flowers** borne in hairy clusters supported by bract-like leaves, on long stems from the leaf axils; white with pale pink blush or sky-blue, slightly translucent; corolla tube slightly star-shaped.

Habitat: Widespread on sandy areas of the floodplain and into the mopane woodland.

Flowering: Throughout the main rains.

Uses and beliefs:
• Subiya use the stems to tie bundles of firewood and sugar cane.

FZ Vol.8 pt1. GM.

Fabaceae (pod-bearing family), Papilionoideae (pea sub-family)

Philenoptera nelsii (Schinz) Schrire subsp. *nelsii*
(syn. *Lonchocarpus nelsii* (Schinz) Schinz)

Common names: Setswana mohata, mohatla, mogata, moponda, mopanda, mhata; **Lozi** mukololo; **Afrikaans** Kalahari-appelblaar; **English** Kalahari apple-leaf, apple-leaf lance-pod.

Derivation: *Phileo*, to love [Greek], *ptera*, winged [Greek]; *nelsii* for L. Nels a German plant collector in the 1880s.

Identification: Deciduous bushy tree with a dense crown, often c. 3 m but up to c. 10 m tall. **Bark** yellow and grey with occasional cracking and peeling and longitudinal striations. Flowering shoots are covered in a white down of **hairs**. **Leaves** usually compound with 1–2 pairs of leaflets and a terminal leaflet, although more often in this area the leaf is reduced to a single oblong leaflet; margins smooth; upper surface of the leaves puckered or quilted whereas the underside has conspicuous **net-veining**; young leaves often densely velvety, becoming dark green and leathery as they age; younger, smaller trees retain their leaves until the new ones come but do not flower. **Flowers** borne in panicles that appear in profusion before the leaves; like the flowers of pea, lilac-purple with a yellow throat. There is an uncommon white variant. Easily confused with *P. violacea* but *P. nelsii* produces flowers before leaves whereas *P. violacea* produces new leaves and flowers at the same time.

Habitat: Widespread throughout the region in deep sand.

Flowering: During the cool dry period.

Uses and beliefs:
- The leaves are excellent fodder.
- The wood is tough and flexible and does not break or shatter easily. In the past, it was often used for wagon wheels and for bearings in agricultural implements.
- Because the wood is very light, it is occasionally used as firewood.

B & P Wyk ZA Trees p.196, Curtis & Mannheimer p.242, Ellery p.125, Palgrave p.327, Setshogo & Venter p.96, Tree Roodt p.59.

Fabaceae (pod-bearing family), Papilionoideae (pea sub-family)

Philenoptera violacea (Klotzsch) Schrire
(syn. *Lonchocarpus capassa* Rolfe)

Common names: Setswana mopororo, mohata, mopanda, mapanda, mokololo, upanda, cawaq, ungqo, mhatla, mhata; **Hmbukushu** mokororo; **Lozi** mukololo; **Shiyeyi** uwara; **Afrikaans** appelblaar; **English** rain tree, apple-leaf tree, lance tree.

Derivations: *Phileo*, to love [Greek], *ptera*, winged [Greek], *violacea*, violet-coloured [Latin]. *Rain tree*, the trees were given this common name because at a certain season they appear to rain! This phenomenon is actually due to the sap-sucking frog-hopper bug *Ptylus grossus*, which takes sugar from the sap and excretes almost pure water, which falls like rain drops.

Identification: Medium-sized deciduous **tree**, c. 10 m tall. **Bark** is grey, fissured and peeling to a light brown below. New wood and flowers covered in **short fine hairs**. **Leaves** alternate, compound with 2–3 pairs of oval leaflets and a larger terminal leaflet, leaf margins slightly wavy; upper surface shiny, lower surface matt and a paler blue-green; net-veined. **Stipules** in pairs at the leaf axils. **Flower buds** dark blue brown with dark grey-brown hairs. **Flowers** in terminal panicles amongst the new growth of leaves, like those of pea, deep purple with white marking on the throat. **Fruit** flat papery pods, creamy-coloured. Easily confused with *P. nelsii* but *P. violacea* produces new leaves and flowers at the same time whereas *P. nelsii* produces flowers before leaves.

Habitat: Widespread and common throughout the region.

Flowering: Late in the cool dry period.

Uses and beliefs:
• Hmbukushu, Wayeyi and Subiya use the wood to make mekoro (dug-out canoes). The wood is yellow to brown in colour and fairly hard. It is used for carving tool handles and small decorative items.
• Hmbukushu pound the roots and throw them into water as a fish poison.
• The trees are very popular with elephant who 'farm' them by browsing them rather as they do mopane.
• The caterpillars of *Charaxes bohemani* and *Coeliades forestan* feed on the foliage.

B & P van Wyk Trees p.458, Curtis & Mannheimer p.244, Ellery p.124, Hargreaves p.30, P Van Wyk Kr Trees p.96, Palgrave p.325, Setshogo & Venter p.96, Tree Roodt p.55, WFNSA p.174. TK, MM, RM, PN.

Lamiaceae (mint family)

Ocimum americanum L. var. *pilosum* (Willd.) A.J.Paton
(syn. *Ocimum canum* Sims)

Common names: Setswana sesunkwane, mogatololo, sebeditona, badingwana, mosukujane; **English** wild basil, hoary basil.

Derivation: *Ocimum*, from *okimon*, the Greek for an aromatic herb; *americanum*, from the Americas.

Identification: Erect **pungent branching herb** up to c. 1 m tall in rainy years. **Stems** buttressed and appearing square. **Leaves** opposite, ovate, curving downwards so that they tend to fold in half at the mid-vein, leaf margins have occasional serrations; leaves dotted with glands that emit a strongly aromatic **perfume** when crushed, a key identifying feature. **Flowers** in long spikes, borne symmetrically in groups around the stem, pale lavender; they emerge between a pair of rounded **sepals**, which persist after the flower has fallen and dry on the plant. **Seeds** 4 nutlets within the calyx.

Habitat: Common and widespread in dry sandy areas in full sun.

Flowering: Throughout the rains and into the cool dry period.

Uses and beliefs:
- The leaves can be dried and used as a substitute for tea. A medicinal tea made by pouring simmering water over the leaves and seed spikes and boiling them for a few minutes is administered at regular intervals to treat colds and fevers.
- Widely used a substitute for basil, especially when barbecuing meat.
- In the Delta, the dried leaves are burnt on hot coals and the smoke inhaled to relieve chesty coughs and asthma.
- When rubbed on the body or hung in the house, the fresh leaves are an efficient insect repellent.
- The powdered dry leaves are used by the San as a deodorant.
- When a person dies, Wayeyi use these plants to make a bed under the body to help keep it fresh until the relatives can assemble for the funeral.
- Larvae of the butterfly *Lepidochrysops van soni* feed on the plant.

Botweeds p.62, Flowers Roodt p.101, Turton p.95. NM, SS.

Nymphaeaceae (water lily family)

Nymphaea nouchali Burm.f.
var. ***caerulea*** (Savigny) Verdc.
(syn. *Nymphaea capensis* Thunb.)

Common names: Setswana makungara, tswii, modidmo, dimhwa, diviya; **Hmbukushu** mahwaa; **Subiya** isiko; **Shiyeyi** sekhumba; **Afrikaans** blouwaterlelie; **English** blue water lily, blue lotus.

Derivations: *Nymphaea* for Nymphe, a water nymph of Greek mythology; *nouchali*, possibly from a wrongly translated English plant name; *caerulea*, blue, the intense blue of the sky at mid-day [Latin]. *Tswii* is the root. *Makungara* is the seed pod. *Sekhumba*, the Seyei word also means the seed pod of the lily.

Identification: Perennial aquatic herb with a submerged **rhizomatous rooting** system. **Leaves** float on the surface of water and are almost round with a narrow v-shaped base, margins smooth, conspicuous veins on the lower surface. **Flowers** floating on the surface of water are large and blue, fading to white as they mature; stamens yellow. The plants easily adapt to changes in water level as they are able to grow very rapidly, as much as a few metres in 24 hours. They quickly colonise newly flooded areas that have been dry for many years, either from seed or from rhizomes lying dormant in the soil.

Habitat: Common and widespread throughout the slow-flowing and still waterways of the Delta.

Flowering: Throughout the year, although the flowers are more plentiful during rainy periods.

Uses and beliefs:
• The plants have a wide variety of uses throughout Africa, but in northern Botswana the tubers are an important element of the diet of local people before and during the rains. Hmbukushu take the thick part of the roots, slice them and either boil or roast them with meat to season and tenderise the meat. The flower is used as a spice and the seeds are also edible.

FZ Vol.1 pt1, Blundell p.20, Ellery p.145, Flowers Roodt p.119, WFNSA p.138. PK, GM, MM, PN.

Rubiaceae (coffee family)

Pentodon pentandrus (Schumach. & Thonn.) Vatke

Derivation: *Penta*, five [Greek], *odontos*, tooth [Greek], referring to the persistent spikes of the calyx on the fruit; *pentandrus*, with 5 stamens.

Identification: Delicate straggly **marginal plant** up to c. 70 cm tall, often found with its **roots** in water. **Leaves** sheathing the stem, opposite, decussate, narrowly elliptical to lanceolate with smooth margins sometimes slightly revolute, c. 40 cm long, bright green above and paler green below with a conspicuous **mid-vein**. **Stipules** are pairs of fine strong hairs at the axils. **Flowers** borne in racemes at the leaf axils, minute, light blue, 5 lobes joined to form a short tube, green towards the calyx; within the flower a ring of hairs. **Fruit** with 5 calyx lobes remaining as small spikes on the rim.

Habitat: Widespread growing in boggy areas on the margin of channels and lagoons.

Flowering: Throughout the rains.

FZ Vol.5 pt1.

Scrophulariaceae (snapdragon or foxglove family)

Aptosimum decumbens Schinz

Derivation: *Aptosimum*, not falling off, persistent [Greek], referring to the leaves that do not fall off when the plant dies; *decumbens*, from the Latin meaning a prostrate plant with the tip of the shoot pointing upwards.

Identification: A mat-forming hemi-parasitic **perennial creeping herb**. **Stems** c. 40 cm long. **Leaves** held rather erect, stalkless, opposite and elliptical, c. 2.5 × 0.8 cm, with smooth margins. **Flower** 5 brilliant blue lobes, c. 1.5 cm dia., fused to form a white corolla tube, c. 1.5 cm long.

Habitat: Found occasionally locally in full sun in open sandy areas on island margins and near pans.

Flowering: If there has been sufficient rain, the plants flower throughout the year.

FZ Vol.8 pt2

Scrophulariaceae (snapdragon or foxglove family)

Craterostigma plantagineum Hochst.

Derivation: *Krater*, a bowl [Greek], *stigma*, stigma [Greek], a bowl or cup-shaped stigma; *plantagineum* refers to the shape and arrangement of the leaves 'like (European) plantain species'.

Identification: Small erect, **perennial herb**, c. 6 cm tall, with a **rhizomatous rooting** system. **Roots** a distinctive brownish-orange. **Leaves** arranged in a basal rosette, broadly elliptical, hairy margins; leaf undersides hairy, occasional bristles on the upper surface. **Flowers** borne in groups of 2–3 on an erect stem, blue and white, with two yellow spots in the throat.

Habitat: Locally rare in damp clayey areas of the mopane woodland.

Flowering: During the main rains.

Uses and beliefs:
• One of the so-called 'resurrection plants': from appearing completely dead, it will put out new green shoots and flowers within hours of rain falling.

FZ Vol.8 pt2, Blundell p.375, Germishuizen p.149, Turton p.53, WFNSA p.366.

Solanaceae (potato family)

Solanum panduriforme Drège ex Dun.

Common names: Setswana morolwane, morolwana, tholwana, tholwane, tholwanakgomo, thontholwana, thulwathulwane, nonura, thola, rutunguza, ntotoba, morola?, hqang'arci, nunukwa; **Hmbukushu** nunura; **Kalanga** thungulu; **Shiyeyi** ntotoba; **Subiya** ntuntulwa; **Afrikaans** gifappel, bitterappel, geelappel; **English** poison apple, bitter apple, Sodom apple, grey bitter apple, inkberry.

Derivation: *Solanum*, the ancient Roman name for one plant in this family, probably this one as it is widespread in Europe, may derive from the Latin *solamen*, which refers to the soothing or narcotic properties of some species; *panduraeformis*, Latin for 'fiddle-shaped', possibly referring to the shape of the leaves on some plants.

Identification: Erect branching **shrubby herb**, c. 80 cm tall. **Leaves** and **stems** hairy, often armed with prickles. Leaves c. 8 cm (inc. c. 1 cm leaf stem) × 2 cm. **Flowers** vary from a brilliant lavender to a soft white with bright yellow anthers, occasional maroon veining, c. 2.5 cm dia. **Fruit** c. 3 cm dia., believed to be highly poisonous, especially when green with lighter markings, becoming bright yellow when ripe.

Habitat: Widespread and common in full sun on the floodplain and in partial shade on islands and in the tree-line.

Flowering: Main rains with sporadic flowering through to the end of the cool dry period.

Uses and beliefs:
- Animals eat the fruit but they are poisonous to humans.
- Subiya burn the fruit and rub the ash on young children's stomachs as a cure for diarrhoea.
- Wayeyi and the Subiya cut the fruit in half and rub it on the scalp to cure rashes and *Mbangash* (Shiyeyi for ringworm).
- In the Okavango Delta, a decoction of the root is applied externally to treat boils and ulcers.
- In Khwai Village, an infusion of the roots is used as a cough mixture. In other parts of Botswana, an incision is made in the chest and the roots are rubbed into the wound to treat breathing difficulties.

B Van Wyk Flowers p.264, Blundell p.190, Ellery p.175, Flowers Roodt p.153, Germishuizen p.172, Hargreaves p.57. TK, CM, GM, BN, SS.

Solanaceae (potato family)

Solanum tettense Klotzsch
var. *renschii* (Vatke) R.E.Gonçalves

Common name: Setswana morolwana.

Derivation: *Solanum* may derive from the Latin '*solamen*', referring to the soothing or narcotic properties of some species; *tettense*, of Tete in Mozambique.

Identification: A strongly growing **annual herb** up to c. 120 cm tall, covered in a tangle of short white **hairs**. **Leaves** alternate, elliptical, asymmetrical at the base, margins wavy, star-shaped **hairs** on leaf surfaces. **Flowers** borne in clusters on the stems opposite the leaf axils; small, mauve, c. 1.2 cm dia., 4–5 recurved twisting narrow lobes, bright yellow protruding anthers. **Fruit** small, spherical, ripening to bright red.

Habitat: Locally uncommon, found growing in medium shade in acacia scrub.

Flowering: During the main rains.

Uses and beliefs:
• Used to treat stomach cramps.

Velloziaceae (blackstick lily family)

Xerophyta humilis (Baker) & Schinz (syn. *Vellozia humilis* Baker)

Derivation: *Xeros*, dry, *phyton*, a plant [Greek], a plant that thrives in dry areas; *humilis*, low-growing [Latin].

Identification: A small c. 8 cm tall, mat-forming, rhizomatous **perennial**. **Rhizomes** are loosely attached to each other and have a heavy rooting system. **Leaves** linear with a heavy central vein at which the leaf folds. **Flowers** pinkish blue, star-like, 6 petals with pointed tips. Stamens yellow, opening to form an inner star as they mature.

Habitat: Locally uncommon in damp clayey areas of mopane woodland.

Flowering: Late in the main rains.

B Van Wyk New p.90, Germishuizen p.155, WFNSA p.80.

Green
or greenish flowers

Amaranthaceae (pigweed or cockscomb family)

Achyranthes aspera L.
var. *sicula* L.

Common names: Setswana molora, motshwarakgano; **Shiyeyi** motweng; **San** xuase or xuamudsa (central Kalahari), gwee (Nxai pan); **English** sickle-fruit, chaff flower.

Derivations: *Achyron*, chaff or husk [Greek], *anthos*, flower [Greek], referring to the dry chaffy nature of the flowers; *aspera*, rough or harsh [Latin], again referring to the flowers. *Molora*, ash or soap. *Motshwarakgano*, the one that catches the slender mongoose. *Motweng*, snuff. *Gwee*, a cow.

Identification: Erect branching **herbaceous annual**, up to c. 50 cm tall. **Stems** square, tinged pink to just below inflorescence, **axils** red. **Leaves** opposite, decussate, ovate, narrowing at the base to form the leaf stalk, c. 8 cm (inc. leaf stalk c. 1.5 cm) × 3.5 cm, margins smooth. **Flowers** in terminal spikes, silver-green, star-shaped, less than 5 mm dia. **Fruit** hooked, and attach themselves to animal fur to aid dispersal.

Achyranthes aspera var. *sicula* may easily be mistaken for *A. aspera* var. *pubescens*, especially in young growth. Great care must be taken if the plant is to be used for medicine or food as var. *pubescens* is highly poisonous. The flowers of var. *sicula* are more widely spaced and less hairy than those of var. *pubescens*. Var. *sicula* usually has green flowers whereas var. *pubescens* has pink flowers. Var. *pubescens* is covered in a white down overall whereas var. *sicula* has sparser hairs.

Habitat: Widespread in shady areas of the riverine forest and tree-line. [Introduced.]

Flowering: From the middle of the main rains into the cool dry period.

Uses and beliefs:
- Leaves are eaten as *morogo*, a vegetable relish.
- The ash of the plant is used to replace salt.
- The ash of the plant is widely used to make snuff. The leaves, stems and flowers are burnt and the ash is ground and mixed with ground tobacco and a little water. San of the central Kalahari use it as a chewing wad like tobacco. They place it behind the lower lip.
- The leaves and roots of *Achyranthes aspera* have many medicinal uses, from preventing miscarriage to stopping wounds bleeding and curing constipation, headaches and toothache. Batswana mix the leaves and roots with *Asparagus africanus* and *Gomphocarpus fruticosus* to improve the metabolic system.

FZ Vol.9 pt 1, Ellery p.153, Flowers Roodt p.13, Blundell p.41, Germishuizen p.73. BB, PN.

Amaranthaceae (pigweed or cockscomb family)

Alternanthera pungens Kunth

Common names: Setswana sepodise; **Afrikaans** kakiedubbeltjie; **English** paper thorn, burweed, khaki weed.

Derivation: *Alternus*, alternate, *anthera*, the anther (part of the stamen) [Latin], referring to alternating sterile and fertile stamens; *pungens*, sharp-pointed [Latin].

Identification: A mat-forming **creeping annual** with stems up to c. 50 cm long. **Stems** tinged with red, **rooting** at each leaf axil as they touch the soil. **Leaves** opposite, oval, with smooth margins, base of the leaf blade is oblique and narrowing to form the leaf stem, up to c. 4 × 2 cm. **Flowers** stalkless and often borne in pairs at the leaf axils, unpleasant prickly balls, c. 8 mm dia., composed of spiny bracts with the flowering parts within. Spiny bracts do not fall until the seeds drop. **Fruit** easily transferred as they stick into feet and shoes.

Habitat: Common and widespread in full sun along roadsides, on waste ground and on overgrazed areas; appears to thrive on being trodden, being generally on compacted sandy soil. [Introduced and invasive.]

Flowering: From the middle of the main rains into the cool dry period.

FZ Vol.9 pt1, Pooley p.528.

Amaranthaceae (pigweed or cockscomb family)

Amaranthus graecizans L.

Common names: Setswana thepe; **San** xhaa; **Hmbukushu** rwedthi.

Derivation: *Amaranthus* from the Greek *amarantos*, unfading, as the flowers of many plants in this genus retain their colour upon drying; *graecus*, Greek, Grecian. *Rwedthi* and *xhaa* mean *morogo*, the vegetable relish eaten with mealie.

Identification: An erect **herb**, branching at base into **stems**, c. 70 cm tall. **Leaves** spathulate and hairless, c. 1.5 × 0.5 cm, leaf margins smooth and sometimes slightly wavy. **Flowers** in clusters on short lateral spikes spiralling up the stems, green and chaff-like.

Habitat: Locally common on the margins of seasonal channels and rivers.

Flowering: During the main rains.

Uses and beliefs:
• Hmbukushu and San collect the leaves and boil them, adding oil and salt to make *morogo* (*rwedthi* in Hmbukushu or *xhaa* in San), a vegetable relish eaten with mealie or porridge. It is considered to be very good to eat.

FZ Vol.9 pt1. MM, RR.

Amaranthaceae (pigweed or cockscomb family)

Amaranthus hybridus L.

Common names: Setswana kgato, mogato, setlepetlepe, rurithi, mojatangombe; **Afrikaans** misbredie; **English** cock's comb, hell's curse, panicled amaranth, common pigweed, prince's feather, red amaranth.

Derivation: *Amaranthus* from *amarantos*, unfading [Greek], as the flowers of many plants in this genus retain their colour upon drying; *hybridus*, mixed, hybrid [Latin].

Identification: An erect **annual herb** up to c. 1 m tall, **hairless**. Veins of the leaves and stems often tinged magenta, the whole plant may occasionally be magenta coloured. **Stems** ribbed. **Leaves** oval with long stems, opposite and decussate, margins entire. **Flowers** form long spikes at the apex of the stems and at leaf axils; small, bristly, green and white.

Habitat: Found occasionally in shaded areas of the river bank. [Introduced]

Flowering: During the main rains.

Uses and beliefs:
• The leaves of this plant are used to make *morogo*, a vegetable relish for mealie.

FZ Vol.9 pt1, B Van Wyk New p.286, Turton p.107, Blundell p.41.

Amaranthaceae (pigweed or cockscomb family)

Amaranthus praetermissus Brenan

Derivation: *Amaranthus* from *amarantos*, unfading [Greek], as the flowers of many plants in this genus retain their colour upon drying; *praetermissus*, neglected, omitted, overlooked [Latin].

Identification: An erect branching **annual herb**, often as little as 20 cm tall but may reach c. 1 m. **Leaves** narrowly oval with a long stalk, folding at the midvein, slightly hairy, uneven margin, occasionally with a hair at the apex. **Flowers** vestigial, growing in groups in leaf axils up the stems, sessile, pinkish; anthers yellow.

Habitat: Found occasionally in light shade on islands in the floodplain.

Flowering: Late in the main rains.

FZ Vol.9 pt1.

Amaranthaceae (pigweed or cockscomb family)

Amaranthus viridis L.

Common names: English waterleaf.

Derivation: *Amaranthus* from *amarantos*, unfading [Greek], some flowers of this genus retain their colour for a long time; *viridis*, green [Latin].

Identification: An erect branching **annual** or **short-lived perennial herb**, c. 50 cm tall. **Stems** with decurrent buttresses from the leaf axils, often suffused with pink in the lower parts of the plant. **Leaves** alternate and deltoid, margins slightly wavy, leaf blades hairless and borne on stems longer than the blade. Small green **flowers** borne in branching spikes at the apex of the stems.

Habitat: Locally uncommon in deep shade in riverine forest. Common and widespread elsewhere.

Flowering: During the main rains.

FZ Vol.9 pt1.

Amaranthaceae (pigweed or cockscomb family)

Cyathula orthocantha (Aschers.) Schinz

Common names: Setswana bogoma; **San** tsakokhee; **English** fertility plant, cyathula.

Derivations: The meaning of *cyathula* is unclear. It may come from *cyathus*, cup or ladle [Greek]; *ortho*, straight, *cantha*, thorn [Greek], referring to the straight spiky bracts surrounding the flowers. *Bogoma*, a generic name for plants of which parts readily attach themselves to passing animals.

Identification: Vigorous branching **annual herb**, up to c. 130 cm in a year of high rainfall. **Stem** ribbed and blotched with red. **New growth** particularly hairy at the leaf axils. **Leaves** opposite, decussate, ovate, up to c. 17 cm (inc. petiole c. 1.5 cm) × 7 cm; leaf margins smooth but margins and the upper leaf surface are covered in long harsh **hairs**, lower surface has hairs along the prominent veins; there is net veining. **Inflorescence** a spike of silver-green spiky bracts surrounding the flowers at the apex of the stems.

Habitat: Widespread, growing in light shade on islands and in the tree-line along the floodplain.

Flowering: From the middle of the main rains into the cool dry period.

Uses and beliefs:
• Batswana men drink an infusion of the leaves and stems to clear the veins when they have sexual problems.
• Batswana mix *Cyathula orthacantha* and *Asparagus africanus* to neutralise snake bites and the stings of scorpions and spiders.
• San use the stems to build roof structures and for fencing.

FZ Vol.9 pt1, Flowers Roodt p.21. BB.

Amaranthaceae (pigweed or cockscomb family)

Pupalia lappacea (L.) A.Juss
var. *velutina* (Moq.) Hook.f.

Common names: Setswana bomama; **Setswana** and **Shiyeyi** bogoma, tedutsabaana; **Subiya** imboke; **English** sweethearts.

Derivation: *Lappaceus*, burr-like [Latin]. *Tedutsabaana*, man's beard. *Bogoma*, a plant, parts of which readily attach themselves to passing animals, various plants of different families have this name. *Imboke* also means that it attaches itself.

Identification: Much branched slightly lax **annual herb**, up to c. 150 cm tall; tall specimens may be supported by other vegetation. **Stems** and **leaves** covered in short **hairs**. **Leaves** elliptical, opposite and decussate with smooth margins. **Flowers** borne in erect spikes, c. 30 cm tall, green and white, arranged alternately, surrounded by hooked burrs. **Fruit** have star-shaped hooked bristles. The plants attach themselves to anything that passes and are a danger to young ground-feeding birds, such as francolin, as the seed heads form rough masses that attach to the birds' legs.

Habitat: Widespread mainly on islands and in woodland with light to heavy shade.

Flowering: During the main rains.

Uses and beliefs:
- San use the roots as an infusion to cure women's pains.
- Subiya and Hmbukushu use the fruits in the same way.
- Much grazed by game when young.

FZ Vol.9 pt1, Blundell p.43, Hargreaves p.9. CM, RM, MT.

Anacardiaceae (mango family)

Rhus tenuinervis Engl.

Common names: Setswana morupaphiri, wanoka, mosasawana, gcan, cau, mugorokoko, modupaphiri; **Afrikaans** krulblaartaaibos; **English** Kalahari currant, commiphora rhus.

Derivations: *Rhus*, red [Greek], the leaves of many plants of this genus turn red in autumn, the wood is also often red, in this case a pinkish yellow; *tenuinervis*, thin-veined [Latin]. *Morupaphiri*, scented by the hyaena.

Identification: Low-growing **shrub**, c. 2 m tall. **Bark** dull dark grey brown, slightly rough. **Leaves** compound, trifoliate, obovate with coarse crenations on the upper 2/3 of the blades, c. 6 cm (inc. c. 1.5 cm leaf stalk) × 4.5 cm; upper surface dark dull green, lower surface paler; both surfaces may be slightly hairy. **Flowers** small, 4 petals, greenish, borne in loose panicles up to 15 cm long at the apex of branches. **Fruit** flattened and purplish black, slightly acidic to taste, c. 5 mm dia. The plant has a distinctive **smell** rather like green apples, and is attractive to flies.

Habitat: Common throughout woodland and islands.

Flowering: Late in the main rains.

Uses and beliefs:
- The fruit are edible.
- Crushed leaves are used to treat stings.
- Popular with San for making bows, although *Grewia flava* is preferred.
- San in the Kalahari add small branches to their drinking water to give it a fresh taste. They say that it cools the water.
- San use it in the same way as *Aerva Leucura* to speak with the ancestors.
- The bark is used for tanning leather.
- Twigs are used to smoke bees.
- The pink to reddish-brown heartwood is excellent for carving.
- Highly palatable to animals and birds: found only where protected by dense acacia trees.

FZ Vol.2 pt2, Ellery p.131, Hargreaves p.49, Palgrave updated p.583, Tree Roodt p.129. RM.

Annonaceae (custard apple family)

Friesodielsia obovata (Benth.) Verdc.

Common names: Setswana mochinga, mokondekonde; **Subiya** mochingachinga; **Afrikaans** noordelike dwababessie; **English** bastard dwababerry, northern dwaba-berry.

Derivation: *Friesodielsia* commemorates the Swedish botanist Elias Magnus Fries (1794–1878) and the German botanist Friedrich Ludwig Emil Diels (1874–1945); *obovata*, inverted ovate shape [Latin], i.e. with the broadest end uppermost, referring to the shape of the leaves.

Identification: Small **shrub**, c. 3 m tall. **Leaves** obovate, c. 9.5 × 6 cm, with net veining, grey-green; leaf margins smooth with a hairy fringe; leaf underside covered in short hairs; leaves have a pleasant citrus **smell** when crushed. Solitary pendant **flowers** borne singly from the leaf axils and occasionally terminally, green; 3 triangular sepals and 2 whorls of 3 petals; the 3 inner petals curve sharply to form a shield over a cushion of cream-coloured ovaries. Leaf-like **bract** below each flower. **Fruit** a cluster of fleshy cylinders, constricted between the seeds, rather like sausages; ripening to reddish brown.

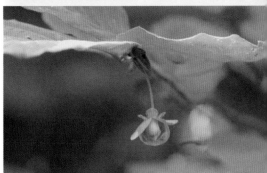

Habitat: Found occasionally locally in riverine bush along the Chobe River.

Flowering: Early in the main rains.

Uses and beliefs:
- The fruit are fleshy and edible with a tart taste, pleasant to eat when stewed.
- The young branches are very strong and flexible and are used for fishing rods.

FZ Vol.1 pt1, Palgrave p.167, Palgrave updated p.205, Setshogo & Venter p.28. BN.

Apocynaceae (oleander family)

Gomphocarpus fruticosus (L.) Aiton subsp. rostratus (N.E.Br.) Goyder & Nicholas

Common names: Setswana mositanokana; **Afrikaans** melkbos; **English** milk weed, wild cotton.

Derivation: *Gomphos*, a bolt or nail; *karpos*, fruit [Greek]; *fruticosus*, shrubby or bushy [Latin]; *rostratus*, a beak [Latin], referring to the long beak on the fruit. *Mositanokana*, the thing that grows along the riverside.

Identification: An erect branching **shrubby herb**, c. 1 m tall, containing latex. **Stems** and **underside** of leaves covered in a tangled mass of short **hairs**. **Leaves** usually opposite and decussate, sometimes alternate; linear on a short stalk, c. 10 × 1 cm; margins smooth; with net-veining and tending to fold at the mid-vein. **Flowers** c. 1.5 cm dia., in hanging clusters on long stems from the leaf axils, pale green to pale cream, 5 petals. **Fruit** often in pairs, pear-shaped with a long sharp 'beak'; sometimes hairless otherwise covered in spines or bristles; bursting when ripe to reveal cotton wool-like **hairs** and black **seeds**.

Habitat: Common and widespread in damp areas on the floodplain and near pans in full sun.

Flowering: During the rains and on into the cool dry period.

Uses and beliefs:
- This plant has many medicinal uses. A cough mixture is made by boiling the roots. A diarrhoea cure is made by boiling a mixture of the leaves, roots and fruit.
- Batswana mix the leaves and roots with those of *Achyranthes aspera* to improve the metabolic system.
- The Wayeyi boil the roots with those of *Pechuel-loeschea leubnitziae* and *Ziziphus mucronata* to make a treatment for gonorrhoea.
- In the Delta, the leaves and latex are taken as an enema.
- *Diphate* (Setswana): if a woman has a miscarriage or an abortion and another person sits where she was sitting, they may develop a stiff neck. Collect and dry the leaves, pound them to a powder and take as snuff. The sneezing will cure the stiff neck.
- The fluffy seed hair is used to stuff pillows.

Blundell p.146, B Van Wyk Flowers p.96, Ellery p.97, Flowers Roodt p.31, Germishuizen p.312, Turton p.87. BB, GM, PN.

Apocynaceae (oleander family)

Orbea caudata (N.E.Br.) Bruyns subsp. *rhodesiaca* (L.C.Leach) Bruyns
(syn. *Orbeopsis caudata* subsp. *rhodesiaca* (L.C.Leach) L.C.Leach

Derivation: *Orbis*, an orb [Latin], referring to the large orb in the centre of the flower; *caudata*, tail-like [Latin], referring to the long points of the lobes; *rhodesiaca*, from Rhodesia.

Identification: Erect leafless **succulent perennial**, c. 20 cm tall. The plant has an extensive **rooting system** and may die back or be completely grazed during dry periods. **Stems** branching, crowded, 4-angled, grey-green with long acute teeth. **Flowers** borne in clusters on the stems, star-shaped, green with brown centres and brown speckles; corolla lobes have long points and a slightly foetid **aroma**.

Habitat: Found occasionally locally in the shade of acacia scrub and in mopane woodland.

Flowering: Middle to end of the main rains.

Apocynaceae (oleander family)

Pergularia daemia (Forssk.) Chiov. subsp. *daemia*

Common names: Setswana lefswe, kgaba, leilane.

Derivation: *Pergularia*, a place where there are vine trellises [Latin]; *daemia*, the Arabic name for this species.

Identification: A rambling **climbing perennial** up to 3.5 m tall with a slightly sticky mass of short **hairs** on almost all parts. **Leaves** opposite, velvety, cordate, with margins entire. **Flowers** c. 1.2 cm dia., in drooping clusters; sepals hairy, green; corolla lobes white, arrow-shaped; flowers open in the evening through to early morning, pollinated at dusk by hawk moths and other insects. Distinctive horse-shoe-shaped double **seed pods** split along one side to reveal wind-borne **seeds**. The plant exudes white **latex** when cut.

Habitat: Common in light shade on islands and in acacia scrub.

Flowering: Main rains through to the end of the cool dry period.

Uses and beliefs:
- The leaves are used in Botswana to make *morogo*, a vegetable relish.
- Used in the Okavango Delta to cure children who do not look after their parents properly.

Turton p.45, Blundell p.148, Fox & Young p.112, Flowers Roodt p.18, WFNSA p.316.

Asteraceae (daisy family)

Acanthospermum hispidum DC.

Common names: Setswana khonkhorose, setlhabakolobe, sephalane; **Subiya** ikunkubo; **Kalanga** ikungubo; **English** upright starbur, bristly starbur.

Derivations: *Akanthos*, thorn or spikes [Greek]; -*spermum* -seeded [Greek], referring to the spiny seeds; *hispidum*, bristly or spiny [Latin]. *Khonkhorose*, like cockroaches because many plants are usually found together.

Identification: Erect branching **annual herb**, up to c. 45 cm tall. **Stems** and **leaves** are **hairy** overall. **Leaves** are sticky, opposite and decussate, obovate with slightly serrated margins on the upper half of the leaf. **Inflorescence** is star-shaped with 7–9 points. **Flowers** c. 2 cm dia., having female ray florets, with single, orange-yellow petals emerging from enlarged spiny phyllaries (the sepal-like structures around composite flowers); a cluster of small orange/green disc florets at the centre. **Fruit** are formed from the individual points of the inflorescence with two spines at the apex; when ripe they separate easily.

Habitat: An invasive plant spreading rapidly. It grows on track sides in full sun in damp areas of the floodplain and in bush near seasonal pans.

Flowering: During the main rains.

Uses and beliefs:
- Listed as a noxious weed and should be controlled by the land user.

Blundell p.159, Botweeds p.16, Turton p.55. BN, KS.

Capparaceae (caper family)

Boscia albitrunca (Burch.) Gilg & Gilg-Ben.

Common names: Setswana motlôpi, makgolela, mareko, monomane, ntopi, mopipi, zang; **Afrikaans** witgat, matopie; **English** shepherd's tree, caper bush.

Derivation: *Boscia*, for Louis A.G. Bosc (1759–1828), a French professor of agriculture; *albi*, white [Latin]; *truncus*, a trunk [Latin], alluding to the white bark of the young branches of the tree.

Identification: Medium-sized, **evergreen tree**, c. 12 m tall, usually a single straight **trunk**. The rounded crown of the tree is almost parasol-shaped as it is very popular with browsers. Young white **branches** particularly distinctive, older **bark** is pale brown, much pitted and often damaged by game as they reach to feed on the leaves. **Leaves** linear, growing in clusters on short stems, margins smooth. **Flowers** on very short stems in clusters at the leaf axils, initially resembling small spherical green berries until the 4 sepals open to reveal a cluster of yellow stamens and later the green ovary. **Fruit** spherical, hairless, yellowish. **Wood** pale yellow-brown.

Habitat: Common and widespread growing in mixed woodland and as single specimens in acacia scrub.

Flowering: During the cool dry period.

Uses and beliefs:
- Dried, roasted and ground roots make a passable substitute for coffee.
- The raw roots are pounded to make a white meal for porridge.
- The hollow trunks hold water, which is tapped by bushmen.
- The buds may be pickled and used as a substitute for capers.
- The ripe fruit are pounded and mixed with milk to make a drink.
- A decoction of the roots is used to treat haemorrhoids.
- Leaves of these evergreen trees are widely used as fodder for cattle; they are also very popular with browsers.
- The fruit is widely eaten by animals, birds and even humans in times of scarcity, although the seeds should not be swallowed because of their high sulphur content.
- A cold infusion of the leaves is used to treat sore eyes in cattle.
- The wood is heavy and fine grained but difficult to work so is used for small domestic objects like bowls and spoons.
- The trees are held in great regard and every effort is made not to destroy them as they are so useful. In some areas, it is believed that the wood should never be burnt as it results in the birth of only bull calves. It is also believed that if the fruit of this tree withers before the sorghum crop is ripe, the crop will be a failure.
- The caterpillars of the Brown-veined white (*Belenois aurota aurota*) and the Queen purple tip (*Colotis regina*) feed on the leaves.

FZ Vol.1 pt1, Ellery p.98, Hargreaves p.19, Palgrave updated p.224, Setshogo & Venter p.6, Tree Roodt p.29. GM.

Caryophyllaceae (carnation family)

Pollichia campestris Aiton

Common names: Setswana gobe-jwatlhoa; **Afrikaans** teesuikerkaroo, suikerteebossie; **English** waxberry.

Derivations: *Pollichia* for J.A. Pollich (1740–80), a German botanist; *campestris*, of the fields or open plains [Latin]. *Gobe*, porridge.

Identification: Climbing erect **semi-perennial herb** from a woody **rooting stock**, up to c. 4 m tall if supported by other vegetation; covered overall in coarse white **down**. **Leaves** grow in whorls and appear paired, oblanceolate, with a pointed tip, smooth margins, no leaf stalks, c. 2.5 × 0.6 cm. **Flowers** minute, c. 1 mm dia., greenish, growing in tight stalkless clusters at leaf axils.

Habitat: Widespread but uncommon, growing in the shelter of acacia scrub and in mixed woodland.

Flowering: During the early rains.

Uses and beliefs:
- The swollen waxy bracts are edible.

FZ Vol.1 pt2, B Van Wyk Flowers p.62, WFNSA p.136. SS.

Celastraceae (spike-thorn or staff-tree family)

Gymnosporia senegalensis (Lam.) Loes.
(syn. *Maytenus senegalensis* (Lam.) Exell)

Common names: Setswana motlhono, mokutemutembuze, mitwa-ya-ntse; **San** xgargum; **Subiya** morwanyeru; **Shiyeyi** rekuXhwa; **Afrikaans** rooipendoring; **English** confetti tree, confetti spikethorn, red spike thorn.

Derivations: *Gymno*, naked, uncovered [Greek]; *spora*, seed [Greek], having naked seeds; *senegalensis*, from Senegal [Latin]. *Morwanyeru*, the teeth of the pike.

Identification: Much branched **shrub or tree**, up to c. 4 m tall. **Bark** light grey and smooth. Side **branches** reddish, growing from leaf axils, spiralling up main stems, often spiked at tip. **Leaves** leathery, obovate, margins finely serrated, c. 7.5 × 2.8 cm (inc. 0.5 cm leaf stem); a pair of leaves or bracts at the base of the flower panicle. **Flowers** in panicles at tips of side branches, c. 4 mm dia., green, 5 petals, yellow stamens; flower **scent** a mixture of honey and feline urine.

Habitat: Widespread and common on islands and in woodland. Often associated with termite mounds.

Flowering: Mainly in the cool dry period but may be found at other times.

Uses and beliefs:
- The bark is used for the treatment of lice.
- The roots are used to treat chest pain.
- The roots are used as an aphrodisiac.
- It is also believed that if the roots are burned with the snake's head, it will cure the snake's bite.
- Highly palatable to animals, much browsed.
- The wood is hard and used to start fires by friction.

FZ Vol.2 pt2, *Ellery p.126*, *Palgrave p.611*, *Tree Roodt p.89*, *WFNSA p.236*. MM, RM, PN.

Celastraceae (spike-thorn or staff-tree family)

Loeseneriella africana (Willd.) N.Hallé var. *richardiana* (Cambess.) N.Hallé
(syn. *Hippocratea africana* (Willd.) Loes. var. *richardiana* (Cambess.) N.Robson)

Common names: Hmbukushu dithyana; **English** paddle-pod.

Derivation: *Loeseneriella*, for Loesener who first described many of the plants in this genus; *africana*, of Africa.

Identification: Climbing twining **woody perennial**, up to c. 10 m tall. Elliptical **leaves** leathery, opposite and decussate, margins becoming become whitened and almost frayed with age. **Flowers** borne in loose panicles at the leaf axils, green, 5 star-shaped sepals, prominent nectaries; 3 orange anthers separate as the flower matures to reveal the stigma. **Fruit** flat, paddle-shaped, splitting into 2 valves that contain flat-winged seeds.

Habitat: Widespread in riverine forest throughout the region and in trees along the spillway margin.

Flowering: Flowers late in the cool dry period.

Uses and beliefs:
- If Hmbukushu are bedevilled by evil spirits, *hathymu* (*dikaba*, Setswana), they make an infusion of the whole plant and bathe in it at dawn before asking help from the ancestors.
- When a cow has no more milk, Hmbukushu make an infusion of the roots by soaking them in water overnight. They then bathe the cow's udder with the water and give it to the cow to drink to encourage it to produce more milk.

Ellery p.121, Palgrave updated p.517. MT.

Combretaceae (bushwillow or combretum family)

Combretum imberbe Wawra

Common names: Setswana motswiri, motswere, madikolo, monyondo, mbgweti, kavimba, movimba, cong, motsore; **Shiyeyi** oyondo; **Kalanga** ngweti; **Afrikaans** hardekool; **English** leadwood, bastard yellow wood, elephant trunk, ivory tree, large fruited bushwillow, large fruited combretum.

Derivation: *Combretum*, the name given by Pliny to a species of climbing plant; *imberbe*, without a beard, hairless [Latin].

Identification: Large, long-lived (up to 1000 years or more) deciduous **tree** with very dense **wood**, up to c. 10 m tall. **Bark** grey with rectangular cracking. **Leaves** opposite, shiny pale green, oval, margin wavy, growing on short twigs that are often tipped with a **spine**. Spikes of small, creamy green **flowers** borne in leaf axils and at the apex of branches; flowers having a fresh smell. **Fruit** 4-winged, c. 1.5 cm dia., straw coloured. The tips of the young shoots emit a citrus **scent**.

Habitat: Common and widespread, especially on damp alluvial soils along water channels. Also common throughout mixed woodland on islands and along the treeline.

Flowering: Early in the rainy season.

Uses and beliefs:
- Batswana and Subiya boil the leaves and inhale the fumes in the evenings to treat coughs.
- Wayeyi use the boiled ashes as a hair relaxant. The mixture is stroked onto the hair for 5 or 6 minutes, then rinsed off. The straightening will last for about a month.
- Herrero and Ovambo never destroy these trees as they are believed to be the ancestors of all their people.
- Much sought for firewood.
- The ashes can be used as a white water-based paint. Traditionally, the houses were painted in patterns of white (from *C. imberbe* ash), black (from charcoal) and greenish brown (from cattle dung, the fresher the vegetation, the greener the dung).

FZ Vol.4 pt0, Ellery p.104, Hargreaves p.33, Palgrave updated p.803, Tree Roodt p.103. PK, GM, PN, BT.

Cucurbitaceae (cucumber or gourd family)

Corallocarpus bainesii (Hook.f.) A.Meeuse

Common names: Setswana moraanoga, pingping-tshegatshega, more-wanoga, coloku; **Hmbukushu** karuarua; **G//ana** orogu?

Derivation: *Corallum*, coral [Latin], *carpus*, fruit [Latin], referring to the bright coral colour of the fruit; *bainesii*, so named for John Thomas Baines, painter, naturalist and explorer, who was part of the Chapman expedition (1861–63) which passed through Namibia to Victoria Falls via Lake Ngami and back.

Identification: Perennial **herbaceous climber**, c. 2 m tall. **Tendrils** from the leaf axils. **Leaves** alternate, palmate, 3 or 5 lobes with a bristle at the point of each lobe, velvety, with smooth margins. Small greenish **flowers** in clusters at the leaf axils; male and female flowers borne on the same plant, female flowers stemless, male flowers borne in long-stemmed racemes. **Fruit** ripen to bright coral-red.

Habitat: Found occasionally locally in light shade on islands and in the tree-line on sandy soils.

Flowering: Throughout the main rains.

Uses and beliefs:
- There are differences of opinion amongst different groups of Subiya about these fruit. Some eat them and report that they are sour tasting, others do not eat them because they believe that snakes eat them and become more venomous. Wayeyi and Hmbukushu also share this belief.
- The leaves and annual stems are cooked and eaten as *morogo*, a vegetable relish for mealie or pap.
- Both Wayeyi and Hmbukushu use a decoction of the roots and fruit to cure a curse, *dungoro* (Shiyei) or *ngoro* (Hmbukushu). For example, a person might say "if you steal my cow, I will curse you... ". The result of the curse is that the victim will split open inside. The decoction will heal the split.

FZ Vol.4 pt0, Ellery p.158, Hargreaves p.18. CM, BN.

Cucurbitaceae (cucumber or gourd family)

Ctenolepis cerasiformis (Stocks) Hook.f.

Common names: English cherry vine.

Derivation: *Cten*, kteis, *ktenos*, comb [Greek], *lepis*, scale or flake [Greek], possibly referring to the bracts with bristled margins; *cerasiformis*, shaped like a cherry [Latin], referring to the brilliant red, cherry-shaped fruit.

Identification: Possibly **perennial** delicate **climbing herb** with **tendrils**, c. 2.5 m tall, depending on height of support. **Leaves** palmate, 3 lobes deeply incised, the lower lobes almost double, margins slightly serrated, with a hairy tip to each serration, c. 12 cm (inc. c. 4 cm petiole) × 1.2 cm. **Flowers** green, minute, male on c. 8 mm stems; female sessile in leaf axils, surmounted by a leafy stipule-like **bract**, with a bristled margin, sheathing the stem. **Fruit** spherical c. 1 cm dia., ripening to red, probably poisonous as they appear to be ignored by birds and mammals.

Habitat: Found occasionally locally but widespread in years of heavy rainfall in light shade on islands and in the treeline of the floodplain.

Flowering: Main rains into the cool dry period.

FZ Vol.4 pt0.

Ebenaceae (ebony family)

Diospyros mespiliformis Hochst. ex A.DC.

Common names: Setswana mokhutsomu, mokutshumo, mokochong, utunda, mochenje, mbiriri; **Hmbukushu** muvichi; **San** suma; **Subiya** muchenje, isuma; **Shiyeyi** oshoma, moktshumo; **Afrikaans** jakkalsbessie; **English** jackal-berry, African ebony, ebony, Transvaal ebony, Rhodesian ebony, Mozambique ebony, ebony diospyros, monkey guava, persimmon, medlar.

Derivations: *Dios*, divine [Greek], *pyros*, wheat [Greek], divine wheat, a name transferred to this genus with edible fruit; *mespilus*, medlar, the Latin name for this fruit, *-formis*, in the form or shape of [Latin], in the form of a medlar. *Jackalberry*, the fruit have been found in the dung of jackals.

Identification: Large, often **deciduous** (probably depending on availability of water) **tree** of mixed riverine woodland, dense crown, 10–25 m tall. **Bark** dark grey, splitting into rectangular blocks. **Leaves** obovate with a wavy margin and net veining; young leaves and shoots are reddish. Male and female **flowers** borne on separate trees, they are rather similar, greenish with 4 or 5 lobes, c. 0.5 cm dia., fused to form a tube; male flowers borne in clusters on short stalks at the leaf axils, with a heavy honey-sweet **perfume**; female flowers borne singly at the leaf axils. **Fruit** are oval with a pointed tip, orange to yellow; sepals remain on the fruit even after they fall.

Habitat: Common and widespread, near water, throughout the area. This species is protected under Botswana's Forest Act, 1968.

Flowering: Before and during the early rains.

Uses and beliefs:
- Batswana usually eat the fruit raw, in preserves or brewed and distilled.
- San dry the fruit, pound them and mix them with milk as a drink.
- Leaves, twigs and bark are used medicinally.
- Wayeyi and the Subiya use the wood to make *mekoro* (dug-out canoes).
- The timber is widely used. It has a reddish-brown to black heartwood when dry.
- Caterpillars of *Charaxes achaemenes* eat the leaves.

FZ Vol.7 pt1, Ellery p.108, Hargreaves p.21, Palgrave updated p.905, Setshogo & Venter p.51, Tree Roodt p.39. PK, TK, RM, BN, PN.

Ebenaceae (ebony family)

Euclea divinorum Hiern

Common names: Setswana motlhakola, nshangule, moshitondo, doboma, motlhakua, motlhahola; **Hmbukushu** moshetondo; **Lozi** mpumutwi; **Shiyeyi** motlhakola; **Afrikaans** towerghwarrie; **English** magic guarri, diamond-leaved euclea, diamond leaf, toothbrush tree.

Derivation: *Eukleia*, good fame, glory [Greek], *divinorum*, the botanist Hiern, who gave the tree its scientific name, noticed that it was being used in 'medicine of the diviners in Batoka country', hence the name.

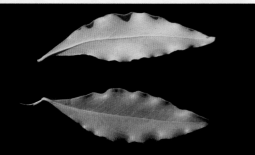

Identification: Locally a **small shrub**, c. 2 m tall but it may grow to c. 8 m. **Bark** usually grey-brown and smooth but it may darken and crack into square sections with age. **Leaves** opposite and decussate, narrowly elliptical, shiny and pale green with strongly wavy margins. **Flowers** pale greeny-yellow, growing in clusters at the leaf axils, with overlapping translucent petals. **Fruit** spherical and thinly fleshy, c. 0.5 cm dia., ripening to black.

Habitat: Widespread and common in shade on islands and throughout the tree-line.

Flowering: During the cool dry period.

Uses and beliefs:
- The fruit are edible although they can have a strong purgative action. Batswana and Hmbukushu boil the roots for use as a purgative. The fruit are also used in brewing.
- Mixed with *Clerodendrum uncinatum* and *Albizia harveyi*, this plant is used to treat sexually transmitted diseases.
- Frayed twigs are used as toothbrushes.
- The branches are used to fight wild fires.
- The bark of the roots is used as a black dye for baskets, if the solution is not strong, it dyes olive green. The fruit are used to make ink.
- The leaves do not appear to be browsed even when there is nothing else available; elephants will walk a long way chewing on a branch; perhaps they are cleaning their teeth?

FZ Vol.7 pt1, Ellery p.110, Hargreaves p.21, Palgrave updated p.888, Setshogo & Venter p.52, Tree Roodt p.109. MM, PN, KS.

Euphorbiaceae (spurge family)

Acalypha fimbriata Schumach. & Thonn.

Derivation: *Akalephe*, a nettle or its sting [Greek], although this particular species does not sting; *fimbriata*, fringed [Latin].

Identification: Erect **annual herb**, occasionally branching from the lower part of the stem, up to c. 120 cm tall. **Stem** faintly ribbed and slightly hairy. **Leaves** alternate, oval, margins serrated with a hair at each serration, 3 main veins, hairy along veins on lower surface, upper surface occasional hairs; leaf stalk is slightly hairy; leaf blade approximately the same length as petiole c. 17 cm (inc. c. 8.5 cm petiole) × 5 cm. **Male and female flowers** grow on the same plant; **female flowers** appearing green, growing on pairs of spikes from the leaf axils, wrapped in a closed **leafy bract** with a deeply toothed margin, producing 3-part seed pods; **male flowers** form a continuation of the flowering spikes, green, c. 0.5 mm dia. May be confused with *Acalypha indica*.

Habitat: Common in deep shade on islands and in the tree-line along the floodplain.

Flowering: During the main rains.

FZ Vol.9 pt4

Euphorbiaceae (spurge family)

Acalypha indica L.

Common names: English poison bush, Indian acalypha.

Derivations: *Akalephe*, a nettle or its sting [Greek]; *indica* means that the plant occurs in India. Poison bush because it is believed to be poisonous to cattle when eaten.

Identification: Erect branching **herb**, c. 75 cm tall, from a woody **root stock**. **Stem** striated. **Leaves** alternate, oval, margins serrated and almost hairless; leaf blade no longer than the petiole, usually c. 10 cm (inc. c. 6 cm petiole) × 3.5 cm. **Male and female flowers** both on same plant; female flowers are usually 3 small hairy balls held in a cup-shaped leafy bract, c. 0.6 cm dia., with a serrated margin, borne either singly at leaf axils or as a spike; male flowers above, cream, c. 1 mm dia.

Habitat: A shade-loving herb, widespread on islands and in the tree-line along the floodplain.

Flowering: Middle to end of the main rains.

Uses and beliefs:
- Used in the treatment of eye complaints.
- Roots used as a purgative, extracted juice used as an emetic.
- Believed to be poisonous to cattle.

FZ Vol.9 pt4, Ellery p.153, Hargreaves p.38.

Acalypha ornata A.Rich.

Common names: Setswana moharatshwene.

Derivation: *Akalephe*, nettle or its sting [Greek]; *ornata*, ornamental or showy [Latin].

Identification: A shrubby **perennial herb**, c. 50 cm tall. **Leaves** ovate, spiralling up the stem, margins serrated, long pointed tip. Separate **male and female flowers** borne on the same plant; **female flowers** bright green, surrounded by conspicuous bracts, growing in tightly packed spikes, c. 6.5 cm tall, at the apex of the plant; **male flowers** creamy white, less conspicuous, growing in c. 4 cm spikes. In other areas, this species may grow to be a small tree, c. 5 m tall, and rarely has the male and female flowers on the same plant.

Habitat: Locally uncommon growing in deeply shaded areas of the riverine forest, preferring the well-drained soil of termite mounds.

Flowering: During the main rains.

FZ Vol.9 pt4

Euphorbiaceae (spurge family)

Croton megalobotrys Müll. Arg.

Common names: Setswana motsebe; **Hmbukushu** ovambenderu omutjivi; **Subiya** morongo; **Shiyeyi** okae; **Afrikaans** grootkoorsbessie; **English** (large) fever-berry (croton), river fever-berry.

Derivation: *Kroton*, tick [Greek], referring to the smooth appearance of the fruit; *megalobotrys* [Greek], a large cluster, referring to the inflorescence.

Identification: Bright green, densely leaved, **deciduous tree** or **shrub**, usually about 4 m tall but occasionally growing to 15 m if ungrazed. **Bark** smooth pale grey. **Leaves** simple, ovate with c. 3 veins from the heart-shaped base of the leaf, net veining, covered in white star-shaped hairs, margins serrated, c. 11 cm (inc. c. 3 cm leaf stalk) × 7 cm but can be much larger; leaves retained almost throughout the year except for a short time towards the end of the hot dry period; turning bright yellow before they fall. **Flowers** in a cream-coloured raceme of separate male and female flowers (which appear similar), c. 8 cm long. **Fruit** 3-lobed, spherical to ovoid, c. 3.5 cm long.

Habitat: Very common as dense under-storey on floodplain margins and in riverine forest, also on island margins.

Flowering: Before and during the early rains.

Uses and beliefs:
- Wayeyi and Subiya use an infusion of bark, leaves and roots to treat influenza.
- Wayeyi use ash from the fruit mixed with vegetable oil to treat wounds.
- More generally, infusions of fruit and bark are used as a purgative, to treat malaria and as a prophylactic.
- Wayeyi and Subiya throw the crushed fruit into a pool to paralyse fish overnight.
- Wayeyi children make spinning tops from the fruit. They push a small stick into the bottom of the fruit.

FZ Vol.9 pt4, Ellery p.106, Palgrave p.496, Roodt p.71. TK, PN, SS, BT.

Euphorbia tirucalli L.

Common names: Setswana sethare-se-tala, motsitsi, moremotala, motsetse, ngocha, ngotza; **Afrikaans** kraalnaboom; **English** rubber hedge euphorbia.

Derivations: *Euphorbia*, in honour of Euphorbus, physician to King Juba II of Mauretania (24 BC). Given the name *tirucalli* by Linnaeus in 1753 because this was the name used by the natives of Malabar. *Motsitsi* means lactating, because of the white latex.

Identification: Large branching **succulent tree**, c. 7 m tall, without spines. **Bark** grey with patches of long cracking. **Branches** often produced in whorls, forming brush-like masses. They are one of the main identifying features. Pairs of small linear **leaves**, c. 2 × 0.25 cm, borne at the ends of the cylindrical smooth green branches; leaves rarely seen as they fall quite rapidly. **Flowers** insignificant, yellowish, borne at the apex of branches. **Fruit** rounded, 3-lobed capsules. When damaged, stems and leaves pour white **latex** that is toxic and caustic.

Habitat: A non-indigenous species but locally common.

Flowering: Late in the main rains.

Uses and beliefs:
- Widely used medicinally but no uses quoted specifically for Botswana.
- Used for hedging around homes and small-holdings, probably because the caustic latex is a deterrent.

FZ Vol.9 pt5, Curtis & Mannheimer p.639, Palgrave p.451, Setshogo & Venter p.57.

Euphorbiaceae (spurge family)

Phyllanthus maderaspatensis L.

Common names: Setswana loetsane, seshangane; **English** creeping milkweed.

Derivation: *Phyllon*, leaf [Greek], *anthos*, a flower [Greek], referring to the fact that some species produce flowers on leaf-like branches; *maderaspatensis* of Madras in southern India.

Identification: Small low-growing prostrate **plant**, c. 20 cm tall. **Leaves** alternate, lanceolate, stalkless, margins smooth to slightly serrated. Green **flowers** with 3 petals and 3 sepals hang on short flower stems at leaf axils below the stalks of the plant. **Seeds** in trilobed capsules. White **latex** when cut.

Habitat: Found occasionally locally in full sun on sandy clay areas of the floodplain.

Flowering: Middle of the main rains.

FZ Vol.9 pt4.

Phyllanthus parvulus Sond. var. *parvulus*

Common names: Afrikaans kleurbossie; **English** dye bush.

Derivation: *Phyllon*, leaf [Greek], *anthos*, a flower [Greek], referring to the fact that some species produce flowers on leaf-like branches; *parvulus*, very small [Latin].

Identification: An unremarkable, small, erect, sparsely branched, **perennial herb**, c. 30 cm tall, growing from a single woody **taproot**. **Leaves** bright green, alternate, borne in 2 rows along the branches, oval and stalkless with smooth margins. **Flowers** borne in pairs of 1 male and 1 female flower, hanging below the leaf axils, c. 1 mm dia.; **male flowers** almost sessile, **female flowers** with short stems; 5 green petals with white margins.

Habitat: Widespread on lightly shaded islands and in sandy mopane woodland.

Flowering: During the main rains.

FZ vol.9 pt4, B Van Wyk Flowers p.296.

Euphorbiaceae (spurge family)

Phyllanthus pentandrus Schumach. & Thonn.

Derivation: *Phyllon*, leaf [Greek], *anthos*, flower [Greek], referring to the fact that some species of the genus produce flowers on leaf-like branches; *pentandrus*, with 5 stamens [Greek].

Identification: Delicate erect branching **annual herb**, up to c. 80 cm tall in years of high rainfall (but more usually c. 25 cm locally). **Leaves** oval with a pointed tip, alternate, folding at mid-vein, c. 2.2 cm (inc. c. 0.2 cm leaf stem) × 1 cm, smooth margins. Female **flowers** minute, with 5 petals with pink centres and a green stripe on the reverse, c. 1.5 mm dia. hanging on short, c. 2 mm, fine stems, below each leaf axil; similar male flowers without stems appear at leaf axils. **Fruit** spherical, c. 2 mm dia., clearly retaining their calyx.

Habitat: Found occasionally locally in open areas of sand and sandy clay throughout the mopane woodland, but also in full sun on the floodplain margins.

Flowering: Middle to end of the main rains.

Euphorbiaceae (spurge family)

Pterococcus africanus (Sond.) Pax & K.Hoffm.
(syn. *Plukenetia africana* Sond.)

Derivation: *Pteron*, wing [Greek], *kokkos*, berry [Greek], referring to the distinctive winged shape of the fruit; *africanus*, of Africa [Latin].

Identification: Climbing, scrambling **perennial herb**, c. 2.5 m tall, from a woody **rootstock**. **Leaves** alternate, hastate, with smooth margins. Minute green **flowers**, c. 1 mm dia., borne in small clusters on fine stems at the leaf axils and at the apex of the plant; buds spherical. **Fruit** distinctive, with 4 segments, square with concave or 'winged' sides.

Habitat: Locally rare in light shade on islands and in mixed woodland.

Flowering: During the early rains.

FZ Vol.9 pt4.

Euphorbiaceae (spurge family)

Tragia okanyua Pax

Common names: Setswana sebabatswane, sebabetsane, sebabatsane, mbabagulo, mbabashulo; **San** tswigho; **Subiya** lovavanzovu; **Shiyeyi** rewawan-jovu; **English** stinging nettle, nettle spurge.

Derivations: *Tragus*, goat [Greek]. *Sebabatswane*, something that burns or itches. *Lovavanzovu* means 'elephant stinger' because the elephant is afraid of nothing except this plant. *Rewawanjovu* also means 'the stinger'.

Identification: A stinging **climber**, c. 2 m tall, covered in highly irritant **hairs**. **Leaves** alternate, bristly, palmate with 3 lobes, c. 17 cm (inc. c. 9 cm petiole) × 5.5 cm, margins serrated. **Male and female flowers** produced separately on the same plant; burr-like **female flowers**, c. 1.5 cm dia., usually growing in pairs below the spikes of **male flowers**, silver green furry calyx holding 3 spherical **fruit; male flowers** green with 3 petals and 3 stamens, c. 2 mm dia.

Habitat: Locally common growing in light shade on islands and in the tree-line along the floodplain.

Flowering: From the middle of the main rains to when the weather dries out.

Uses and beliefs:
- Hmbukushu make a cut on the boys' arms and rub the seed heads of this plant into the wound to make their punch a 'real stinger'.
- San soak this plant in water, which they then use to wash the genitals of bulls to increase their potency.
- Young Subiya and Wayeyi girls rub their breasts with the fruit to enlarge them.
- It is also said to be used to treat sexually transmitted diseases, itching, heart palpitations and to prevent witchcraft.

FZ Vol.9 pt4, Hargreaves p.43. CM, PM, RM, BN, PN.

Hyacinthaceae (hyacinth family)

Dipcadi glaucum (Burch. in Ker-Gawl.) Baker

Common names: English poison onion, wild onion.

Derivation: *Dipcadi*, possibly from an oriental name for grape hyacinth; *glaucum*, having a fine whitish waxy coating that appears on some leaves [Latin].

Identification: Perennial herb with a **bulbous root**, erect annual growth, c. 80 cm tall. **Bulb** c. 4 cm dia., has a mahogany brown covering. **Leaves** strap-like, up to c. 4 cm wide and c. 30 cm long, U-shaped cross-section, glaucous, covered in a fine, white, waxy powder. **Flowers** green, c. 1.5 cm long, borne on long stems, spiralling up the stem, with 2 whorls of 3 petals, 3 pointed tepals aligned with the lower whorl of petals, tepals recurved, each flower supported by a small bract. **Fruit** trilobed capsules.

Habitat: Widespread locally in light shade on well-drained soil.

Flowering: From the first good rainfall into the early part of the main rains.

Uses and beliefs:
• The caterpillars of the genus *Colotis* (Orange tips) feed on this plant.

Hyacinthaceae (hyacinth family)

Dipcadi longifolium (Lindl.) Baker

Derivation: *Dipcadi*, a Turkish name for the grape hyacinth family; *longus*, long [Latin], *folium*, leaf [Latin], long-leaved.

Identification: Perennial herb with a **bulbous root**, c. 70 cm tall. **Leaves** grass-like, no mid-rib but deeply recurved along their length. **Flowers**, c. 1 cm long, hanging in a single row below the stem, pale green, 3 offset petals and tepals, each flower supported by a small linear bract. **Fruit** erect trilobed pods.

Dipcadi longifolium may easily be confused with *D. marlothii* but is distinguished by its more pendulous flowers, which are paler in colour and do not have such recurved petals. The bracts supporting the flowers of *D. longifolium* are smaller and less papery than those of *D. marlothii*.

Habitat: Found in light shade in mixed woodland areas of islands and river bank.

Flowering: Early in the main rains.

Uses and beliefs:
- Lozi crush and then smoke the empty seed pods as a substitute for tobacco.

KS.

Hyacinthaceae (hyacinth family)

Dipcadi marlothii Engl.

Derivation: *Dipcadi*, a Turkish name for the grape hyacinth family; *marlothii* for the South African botanist Rudolph Marloth.

Identification: Perennial herb with a **bulbous** root, c. 70 cm tall. **Leaves** grass-like, no mid-rib but deeply recurved along their length, hairy at the base. **Flowers**, c. 1 cm long, hanging in a single row below the stem, green with 3 offset petals and tepals, often recurved; each flower is supported by a **bract**. **Fruit** erect trilobed pods.

Dipcadi marlothii may easily be confused with *D. longifolium* but is distinguished by its more erect flowers, which are brighter green in colour and have distinctly recurved petals. The bracts supporting the flowers in *D. marlothii* are larger than those in *D. longifolium*.

Habitat: Found in light shade in grassy areas of the floodplain margin.

Flowering: Early in the main rains.

Uses and beliefs:
- Lozi crush and then smoke the empty seed pods as a substitute for tobacco.

B Van Wyk Flowers p.298, Turton p.128. KS.

Hyacinthaceae (hyacinth family)

Ledebouria revoluta (L.f.) Jessop

Common names: Setswana segokwe, kobo; **Hmbukushu** ngarakashe.

Derivation: *Ledebouria*, for the German botanist Carl Friedrich von Ledebour (1785–1851), professor of botany, traveller and plant collector; *revoluta*, rolled backwards [Latin], alluding to the lobes of the flower.

Identification: Bulbous perennial with an annual flowering spike, c. 10 cm tall. **Leaves** forming a basal rosette, grey-green with purple spots, linear, slightly recurved with a longitudinal fold in the middle of the leaf; produced at the same time as the flowers. **Inflorescence** c. 5 cm tall, a tight spike of greenish-white flowers. **Flowers** with 6 revolute lobes having a blue or mauve stripe on the inside; stamens indigo with yellow anthers.

Habitat: Found occasionally in light shade in mopane woodland.

Flowering: During the early rains.

B Van Wyk New p.258. MT.

Loranthaceae (mistletoe family)

Erianthemum ngamicum (Sprague) Danser

Common names: Setswana palamêla; **English** Ngami mistletoe.

Derivations: *Eri*, woolly [Greek], *-anthus*, flowered [Greek], having woolly flowers; *ngamicum*, from Ngamiland. *Palamêla*, the one that climbs, referring to the fact that the plant grows perched in trees.

Identification: Shrubby **parasitic plant** found growing on *Dicrostachys cinerea*. The plant is **hairy** overall with star-shaped hairs on both sides of the leaves. Silver-green **leaves** elliptical with wavy margins; leaf arrangement very variable, opposite on main stems and alternate on new growth, otherwise in clusters. **Flowers** with a yellow corolla tube and green lobes; hairy, especially the corolla tube; the flowers spring open and fling out the enclosed pollen at the slightest touch (an evolutionary adaptation to pollination by birds or large insects).

May be confused with *Erianthemum virescens*, but *E. ngamicum* flowers are larger and more hairy and its leaves have wavy margins.

Habitat: Found occasionally locally growing on *Dichrostachys cinerea*.

Flowering: Mainly during the early rains but may flower throughout the year.

Loranthaceae (mistletoe family)

Erianthemum virescens (N.E.Br.) Wiens & Polhill

Common names: Setswana palamêla; **English** green mistletoe.

Derivations: *Eri*, woolly [Greek], *-anthus*, flowered [Greek], having woolly flowers; *virescens*, becoming green [Latin]. *Palamêla*, the one that climbs, referring to the fact that the plant grows perched in trees.

Identification: Woody **parasitic plant** found growing in mopane woodland. Creamy-green **leaves** oval with a rounded tip; arranged in pairs on the main stems or occasionally in clusters; covered in a thick down of **hairs**. **Flowers** have a cream corolla tube with greenish-cream lobes; slightly hairy overall; the flowers spring open and fling out the enclosed pollen at the slightest touch (an evolutionary adaptation to pollination by birds or large insects).

May be confused with *Erianthemum ngamicum* but *E. virescens* flowers are narrower and less hairy and its leaves do not have wavy margins.

Habitat: Locally uncommon growing in mopane woodland.

Flowering: Mainly during the early rains but may continue into the main rains.

Meliaceae (mahogany family)

Trichilia emetica Vahl

Common names: Setswana mosiki, moshikiri, mosikili; **Lozi** musikili; **Afrikaans** rooiessenhout; **English** Natal mahogany.

Derivation: *Trichilia*, 3 parts [Greek], alluding to the ovary and ensuing seed capsules that usually have 3 segments; *emetica*, an emetic, causing vomiting [Latin].

Identification: Medium to large **evergreen tree** with a spreading dense crown. **Bark** smoothish, dark brownish grey. **Wood** light coloured, pinkish-grey. Shiny dark green **leaves** compound, 4–5 pairs of leaflets and a terminal leaflet; upper surface of leaflets shiny, dark and leathery, lower surface densely covered in short curly hairs and with prominent veins; swollen joint at the leaf axils. **Flowers** pale grey-green with long rounded lobes, c. 2 cm dia., fused to form a tube; borne in clusters at the apex of branches. A fragrant lemony **perfume**. **Fruit** almost spherical, splits open to reveal black seeds with a scarlet cover.

Habitat: Widespread on islands and along river banks in NE Botswana.

Flowering: During the cool dry period.

Uses and beliefs:
- An infusion of the bark is used as an emetic or an enema.
- Oil from the seeds is widely used for cooking, for soap-making, as an anointing oil, as a hair oil and as a treatment for the broken skin when a bone has been fractured.
- A hot infusion of the leaves is applied to bruises as a soothing lotion.
- In the Zambezi Valley, a root infusion is drunk to facilitate labour in pregnancy.
- The seeds are edible.
- Placing leaves in the bed at night is said to induce sleep.
- The bole is used for making *mekoro* (dug-out canoes). The timber makes excellent furniture, planking and boat material.
- Frequently planted as a shade tree in towns.

FZ Vol.2 pt1, Palgrave updated p.454, Setshogo & Venter p.102, WFNSA p.214, Plowes & Drummond.

Menispermaceae (monkey vine or curare family)

New growth on mature plant

Young plant

Cocculus hirsutus (L.) Diels

Common names: Setswana motsweketsane, mokau, palamêla; **Shiyeyi** diangamoti; **English** python vine, python climber.

Derivations: *Cocculus*, the diminutive of *kokkos*, a berry [Greek]; *hirsutus*, hairy [Latin]. *Diangamoti*, twisting the tree. *Motsweketsane*, veins, referring to varicose veins, describing the appearance of the plant as older stems wind round the trunks of trees. *Palamêla*, the one that climbs.

Identification: Large **perennial climber**, c. 10 m tall. Young plants dark green, hairless, with palmate leaves with up to 7 lobes. **New growth** on mature plants very hairy. **Mature leaves** blue-green especially on underside, shape varies from ovate to palmate or entire, 7-lobed with 3–7 main veins and net veining, apiculate with smooth margins, tending to be deltoid c. 8 cm (inc. 1.5 cm petiole) × 4.5 cm, petioles often bent, spiralling up the stem but all facing the same way. **Flowers** yellow-green, c. 3 mm dia., borne in groups on short stems at the leaf axils; 3 sepals and 6 petals; pleasant slightly astringent **perfume**. **Fruit** borne in threes but often only 2 survive.

Habitat: Common throughout all the woodlands of the area but less so in the mopane.

Flowering: Flowers early in the cool dry period.

Uses and beliefs:
- The huckleberry-like fruit are edible.
- San use the young stems as strings for bows and to tie spearheads to shafts.
- The rope-like stems are used to tie grass bundles in the Okavango.

FZ Vol.1 pt1, Hargreaves p.4, Tree Roodt p.35. MMo, RM.

Molluginaceae (mollugo family)

Glinus lotoides L.

Derivation: *Glinus*, origin obscure as '*glinos*' is Greek for the maple tree; *lotus* is Greek for a wide variety of plants; *-oides*, like or similar to, possibly meaning similar to the *Fabaceae* genus *Lotus*.

Identification: Prostrate, creeping **annual herb** with red main **stems** and red-green side stems forming mats up to c. 1 m dia.; covered in dense short stellate **hairs**. Grey-green **leaves** lanceolate to obovate and borne in whorls. Minute **flowers**, c. 2 mm dia., borne in clusters in the leaf axils; 3 or 5 petals, green-white with a pink tinge to the sepals, sepals alternate to the petals; flowers and buds covered in down.

This plant frequently hybridises with *Glinus oppositifolius* and a variety of forms may be found if both plants exist in an area.

Habitat: Found occasionally locally in the clay of dried out seasonal pans.

Flowering: During the early rains.

FZ Vol.4 pt0.

Molluginaceae (mollugo family)

Limeum argute-carinatum Wawra & Peyr. var. *kwebense* (N.E.Br.) Friedr.

Derivation: *Limeum*, meaning pest, to the point of ruin [Greek], referring to the toxicity of this genus; *argute*, sharply toothed or notched; *carinatus*, having a keel [Latin], possibly alluding to the shape of the fruit; *kwebense* for the Khwebe Hills, south of Maun.

Identification: A low-growing, branching, **annual herb**, c. 15 cm tall, **hairless** overall. **Leaves** linear, arranged in clusters, spiralling up the stem, smooth margins. **Flowers** greenish yellow with 5 petals, borne in clusters on short fine stems at the leaf axils.

Habitat: Found occasionally locally in full sun on the floodplain.

Flowering: During the main rains.

FZ Vol.4 pt0.

Moraceae (fig family)

Ficus thonningii Blume
(syn. *Ficus burkei* (Miq.) Miq.)

Common names: Setswana mmumo, moumo, utata; **Shiyeyi** ovumo; **Afrikaans** gewone wildevy, petersse-vy; **English** common wild fig, strangler fig, Burkis fig, bark cloth fig, Peters fig.

Derivation: *Ficus*, the Latin name for the edible fig tree.

Identification: A very variable species as may be seen by all the names that it has been given in different parts of southern Africa. Medium-sized **evergreen tree**, c. 10 m tall, with a dense dark green crown. Germinates in the forks of the branches of its host tree, which is eventually encircled and strangled by **aerial roots** growing down to the ground. Trunks and roots may be seen forming a lattice on the trunk of the host tree. This tree can also grow by itself as a free-standing specimen. **Bark** is smooth and pale grey. **Leaves** elliptical, alternate, spiralling up the stems, dark green, leathery with net veining, smooth margin; lower surface is paler green. **Flowers**, like those of all figs, are borne inside 'fruit' which are actually enclosed inflorescences; fertilised by a tiny parasitic wasp that enters the fig through the minute hole at its base. **Fruit** small, c. 1 cm dia., spherical, covered in a velvety down, growing either singly or in pairs at the leaf axils, becoming yellowy-green as they ripen. White **latex** present in leaves and stems.

Habitat: Widespread, found occasionally on host trees and rarely free-standing, on termite mound islands in the floodplain or in mixed woodland.

Flowering: May produce figs at any time of the year but they are more common during the main rains.

Uses and beliefs:
- The roots of the fig are used to treat colic, syphilis and certain skin conditions. Mixed with the grass *Sporobulus indicus*, the roots are used to treat snakebites.
- The adhesive latex from the roots is used to trap birds and hares.
- An extract of the bark is used to treat influenza.
- The bark fibre is used to make fabric, cord and mats.
- Fruit eaten by birds, bats and antelope.
- The caterpillars of *Myrina dermaptera* and *M. silenus* feed on the leaves.

FZ Vol.9 pt6, Ellery p.113, Hargreaves p.4, Palgrave updated p.146, Setshogo & Venter p.104, Tree Roodt p.25. PN.

Rhamnaceae (buckthorn or buffalo thorn family)

Berchemia discolor (Klotzsch) Hemsl.

Common names: Setswana motsentsela, mutsintsila, motsintsila, mokerete, mozinzila, lole; **San** tsizyna; **Subiya** iziye; **Afrikaans** bruin-ivoor; **English** bird plum, brown ivory.

Derivation: *Berchemia* for M. Berchem, a French botanist of the 17th Century; *discolor*, of two different colours [Latin], referring to the difference in the colour of the two leaf surfaces.

Identification: Slender **deciduous tree** with a dense crown, bright green in spring, c. 15 m tall. Both leaves and flowers are borne on **new growth**. **Bark** grey with longitudinal fissures. **Leaves** alternate, elliptical with a slightly wavy margin towards the tip; paler green on the lower surface; veins form a distinct herring-bone pattern on the lower surface. **Flowers** green; 5 conspicuous triangular sepals; stamens alternate with the sepals and folded back behind the flower, probably to ensure that the flower is not self-pollinated; delicately **scented**. **Fruit** fleshy, single-seeded, becoming pale yellowy orange when ripe.

Habitat: Widespread found occasionally locally.

Flowering: During the early rains.

Uses and beliefs:
- San pound the bark and place a wad on painful teeth. It acts as an anaesthetic and antibiotic.
- The juice from the crushed fruit is mixed with water to make an alcoholic beverage *irimba* (Subiya).
- The crushed fruit are mixed with mealie meal, sugar and water to make scones or bread rolls, *mukamu* (Subiya). Reeds are laid in the bottom of a pan, balls of the dough are placed on them and a little water added. The dough is then steamed until cooked.
- The bark is used as an orange/brown dye for basket making. The dried roots are boiled with charcoal to make dark brown dye.
- The hard wood has an attractive colour and is used to make furniture.

FZ Vol.2 pt2, Ellery p.98, Hargreaves p.44, Palgrave updated p.668, Setshogo & Venter p.111, Tree Roodt p.111, WFNSA p.240. TK, CM, GM, RM, BN, KS.

Rhamnaceae (buckthorn or buffalo thorn family)

Ziziphus mucronata Willd. subsp. *mucronata*

Common names: Setswana mokgalo, jujube, monganga, nchecheni, moketekete; **Kalanga** ntjetjeni; **Subiya** mokalu; **Hmbukushu** mokeketi; **Afrikaans** blinkblaar-wag-'n-bietjie; **English** buffalo thorn, shiny-leaf-wait-a-bit.

Derivations: *Ziziphus* from the Arabic *zizouf*, the name for the jujube tree (*Z. lotus*); *mucronata*, each leaf having a single short hair or bristle at its tip [Latin] (a mucronate tip). *Blinkblaar-wag-n-bietjie*, shiny-leaf-wait-a-bit because of the pairs of hooked and straight thorns, which having hooked a victim, make it difficult for them to extract themselves.

Identification: Spiny **shrub** or **small tree**, up to c. 6m tall. **Bark** grey to dark grey with small fissures making rectangular sections that peel off occasionally. The young stems and flower buds are hairy. **Leaves** shiny, leathery, alternate on a spiny zigzag stem; oval with a slightly cordate base, c. 6 cm (inc. c. 0.8 cm leaf stem) × 3.5 cm, margins serrated. **Flowers** small, c. 4 mm dia., green; 5 sepals alternate with the stamens. **Fruit** shiny reddish to yellowish-brown, single-seeded.

Habitat: Widespread growing in full sun on the floodplain.

Flowering: Throughout the rains.

Uses and beliefs:
- Hmbukushu and Batswana eat the fruit, Batswana also use them for brewing.
- Hmbukushu use *lefetho* (a wooden whisk) to mix the fruit with *madila* (sour milk) until it thickens. The resulting drink acts as a stimulant or tonic.
- Subiya chew a leaf to make a poultice for infected insect bites and boils, after a new leaf has been applied each day for 5 days, the infected tissue will come away but this can be painful.
- Young Wayeyi men take an infusion of the roots of young plants to make them sexually strong.
- The wood is used for tool handles (including the stocks of whips), yokes, spoons and as general-purpose timber.
- The sap is used to mix arrow poison from beetle grubs.
- It is believed that the tree is immune to lightning.
- Highly palatable and widely browsed by animals. The fruit is much sought by birds, especially the grey lourie.
- Caterpillars of *Tuxentius calice*, *T. meleana*, *T. hesperis* and *Tarucus syberus* feed on it.

FZ Vol.2 pt2, Ellery p.138, Hargreaves p.449; Palgrave p.666, Tree Roodt p.91, WFNSA p.240. TK, GM, MM, BN, KN, SS, MT.

Rubiaceae (coffee family)

Kohautia subverticillata (K.Schum.) D.Mantell
subsp. *subverticillata*

Common names: Setswana mollo-wa-badimo.

Derivation: *Kohautia* in honour of Francis Kohaut who collected plants in Senegal during the 19th Century; *verticillatus*, whorled [Latin], referring to the lower part of the inflorescence.

Identification: Slim erect **annual herb**, c. 45 cm tall. **Leaves** borne in whorls, linear to elliptical, sessile (lacking leaf stalks), margins smooth; leaf surfaces **slightly hairy**. **Flowers** small, borne in pairs at the nodes, greenish, paler above and darker below; usually 4 lobes, c. 3 mm dia., fused to form a long corolla tube, c. 1.5 cm long, which is often covered in sparse long hairs. **Seed capsule** retains long points on the calyx.

This plant is easily confused with *K. caespitosa* but the retained sepals on the seed capsule are longer and narrower in *K. subverticillata*. It is also a brighter green overall and generally slightly smaller than the more common *K. caespitosa*.

Habitat: Widespread, found occasionally locally growing in full sun on the floodplain.

Flowering: Early in the main rains.

FZ Vol.5 pt1.

Solanaceae (potato family)

Withania somnifera (L.) Dunal

Common names: Setswana moarasope, morarasope, morararupe, moherasope, modikaseope; **English** winter cherry, Indian ginseng, poisonous gooseberry.

Derivation: *Withania*, possibly named (misspelling included!) after the English palaeobotanist Henry Thomas Maire Witham (1779–1844), geologist and author of *Observations on Fossil Vegetables*; *somnifera*, sleep-inducing [Latin].

Identification: Shrubby branching **plant**, up to 1.5 m tall, potentially taller. New leaves and buds are covered with a white **down**. **Leaves** oval and hairless, margins wavy; growing in pairs, 1 leaf is often slightly larger than the other, older leaves may be single. Green **flowers** with 5 petals in clusters in the leaf axil.

Habitat: Found only once on a termite mound (recently naturalised).

Flowering: During the main rains.

Uses and beliefs:
• Believed to be an excellent adaptogenic tonic; it helps to regulate the body's major systems.
• The caterpillars of the Death's Head Hawk Moth, *Acherontia atropos*, feed on it.

Hargreaves p.57.

Orange
brown or red-brown flowers

Acanthaceae (acanthus or spinyflower family)

Ruspolia seticalyx (C.B.Clarke) Milne-Redh.

Derivation: *Ruspolia* in honour of an Italian explorer of Somaliland, Prince Eugenio Ruspoli, who was killed by an elephant; *seti*, bristled [Latin], referring to the bristly calyx.

Identification: Vibrant salmon-red flowered **herbaceous perennial**, up to 1.8 m tall but usually much smaller (c. 50 cm). **Stems** woody. **Leaves** opposite and decussate, oval with a slightly elongated point, up to c. 22 cm (inc. leaf stem c. 3 cm) × 10 cm, narrowing to form the leaf stalk, margins smooth, net veining; **hairs** along the veins on lower leaf surface of the leaves, occasional tufts and stellate hairs on the upper surface. **Flowers** with 5 lobes, c. 1.8 cm dia., corolla tube c. 2.5 cm long; borne in groups at the apex of the plant. **Seed pods** fiddle-shaped.

Habitat: Locally uncommon, growing in deep shade in the riverine forest.

Flowering: During the main rains.

Alliaceae (onion family)

Scadoxus multiflorus (Martyn) Raf.
(syn. *Haemanthus multiflorus* Martyn)

Common names: Setswana phalatsi, lephutse; **San** bamotshai; **Shiyeyi** sexhodo; **English** fireball lily, pincushion lily, royal shaving brush lily.

Derivations: *Skiadion*, abbreviated to *Sca*, parasol or umbel [Greek], *doxa*, repute or glory [Greek], a glorious parasol; *multi*, many [Latin], *florus*, flowers [Latin], abounding in flowers.

Identification: Deciduous **perennial bulb** produces a plant c. 50 cm tall. **Stems** with purplish spots and a spiral ruff at the base, 'D'-shaped in cross-section. **Leaves** large, bright green, shiny, ovate with a wavy margin, c. 35 cm long. The spectacular red **inflorescence**, c. 20 cm dia., appears (before the leaves) at the apex of the stem, very variable consisting of 10 to more than 150 florets. Scarlet star-shaped **florets** each with 6 linear petals and 6 stamens, anthers bright yellow. **Fruit** bright red and fleshy.

Habitat: Widespread, becoming less common as they are poached for garden decoration. In grassland and scrub, usually in the shelter of bushes.

Flowering: During and after the early rains.

Uses and beliefs:
- This plant is highly toxic.
- Batswana mix a powder of the dried bulbs with water and sprinkle the liquid around the house to prevent witchcraft.
- Wayeyi burn the flowers and mix the ash with vegetable oil or petroleum jelly to treat wounds. They also drink an infusion of the roots to improve the skin.

Flowers Roodt p.27, WFNSA p.76, Blundell p.426. BB, SS, BT.

Apocynaceae (oleander family)

Orbea huillensis (Hiern) Bruyns subsp. *huillensis*

Derivation: *Orbis*, a circular shape or disc [Latin], referring to the disc in the centre of the flower; *huillensis*, first described by Huilla in Angola.

Identification: Erect, leafless, **perennial succulent** with 4 rows of triangular spiky blunt outgrowths. **Stems** branching and crowded, c. 20 cm tall (significantly larger than those of other species of this genus), pale grey-green with red flecks. **Flowers** dark red, star-shaped with 5 elongated, pointed lobes, c. 3 cm dia., borne in clusters on the stems.

Habitat: Rare, growing in exposed sunny positions on slight mounds in flat rather impermeable areas or in deep sand.

Flowering: Late in the cool dry period.

Apocynaceae (oleander family)

Orbea lugardii (N.E.Br.) Bruyns
(syn. *Caralluma lugardii* N.E.Br.)

Derivation: *Orbis*, an orb [Latin], referring to the large orb in the centre of the flower; *lugardii*, Edward James and Charlotte Eleanor Lugard visited Ngamiland in the late 1890s, many of the plants they collected were named in their honour.

Identification: Erect, leafless, **succulent perennial**, c. 20 cm tall, with an extensive **rooting system**, may die back or be completely grazed during dry periods. **Stems** branching, grey-green with small short acute teeth. **Flowers** borne singly or in small clusters on stems, star-shaped, dark red with tips of long pointed lobes paling to brown, centre yellow, with a very strong carrion **odour**.

Habitat: Locally uncommon, found in light shade in mixed riverine woodland.

Flowering: During the early rains.

Apocynaceae (oleander family)

Orbea schweinfurthii (A.Berger) Bruyns
(syn. *Caralluma schweinfurthii* A.Berger)

Common name: Setswana seboka.

Derivations: *Orbis*, an orb [Latin], referring to the large orb in the centre of the flower; *schweinfurthii*, for G.A. Schweinfurth, who collected plants in Central Africa in the 1860s. *Seboka*, describes the way the plant grows looking like a crowd of people.

Identification: Leafless **perennial succulent**, stems branching and crowded, lying on the ground with tips ascending (decumbent). **Rooting system** extensive, may die back or be completely grazed during dry periods. **Stems** grey-green with short irregularly placed acute teeth. **Flowers** borne in small clusters or singly on stems, star-shaped, yellow with brown spots with a yellow centre, c. 1 cm dia. **Fruit** a pair of green splotched brown tubular pods (follicles), opening along one side to release seeds with long hairs (pappi) for wind distribution.

Habitat: Locally uncommon, found in light shade in the tree-line on the spillway.

Flowering: Middle to end of the main rains.

KS.

Apocynaceae (oleander family)

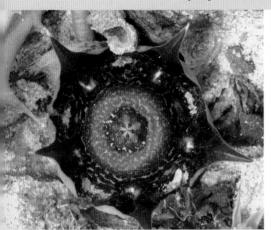

Duvalia polita N.E.Br.

Common name: English polished star.

Derivation: *Duvalia*, for Henri Auguste Duval (1777–1814), a French medic who wrote a pamphlet about succulents; *polita*, polished, elegant [Latin].

Identification: Upright, branching, leafless **perennial succulent**, c. 10 cm tall, with 6 rows of short teeth on the angles of its **stem**, each tooth with a pair of small spines on its upper surface. **Flowers** c. 3 cm dia., 5 pointed lobes which are green externally and dark red-brown internally, blotched on white towards the centre; flowers towards the base of the plant on flower stalks (c. 2.5 cm long); with a strong **smell** of carrion. Clear **latex** is present at breaks in the stems.

Habitat: Found occasionally locally in good-sized growths, in sandy clay areas of mopane woodland near seasonal pans.

Flowering: Late in the main rains.

Uses and beliefs:
- The stems are edible, raw or roasted, but are bitter.

WFNSA p.314.

Asteraceae (daisy family)

Dicoma schinzii O.Hoffm.

Common names: Setswana pelobotlhoko, makue; **San** chotho.

Derivations: *Di*, two [Greek], *choma*, tuft at the summit [Greek], referring to the double seed pappus of the first species described; *schinzii* for the Swiss botanist Professor Schinz. *Pelobotlhoko* and *chotho* mean 'painful heart' in the sentimental sense.

Identification: Lax-branching **perennial herb**, c. 35 cm dia. **Stems**, leaf margins and veins covered in golden **hairs**, which occasionally spread onto the leaf blade. **Leaves** alternate, elliptical, covered in a heavy white velvet of hairs, c. 5 cm (inc. c. 1 cm leaf stalk) × 1.5 cm. **Inflorescence** a composite of disc florets, c. 1.5 cm dia., appearing mainly as a mass of golden bristly hairs supported by a whorl of leaves. **Seed heads** become harsh and sharply spiked.

Habitat: Found occasionally locally in sandy places throughout the area.

Flowering: Main rains into the cool dry period.

Uses and beliefs:
- San soak one plant in water and bathe at dusk (after sunset) and again at dawn (before sunrise). They then speak to their ancestors about their problems as a method of curing bad luck.

FZ Vol.6 pt1. RM.

Campanulaceae (bluebell family)

Gunillaea emirnensis (A.DC.) Thulin

Identification: Erect **annual herb** with branching stems, c. 60 cm tall, covered in fine, short, white **hairs**. **Stems** slightly red-coloured. **Leaves** lanceolate, without stalks, wavy margins, often in a basal rosette and spiralling up the stems. **Flowers** pale orange with bright yellow stamens; 4 recurved sepals. **Seed capsules** erect, slightly ribbed, cylindrical pods, c. 2 cm long.

Habitat: Locally rare, growing in seasonally damp areas of the Khwai River floodplain.

Flowering: Early in the main rains.

FZ Vol.7 pt1.

Colchicaceae (colchicum family)

Gloriosa superba L.

Common names: Setswana sehokgwe, nachwa, mokuku, gondovoro; **Hmbukushu** ngondovuro, mayanga-ombwa; **San** kobo; **Shiyeyi** letakana; **Afrikaans** geelboslelie, rooiboslelie; **English** flame lily, superb lily, climbing lily.

Derivations: *Gloriosa*, glorious [Latin], *superba*, superb [Latin], both referring to the brilliant showy flowers of this plant. *Ngondovuro*, cockerel.

Identification: One of the most beautiful plants in the area but **very poisonous**. Climbing to c. 2.5 m, free-standing up to c. 60 cm. **Tuberous root stock**. **Leaves** alternate, sessile, narrowly oval, margins entire; **tendril** at the leaf tip used to climb and clasp other vegetation. **Flowers** borne on long stems at the leaf axils, large, brilliant red, 6 wavy recurved petals with yellow-green margins that disappear with age. **Fruit** ribbed, roughly heart-shaped although often distorted by insect damage. Plant produces a **yellowy latex** when cut.

Habitat: Widespread but not common throughout the area, more prolific in years of heavy rainfall. In the shade and protection of other trees and shrubs, on islands, in the tree-line and even in acacia scrub.

Flowering: Throughout the main rains, flowering stops immediately there is a long gap in the rains.

Uses and beliefs:
- Although Roodt lists the plant as highly toxic (it can kill a dog in 15 mins), it has many medicinal uses.
- Hmbukushu men use it to improve poor erections. The root is boiled with milk from a cow whose calf is almost weaned and the infusion drunk once or twice a day. Wayeyi eat the raw roots to enhance penis growth.
- Batswana treat open wounds with slices of the root bound in a cloth. They also boil and drink an infusion from the roots of the green plant to neutralise the effect of snake, spider or scorpion bites.
- In the Delta, the sap of the leaves is used to treat pimples and acne.
- Wayeyi use a mixture of the roots of this plant with those of wild sage (*Pechuel-loeschea leubnitziae*) and buffalo thorn (*Ziziphus mucronata*) as an internal treatment for gonorrhoea.
- San mix it with a soft porridge to improve the metabolic system.
- Hmbukushu mix pounded fruit with the food of lazy hunting dogs, who vomit and then will avidly follow the scent of any animal.
- Subiya call it the 'Christmas flower'. They cut it and place it in their houses to show happiness and peace.

Roodt p.105, WFNSA p.36. BB, CM, MM, SS, MT.

Commelinaceae (spiderwort family)

Commelina subulata Roth
(This family is currently under revision and consequently these scientific names may change.)

Derivation: *Commelina*, for the Dutch botanists Johan Commelijn (1629–92) and his nephew Caspar (1667–1731); *subulata*, awl-shaped [Latin], referring to the spathes.

Identification: Erect rangy **annual herb**, c. 25 cm tall but may grow taller if supported. **Leaves** alternate, linear, folding slightly at the mid-vein, margins decurrent and forming a short stem sheath from which the spathes emerge. There may be long **hairs** at the lower leaf axils. **Flowers** dull orange, c. 1 cm across, emerging from a crescent-shaped folded **spathe** that is c. twice as long as it is wide; spathe with a margin of **hairs**, no fluid present in the spathe.

Most *Commelina* species flower at the apex of the plant but *C. subulata* flowers at the leaf nodes along the stem.

Habitat: Found occasionally locally in large quantities beside seasonal pans in mopane woodland.

Flowering: During the main rains.

B Van Wyk Flowers p.124, WFNSA p.32.

Crassulaceae (stonecrop or jade tree family)

Kalanchoe lanceolata (Forssk.) Pers.

Common names: Setswana semonye, moithimodiso; **English** kalanchoe.

Derivations: *Kalanchoe*, derived from the Chinese name for one species of this genus; *lanceolata*, lanceolate, spear-shaped [Latin], referring to the leaves being of long narrow shape tapering to a point. *Moithimodiso*, 'the thing that makes you sneeze', referring to its use as a medicinal snuff.

Identification: Semi-perennial, erect, single-stemmed, **succulent**, up to c. 1.5 m tall. **Stem** round with 4 buttress ribs. **Leaves** silver-grey, sessile, lanceolate, irregularly toothed, opposite, decussate, c. 20 × 6 cm, growing at right-angles to the stem. **Flowers** brilliant orange, 4 rounded petals with distinct points; growing in scorpioid racemes from the leaf axils and at the apex of the plant. The plant has a strong green-vegetable **smell** when crushed.

Habitat: Widespread, found occasionally locally in light shade near seasonal pans and in mopane woodland.

Flowering: From late in the main rains well into the cool dry period.

Uses and beliefs:
- The plant is dried, ground and mixed with snuff for use as a cold remedy. It causes violent sneezing and is said to clear the head. San from the Ghanzi district dry and grind the roots for snuff as a treatment for headache.
- The same bushmen also chew the roots and swallow the saliva or make a boiled infusion from the roots, if there is water available, as a tonic for general ill-health or loss of appetite.

FZ Vol.7 pt1, Ellery p.166, Flowers Roodt p.67, Germishuizen p.154, Hargreaves p.22, Blundell p.34.

Elatinaceae (waterwort family)

Bergia ammannioides Roth

Derivation: *Bergia* for Peter Jonas Bergius (1730–90), Swedish botanist and physician, pupil of Linnaeus; *ammannioides*, similar to the Lythraceae genus *Ammannia*.

Identification: Small erect **annual herb**, c. 15 cm tall, branching from low down. An overall red-green colouration but collects sand in its hairs to give it a dirty unkempt appearance. **Leaves** are opposite, lanceolate, with serrated margins, c. 2 cm long. **Flowers** borne in stalkless clusters in leaf axils and at the apex of the plant; flower buds red opening to pale pink cup-shaped flowers.

Habitat: Found occasionally in damp clayey areas of the floodplain.

Flowering: During the early rain.

FZ Vol.1 pt2.

Euphorbiaceae (spurge family)

Acalypha vilicaulis Hochst. ex A.Rich.

Common names: Setswana chaodabi, makgonatshe, makgonatsotlhe, mogalori-kodumela; **English** hairy stemmed acalypha.

Derivation: *Akalephe*, a nettle or its sting [Greek]; *villus*, hairy [Greek], *kaulos*, stem [Greek], hairy-stemmed.

Identification: Shrubby branching **herb** from a woody rootstock, up to c. 50 cm tall, densely covered in fine silky **hairs** overall. **Leaves** alternate, narrowly cordate with sharply toothed margins; leaf blades hang down almost against the stem; pairs of small bristle-like **stipules** at leaf axils, c. 3 mm long. **Male and female flowers** are borne separately on the same plant; **female flowers** are terminal, c. 2 cm tall, a mass of striking red styles emerge from a green receptacle; **male inflorescences** are spikes, c. 4 cm long, of c. 1 mm dia. cream flowers, borne at the leaf axils.

Habitat: Locally rare on eroded termite mounds on the spillway margin.

Flowering: During the main rains.

Uses and beliefs:
- The roots are used for a variety of conditions from heart problems and coughs to male or female infertility, and as a contraceptive.

FZ Vol.9 pt4, B Van Wyk New p.204, Hargreaves p.38, WFNSA p.224.

Euphorbiaceae (spurge family)

Euphorbia crotonoides Boiss.

Derivation: *Euphorbia*, in honour of Euphorbus, physician to King Juba II of Mauretania (24 BC); *croton*, *-oides*, croton-like [Greek], similar to the genus *Croton*.

Identification: Erect, branching, **annual herb**, up to c. 1 m tall. **Stems** 5-ribbed, almost winged, a thin milky **latex** pours out when cut. **Leaves** opposite, ovate with serrated margins; lower surface has a prominent, keeled main vein; leaf stalk flattened, c. 4 mm wide; leaf axils are stained red and there are pairs of glandular stipules. **Flowers** c. 1.5 mm, vestigial, borne at the apex of the plant and of the side branches, hairy; yellow stamens surrounded by 4 yellow or dark red nectariferous glands. **Fruit** pear-shaped, hairy, with a prominent retained stigma at the apex.

Habitat: Found rarely, but in considerable numbers when found, by sandy-clay track-side in mopane woodland.

Flowering: During the main rains.

FZ Vol.9 pt5.

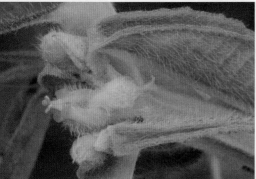

Euphorbiaceae (spurge family)

Euphorbia hirta L.

Common names: English dove milk.

Derivation: *Euphorbia*, in honour of Euphorbus, physician to King Juba II of Mauretania (24 BC). *hirta*, hairy [Latin], with long distinct hairs.

Identification: A semi-recumbent **annual herb**, c. 20 cm tall, covered in short, flat, white **hairs**. The whole plant is suffused with red, although the underside of the leaves are a lighter green. **Leaves** opposite, oval, slightly asymmetrical, stalkless, with serrated margins. **Inflorescences** tight, c. 1 cm, balls of separate **male and female flowers** in the leaf axils and at the apex of the stems; **female flowers** have pale pink petals.

Habitat: Found occasionally locally in areas of mixed riverine woodland.

Flowering: During the main rains.

FZ Vol.9 pt5, Turton p.22.

Euphorbiaceae (spurge family)

Ricinus communis L. var. *communis*

Common names: Setswana mokhure, moono, mono, mfuthe; **English** castor oil plant, castor bean, castor-oil bush.

Derivation: *Ricinus*, a tick [Latin], referring to the appearance of the seeds; *communis*, common, general [Latin].

Identification: Large vigorously growing **perennial herb**, c. 2 m tall. **Leaves** peltate and deeply 8-lobed with serrated margins; leaf stalks long, blotched grey-green, with small spines on the underside; leaf blade dark green, shiny. **Male and female flowers** borne in the same raceme; **male flowers** towards the base, small balls of yellow stamens emerging from a pair of sepals; **female flowers** have no visible petals or sepals, bright orange-red stigma emerging from a large spherical ovary with many protuberances. Ripe **fruit** dehisce abruptly shooting seeds several metres.

Habitat: This introduced, potentially invasive, herb is locally uncommon growing in light shade near water.

Flowering: Appears to flower and seed throughout the year.

Uses and beliefs:
- Introduced for medicinal purposes. The seeds are very poisonous and must be harvested with care. They are crushed and the oil is used to make the well-known purgative.
- Unattractive to animals, never browsed.

FZ Vol.9 pt4, Hargreaves p.42.

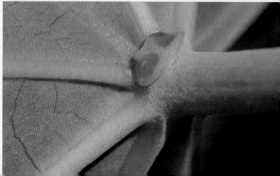

Fabaceae (pod-bearing family), Caesalpinioideae (cassia sub-family)

Chamaecrista absus (L.) H.S.Irwin & Barneby

Common names: English pig's senna.

Derivation: *Chamae*, dwarf or low-growing [Greek], *crista*, crest or terminal tuft [Latin]; the derivation of *absus* is obscure, but possibly *ab*, away from, out of, *sub*, somewhat, a little [Latin], meaning that it is a little different from other plants of the genus, referring to the colour of its flowers.

Identification: Erect branching **annual herb**, c. 40 cm tall. **Leaves** compound, spiralling up the stem, with 2 pairs of broadly oval leaflets with blunt tips that fold together; leaflet margins entire. **Flowers** orange-red, borne in racemes from the leaf axils, the 4 main petals have a claw. **Fruit** in the form of a pod, c. 8 × 0.8 cm, flat with clearly delineated seeds.

Habitat: Widespread in years of high rainfall in grassy damp areas of seasonal pans, not actually on the pan margin.

Flowering: During the main rains.

Fabaceae (pod-bearing family), Papilionoideae (pea sub-family)

Crotalaria podocarpa DC.

Derivation: *Krotalon*, rattle [Greek], as the seeds of many species rattle in the pod; *podos*, foot [Greek], *karpos*, fruit [Greek], alluding to the shape of the fruit.

Identification: Annual prostrate herb with stems c. 25 cm long. **Leaves** alternate, trifoliate; leaflets linear with a rounded tip with a bristle, margins smooth. **Flowers** like those of pea, an orange standard with a yellow centre and yellow wings with a red blush, all slightly variable; usually borne in pairs on stems at the leaf axils; 2 large leafy **stipules** at the axils. **Fruit** short inflated cylindrical pods with rounded ends, c. 2.5 × 1.2 cm; initially pinky-green turning to green with a maroon seam between the 2 halves.

Habitat: Locally uncommon in full sun on open sandy ridges in mopane woodland.

Flowering: Late in the main rains.

FZ Vol.3 pt7.

Indigofera astragalina DC.

Derivation: *Indigo*, the colour indigo, *fera*, bearing [Latin], referring to plants of this genus that are used to obtain indigo dye; *astragalus* is the genus of milk vetch. This particular *Indigofera* resembles a milk vetch.

Identification: A strong branching **annual herb**, usually c. 30 cm but up to c. 60 cm in years of high rainfall. All **stems** and **leaves** hairy; the **hairs** on the stems and leaf stems are dark red. **Leaves** compound imparipinnate, usually 4 pairs plus terminal leaflet, oval with blunt tips, margins entire, net veining, each leaflet c. 4.5 × 2 cm on a c. 2 mm leaf stalk; a pair of long (c. 2 cm) pale-coloured **stipules** at each axil. **Flowers** pale salmon pink, c. 3 mm across, with long pointed sepals, in a spike up to c. 8 cm long, emerging from leaf axils. **Fruit** short, ribbed cylinders with rounded ends, the style is often retained.

Habitat: Widespread but not common in mopane woodland and along spillway margin. Appears to need a certain amount of leaf litter in the sand to thrive.

Flowering: Middle to end of the main rains

Uses and beliefs:
- Much loved by black rhinoceros as a food plant.

MI.

Fabaceae (pod-bearing family), Papilionoideae (pea sub-family)

Indigofera colutea (Burm.f.) Merr.

Derivation: *Indigo*, the colour indigo, *fera*, bearing [Latin], referring to plants of this genus that are used to obtain indigo dye; *kolutea*, the Greek name for bladder senna, another member of the Fabaceae.

Identification: Small erect **annual herb**, c. 40 cm tall; **sticky** and **hairy** overall, with pairs of c. 3 mm stipules in the leaf axils. **Leaves** compound with up to 5 pairs of leaflets plus a terminal leaflet, c. 3.5 × 1 cm. **Flowers** c. 3 mm long with strong salmon pink standard, wings and throat of the standard a darker salmon-red. **Seed capsules** are tubular pods that give off a pungent spicy **smell** when crushed.

Habitat: Found occasionally in dry years but widespread in years of heavy rains growing in full sun on the floodplain and on island margins.

Flowering: During the later part of the main rains.

Fabaceae (pod-bearing family), Papilionoideae (pea sub-family)

Indigofera filipes Harv.

Derivation: *Indigo*, the colour indigo, *fera*, bearing [Latin], referring to plants of this genus that are used to obtain indigo dye; *fili-* as a prefix means thread-like [Latin], referring to the flower and leaf stalks.

Identification: A delicate, erect, branching, **annual herb**, c. 30 cm tall. **Leaves** alternate, compound, with fine thread-like stalks, up to 5 pairs of leaflets plus a terminal leaflet; leaflets linear, very narrow relative to their length. **Flowers** pea-like, deep salmon red, usually borne in pairs on fine thread-like stems. **Fruit** small, almost cylindrical pods.

Habitat: Locally uncommon, growing in full sun in deep sand.

Flowering: During the main rains.

B Van Wyk Flowers p.208, WFNSA p.190.

Fabaceae (pod-bearing family), Papilionoideae (pea sub-family)

Indigofera trita L.f.
subsp. *subulata* (Vahl ex Poir) Ali
var. *subulata*

Derivation: *Indigo*, the colour indigo, *fera*, bearing [Latin], referring to plants of this genus that are used to obtain indigo dye; *trita*, either common-place, ordinary [Latin] or *tri-* three, referring to the number of leaflets [Latin]; *subulata*, awl-shaped [Latin].

Identification: Small **semi-recumbent herb**, c. 40 cm tall, may be either **annual or perennial**; covered in fine silky white **hairs** overall. **Leaves** alternate, compound, leaf tips rounded, smooth margins, 3 oval leaflets; with pairs of, often brown, bristle-like **stipules** at the leaf axils. **Flowers** borne in close spikes, soft orange with yellow markings in the throat of the standard.

Habitat: Found occasionally locally in lightly shaded mopane woodland.

Flowering: During the main rains.

Ellery p.164.

Zornia glochidiata DC.

Derivation: *Zornia*, for Johannes Zorn, a German pharmacist and botanist, 1739–99; *glochidiata*, provided with barbs [Latin], referring to the spiny seedpods.

Identification: Erect branching **annual herb**, up to c. 30 cm tall. **Stems** covered in short fine hairs. **Leaves** c. 3.5 × 0.9 cm, alternate, compound, smooth margined, with sparse long hairs and 2 narrowly ovate hairless leaflets; pairs of diamond-shaped stipules attached at the mid-point at the leaf axils. **Inflorescence** a spike of **flowers**, c. 3 mm dia., borne between 2 leafy bracts with hairy margins; keel and wing petals are red with a yellow margin. **Fruit** are pods covered with barbed bristles, a constriction between each seed.

Habitat: Widespread in full sun in open sandy areas of mopane woodland or the edge of the tree-line.

Flowering: During the main rains.

FZ Vol.3 pt6, B Van Wyk Flowers p.144.

Iridaceae (iris family)

Ferraria glutinosa (Baker) Rendle

Derivation: *Ferraria* commemorates the Italian botanist Giovanni Battista Ferrari (1584–1655), who was horticultural advisor to the papal family; *glutinosa*, sticky, glutinous [Latin].

Identification: Erect, branching **herb**, c. 1 m tall, growing from a **corm**, completely **hairless**. **Stem** very sticky for c. 2 cm below the leaf sheaths, more yellow in colour in those areas. **Leaves** linear and grass-like, up to c. 60 cm long and 1.2 cm wide, tightly folded at the mid-rib, mainly at the base of the plant. **Flowers** with 6 brown petals edged with yellow, also splotched yellow in the centre, emerging in pairs from a cylindrical pointed sheath; stamens orange. **Fruit** erect and globose, with 6 segments.

Habitat: Locally rare in sandy mopane woodland.

Flowering: Early in the main rains.

FZ Vol.1 pt2.

Lamiaceae (mint family)

Clerodendrum uncinatum Schinz
(syn. *Rotheca uncinata* (Schinz) Herman & Retief)

Common names: Setswana thiba-di-molekane, mokhesa; **English** cat's claw.

Derivation: *Kleros*, chance [Greek], *dendron*, tree [Greek], supposedly an allusion to the variable medicinal qualities of these trees, shrubs and climbers; *uncinatus*, hooked at the end [Latin], referring to the vicious thorns on this plant.

Identification: Low-growing, much-branched **shrubby perennial**, up to c. 1 m tall but usually smaller. Exceptionally long **roots** because these plants flower in the cool dry period and need to go deep in search of moisture. **Stem** square and buttressed. Stems and leaves covered in fine white **hairs**. **Leaves** dull blue-green, c. 5 × 2.2 cm, opposite and decussate, folding up against stem to reduce transpiration, oval with smooth margins and net veining; pairs of hooked **thorns** at the leaf axils. **Flowers** scarlet with yellow streaking towards centre, c. 2.5 cm dia., borne singly in the leaf axils, 5 asymmetrical petals, 4 stamens. **Fruit** have 4 segments.

Habitat: Widespread but not common locally in deep sand in full sun.

Flowering: During the cool dry period.

Uses and beliefs:
- Kade bushmen eat the berries when other foods are scarce.
- Mixed with *Euclea divinorum* and *Albizia harveyi*, this plant is used to treat sexually transmitted diseases.
- A sign of deep sand in trackways.

Flowers Roodt p.156, Hargreaves p.21.

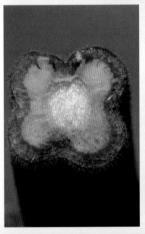

Lamiaceae (mint family)

Leonotis nepetifolia (L.) R.Br.

Common names: Setswana semonamone, misnamed as dagga or motokwane (see derivation); **San** oi?; **English** wild dagga.

Derivations: *Leon*, lion [Greek], *otis*, ear [Greek], referring to the resemblance of a single flower to a lion's ear; *nepeta*, catnip, *folia*, leaves [Latin], having leaves like catnip. *Semonamone*, something you suck, referring to the sweet nectar in the flowers. *Dagga* and *motokwane*: the plant is misnamed because the young plants resemble those of *Cannabis sativa* and it is erroneously supposed to have the same psychoactive properties.

Identification: Erect, branching, **annual herb**, up to c. 2 m tall. **Stem** buttressed and square. **Leaves** opposite and decussate, velvety and broadly ovate, with toothed margins; in dry weather, leaves often hang almost against the stem. **Flowers** brilliant orange, grouped in spherical clusters at the upper leaf axils, tubular and softly hairy, upper lip forms a hood over the lower, opening in rows down the ball, emerging from a spiny calyx; mainly pollinated by sunbirds.

Habitat: Common and widespread on islands and in riverine woodland, usually in light shade but occasionally in full sun.

Flowering: During the main rains.

Uses and beliefs:
- Locally known as *dagga*, the leaves are smoked in place of cannabis but have been proven to have no psychoactive effect.
- Batswana women believe this plant to be a contraceptive: drinking water boiled with 2 seeds gives 2 years without pregnancy, 3 seeds, 3 years and so on.
- An infusion of pounded leaves is used as an antiseptic wash that also helps counteract insect bites and heat rash.

Ellery p.168, Flowers Roodt p.97, Turton p.101, Blundell p.403. CM, RR, KS.

Lythraceae (bloodflower family)

Ammannia baccifera L. subsp. *baccifera*

Derivation: *Ammo-*, sand- [Greek], possibly referring to the sandy habitat of this species; *bacca*, berry [Latin], *fera*, bearing [Latin], berry-bearing.

Identification: Small, erect, **annual**, pink-stemmed **herb**, branching from the base, c. 25 cm tall. Dark red-green appearance overall. **Stems** 4-angled. **Leaves** c. 1 cm long, dark green, opposite, decussate, linear with smooth margins. **Flowers** c. 1 mm dia., pinky-red, with 4 lobes, borne on short stalks in clusters at the leaf axils.

Habitat: Found occasionally on the floodplain.

Flowering: During the early rain.

FZ Vol.4 pt0.

Lythraceae (bloodflower family)

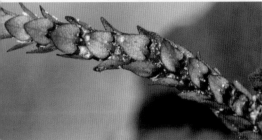

Rotala filiformis (Bellardi) Hiern
(syn. *Rotala heterophylla* A.Fern & Diniz)

Derivation: *Rotalis*, wheeled or wheel-like [Latin], a reference to the whorled leaves of many species of this genus; *hetero*, diverse [Greek], *phyllon*, leaf [Greek], diversely leaved.

Identification: Small erect herb, c. 6 cm tall, growing in a square cross-section, columnar form. The red-green colouring of the plant allows it to absorb light across a wide spectrum. **Leaves** deltoid in shape and scale-like, tightly packed against the stems in opposite and decussate arrangement. Minute pinkish-red **flowers** appear in the axils of the leaves.

Habitat: Found occasionally in damp areas of the floodplain.

Flowering: During the early rain.

FZ Vol.4 pt0.

Portulaca kermesina N.E.Br.

Derivation: *Portare*, to carry [Latin], *laca*, milk [Latin], i.e. milk carrying, originally referring to the milky latex in *Portulaca oleracea* and then applied to other plants of the genus; *kermesina*, carmine [Latin].

Identification: A fairly erect, branching, **annual or perennial succulent**, c. 28 cm tall. **Leaves** long, narrow and almost cylindrical, c. 2 × 0.3 cm, spiralling up the stem; long white **stipular hairs** at the **leaf axils**; a whorl of leaves below the cluster of flowers at the apex of the stem. **Flowers** with 5 overlapping, oval, carmine petals, a point at the apex of each; at least 10 stamens with golden yellow anthers and carmine filaments.

Habitat: Locally rare in sandy clay areas of the floodplain and beside tracks in clearings in mopane woodland.

Flowering: Towards the end of the main rains.

FZ Vol.1 pt2, B Van Wyk New p.230, WFNSA p.134.

Sterculiaceae (cacao or cola family)

Hermannia eenii Baker f.
(syn. *Hermannia angolensis* K.Schum.)

Derivation: *Hermannia*, in honour of Paul Hermann (1646–95), a German botanist, professor of botany at Leiden in Holland, who visited southern Africa; *angolensis*, from Angola.

Identification: A prostrate mat-forming **perennial herb** with stems c. 50 cm long; coarse yellowy stellate **hairs** overall. **Leaves** bright green, alternate, oval, with coarsely serrated margins; pairs of long **stipules** at each leaf axil. **Flowers** borne mainly in the leaf axils of side shoots, salmon pink, c. 8 mm dia., both the calyx and the stamens markedly longer than the petals.

Habitat: Widespread in full sun on sandy areas of the floodplain and spillway.

Flowering: From the middle of the main rains into the cool dry period.

FZ Vol.1 pt2.

Sterculiaceae (cacao or cola family)

Hermannia glanduligera K.Schum.

Common names: Setswana masigomabe, masogomabe; **Subiya** masiku maave; **Afrikaans** gombossie.

Derivations: *Hermannia*, in honour of Paul Hermann (1646–95), a German botanist, professor of botany at Leiden in Holland, who visited southern Africa; *glanduligera*, gland-bearing. *Masiku maave*, sleepless night.

Identification: Erect branching **woody herb**, c. 2 m tall if supported by other vegetation, covered with sticky glandular hairs overall. **Leaves** ovate with serrated margins, up to c. 4 × 2.5 cm but usually smaller, spiralling up the stem. **Flowers** c. 7 mm dia., borne singly in the leaf axils on pendant stems, with 5 salmon pink petals (there is a white variant), a pair of small bracts (c. 3 mm) behind flower head; rolled-back sepals longer than the petals; stamens white, almost twice the length of the petals. **Seed capsules** have 5 segments with a single erect spike on each segment. A spicy pungent **aroma** when the plant is crushed.

Easily confused with *Hermannia kirkii* but the lengths of the sepals and stamens, and the single spikes on the seed capsule segments, are diagnostic.

Habitat: Widespread on lightly shaded islands on the floodplain.

Flowering: During the main rains.

Uses and beliefs:
• The roots are used by San to burn decorative marks on the skin.

FZ Vol.1 pt2, Ellery p.162, Germishuizen p.83, Turton p.77, WFNSA p.260. GM.

Sterculiaceae (cacao or cola family)

Hermannia kirkii Mast.

Common names: Setswana matsogomabe.

Derivation: *Hermannia*, in honour of Paul Hermann (1646–95), a German botanist, professor of botany at Leiden in Holland, who visited southern Africa; *kirkii* in honour of Sir John Kirk (1832–1922), British Consul at Zanzibar, who accompanied David Livingstone on his Zambezi expedition of 1858.

Identification: An erect **annual herb** with branching stems, c. 50 cm tall but occasionally up to c. 1.2 m, covered with short sticky glandular **hairs**. **Leaves** narrowly elliptical to linear with serrated margins, spiralling up the stem. **Flowers** c. 1 cm dia., borne singly at the leaf axils on long almost straight stems which are often jointed just behind the flower, 5 salmon red petals (there is a rare white variant), petals almost twice as long as the sepals and the stamens; stamens blue-mauve. **Seed capsules** 5 segments with a pair of spikes radiating from each segment.

This species is easily confused with *Hermannia glanduligera* but the lengths of the sepals and stamens and the double spikes on the seed capsule segments are diagnostic.

Habitat: Locally widespread, found in full sun or very lightly shaded areas of floodplain and islands. It thrives on eroded termite mounds.

Flowering: During the main rains.

FZ Vol.1 pt2, *Blundell p.72*.

Turneraceae (turnera family)

Streptopetalum serratum Hochst.

Derivation: *Streptos*, twisted [Greek], *petalon*, petal [Greek]; *serratum*, having serrated leaf margins, like the teeth of a saw [Latin].

Identification: Erect branching **annual herb**, c. 40 cm tall, all parts **hairy** and rough. **Leaves** alternate, narrowly ovate with serrated margins. **Flowers** c. 1 cm dia., borne in spikes from the leaf axils, bright apricot-yellow with 5 overlapping blunt petals emerging from a c. 1.5 cm calyx tube. **Fruit** ellipsoid, dehiscent, opening into 3 sections.

Habitat: Found occasionally locally in damp degraded areas of mopane woodland, in full sun.

Flowering: During the main rains.

FZ Vol.4 pt0, Blundell p.254, Turton p.77.

Turneraceae (turnera family)

Tricliceras glanduliferum (Klotzsch) R.Fern.

Common names: Setswana monontshane, tobi.

Derivation: *Tricliceras*, 3-chambered, referring to the fruit; *glans*, *glandis*, gland [Latin], *fero*, *ferre*, to bear or carry [Latin], gland-bearing.

Identification: Erect **annual herb**, up to c. 45 cm tall. **Stems** covered in long erect **hairs**. **Leaves** lanceolate with deeply serrated margins. **Flowers** borne in groups of 2 or 3 on long stems from the leaf axils, c. 1.5 cm dia., bright orange, a flat disc of 5 blunt overlapping petals. **Fruit** thin cylinders up to c. 8 cm long with restrictions between the seeds, fruit stalks are as long as the fruit.

Habitat: Locally uncommon in damp clayey areas and wallows in the mopane woodland.

Flowering: Flowers during the main rains.

FZ Vol.4 pt0.

Vitaceae (grape family)

Ampelocissus africana (Lour.) Merr.

Derivation: *Ampelos*, vine [Greek], *kissos*, ivy [Greek], an ivy-like climbing vine; *africana*, of Africa.

Identification: Herbaceous climber, c. 3 m tall, scrambling through other vegetation, may be quite **hairy**. **Immature leaves** oval with palmate veining and serrated margins. **Mature leaves** palmate with 5 lobes, margins incised and irregularly toothed, up to c. 30 × 22 cm; branching **tendrils** from the leaf axils. **Flowers** dark red, c. 5 mm dia., in branching clusters at the leaf axils, star-shaped with 5 pointed petals; stamens opposite the petals, anthers yellow. **Fruit** small, spherical.

Habitat: Found occasionally locally in riverine bush along the Chobe River.

Flowering: During the early rains and into the main rains.

FZ Vol.2 pt2.

Pink
red or red-purple flowers

Aizoaceae (carpetweed family)

Gisekia africana (Lour.) Kuntze

Common names: Setswana motlhabana; **Shiyeyi** onshabwe; **English** gisekia.

Derivations: *Gisekia* for P.D. Giseke, a German botanist, pupil of Linnaeus; *africana*, from Africa. *Onshabwe* is a nickname meaning 'the buffalo' because they are everywhere.

Identification: Prostrate to procumbent hairless **annual herb**, spreading to c. 0.8 m. **Stems** often pinkish or red. **Leaves** opposite, linear, smooth margined, stalkless. **Flowers** with 5 petals, greenish yellow, pink or white, c. 3 mm dia.

Easily confused with *Gisekia pharnacioides* but it has 5 stamens whereas *G. africana* has 10–15.

Habitat: Locally common in open sandy locations in the floodplain in full sun, also on disturbed sandy soils.

Flowering: After the early rains and during the main rains.

Uses and beliefs:
• Unmarried Subiya women are not supposed to cut this plant or grow it in their homes because it is believed that it will prevent them from finding the right husband.

FZ Vol.4 pt0, Botweeds p.2, Hargreaves p.8, WFNSA p.132. CM.

Aizoaceae (carpetweed family)

Sesuvium hydaspicum (Edgew.) M.L.Gonç.

Identification: Low-growing **succulent**, c. 20 cm tall. **Stems** pinkish, hairy. **Leaves** c. 2.2 cm (inc. c. 0.5 cm stalk) × 0.5 cm, opposite, narrowly oval with smooth margins; lower surface hairy, upper surface rough. **Flowers** borne at the leaf axils, c. 0.7 cm dia., pinkish-purple, 5 petals, 5 long sepals that appear as points behind the petals.

Habitat: Locally uncommon found in full sunshine on lightly shaded islands.

Flowering: During the main rains.

Amaranthaceae (pigweed or cockscomb family)

Achyranthes aspera L. var. ***pubescens*** (Moq.) Townsend

Common names: Setswana molora, motshwarakgano; **Afrikaans** grootklits, kafblom; **English** (rough) chaff flower, burweed, devil's horsewhip.

Derivations: *Achyron*, chaff or husk [Greek], *anthos*, flower [Greek], referring to the dry chaffy nature of the flowers; *pubescens*, downy [Latin]. *Molora* is Setswana for both ash and soap. *Motshwarakgano*, 'the one that catches the slender mongoose'.

Identification: Vigorously growing **herbaceous annual**, up to c. 2 m tall if supported by other vegetation. **Leaves** c. 12 cm (inc. c. 0.8 cm leaf stalk) × 6 cm, elliptical, opposite, decussate, with decurrent and slightly wavy margins, downy; leaf narrows at the base to form a leaf stalk. The decurrent leaf margins form the buttresses of the square **stem. Flowers** pink, star-shaped, c. 0.6 cm dia. **Fruit** readily attach themselves to animal fur to aid seed dispersal.

Achyranthes aspera var. *pubescens* can easily be mistaken for *A. aspera* var. *sicula*, especially in young growth; great care must be taken if the plant is to be used for medicine or food as var. *pubescens* is very poisonous. The flower spike of var. *pubescens* is larger and more heavily pubescent than that of var. *sicula*. Var. *pubescens* often has pink flowers whereas var. *sicula* usually has green flowers. Var. *pubescens* is covered in a white down.

Habitat: Locally common and widespread in the shelter of heavy vegetation on decaying termite mounds and islands. [Introduced.]

Flowering: From the middle of the main rains into the cool dry period.

Uses and beliefs:
• This introduced variant of *Achyranthes aspera* is highly poisonous. Care must be taken not to confuse it with the indigenous variants, which have a multitude of beneficial nutritional and medicinal uses.

FZ Vol.9 pt1, Ellery p.153.

Amaranthaceae (pigweed or cockscomb family)

Hermbstaedtia linearis Schinz

Common names: Setswana mosiama; **San** xao, xummu; **Subiya** muchira ukaza; **Afrikaans** katstertjie, rooiaarbossie; **English** cat's tail, woolflower.

Derivations: *Hermbstaedtia* for S.F. Hermbstadt, Professor of Chemistry and Pharmacy at Berlin University; *linearis*, narrow, with sides almost parallel [Latin]. *Muchira ukaza*, tail of the cat.

Identification: Annual (occasionally perennial) erect woody **herb**, branching at the axils, c. 90 cm tall, covered in fine short **hairs**. **Stems** ribbed. **Leaves** dark green to red, alternate, linear, margins smooth; leaf blade narrows to form a leaf stalk. **Flowers** a tight spike of papery pink blooms, up to c. 25 cm long.

Hermbstaedtia linearis and *H. odorata* can easily be confused, but *H. odorata* has a short (or obsolete) style, whereas *H. linearis* has a distinct, slender style protruding beyond the flower. *H. linearis* is also a less vigorous plant. In *H. odorata*, the seed spike is covered with a mass of soft hairs.

Habitat: Widespread but locally uncommon on heavily trampled sandy areas of the floodplain in full sun.

Flowering: From the middle of the main rains into the cool dry period.

Uses and beliefs:

Local people do not necessarily differentiate between *Hermbstaedtia linearis* and *H. odorata* at the flowering stage but may do so in seed and may use the plants in similar ways.
- The burnt flowers are mixed with fat and used to treat depressed fontanelles in babies.
- Wayeyi men drink an infusion of the roots, boiled with fresh milk, to improve their potency.
- San take a strong infusion of the roots as a cure for gonorrhea for both men and women. The roots are soaked in one cup of water and taken 3 times per day until cured. It is said to be very effective. The roots may also be chewed and the saliva swallowed if water is not available.

FZ Vol.9 pt1, Germishuizen p.73, WFNSA p.130. BB, BN, KN, MT.

Hermstaedtia odorata

Asteraceae (daisy family)

Dicoma tomentosa Cass.

Common names: Setswana pelobotlhoko, tsetwane; **San** xaoqoo; **Shiyeyi** moXhXo; **English** doll's protea, hairy dicoma.

Derivations: *Dicoma*, *di*, two [Latin], *choma*, tuft at the summit [Latin], referring to the double seed pappus of the first species to have been described. *Tomentosa* refers to the matted hairiness of the plant. *Pelobotlhoko* means 'sore heart', referring to the use of the plant to cure heartache.

Identification: Erect **annual herb**, up to c. 1 m tall. **Stems** dark red. **Leaves** linear and sessile with smooth margins, folding along the mid-vein, covered in fine **hairs**, c. 5.5 × 0.5 cm. **Inflorescences** a tight group of spiny bracts, c. 1.5 cm dia. × 2 cm tall; 2 leafy bracts at the base of the flowers in the leaf axils.

Habitat: Widespread in the grassland and scrub surrounding seasonal pans where the soil is much trampled by animals grazing and coming to drink.

Flowering: From the middle of the main rains into the cool dry period.

Uses and beliefs:
- In Khwai Village, an infusion of boiled roots is drunk at regular intervals to cure heartache.
- Wayeyi and Hmbukushu use this species to treat *dikgaba*. If a child does not look after his parents, he may be bedevilled by *tokolosi*, evil spirits. The plant is soaked in water overnight and very early the next morning the parents go with the child to the place where the plants grow. They wash him with the infusion. They also take it in the mouth, swallow some and spray him with the rest. Alternatively, the plant may be burned on hot coals. If the problem is severe, it is mixed with *Aerva leucura*.
- *Dicoma tomentosa* is also used to cure the 'curse of raptors'. An infant is cursed if either the Bateleur or the Martial eagle flies over and casts its shadow on the child. The symptoms are sunken fontanelle and skew eyes. To cure the curse, the plant is cut into small pieces and placed on red hot coals and the baby is placed nearby to inhale the vapour. San in Ghanzi district have a similar belief concerning the Black-shouldered kite. Subiya use the ash of the plant to make a cross on the fontanelle, both on front and back of the head, thereby curing the curse invoked by the Bateleur.

FZ Vol.6 pt1, Flowers Roodt pp.18 & 45. BB, BN, SS.

Asteraceae (daisy family)

Sphaeranthus peduncularis DC.

Common names: English purple pan weed.

Derivation: *Sphaira*, globe, *anthos*, flower [Greek], referring to the globose flower heads; *peduncularis*, with a well-developed flower stalk or peduncle [Latin].

Identification: A densely growing branching **annual herb**, 20–50 cm tall and forming clumps up to c. 1 m across. **Stems** harshly hairy. **Leaves** decurrent forming 'wings' along the length of the stem, alternate, lanceolate, with serrated margins, slightly **sticky** and covered with soft **hairs**. **Inflorescences** composite, spherical, purple balls, c. 1 cm dia.; stigmas protrude from the florets. **Seed heads** retain the shape of the inflorescence. The plant has a spicy **aroma** when crushed.

Habitat: Although on the international list of endangered species, this herb is locally common and widespread in drying seasonal pans and surrounding areas.

Flowering: Almost throughout the year.

Bignoniaceae (jacaranda family)

Kigelia africana (Lam.) Benth. subsp. *africana*

Common names: Setswana moporota, kazungula, uvunguvungu, mosungula, mozungula; **Hmbukushu** movunguvungu; **Subiya** izungwe; **Shiyeyi** uxhoro; **Afrikaans** worsboom; **English** sausage tree.

Derivations: *Kigelia*, a Latinised version of the Mozambique vernacular name, *kigeli-keia*; *africana*, of Africa [Latin]. *Kazungula* is the name of the settlement in Botswana where Namibia, Zambia, Zimbabwe and Botswana meet at a point on the Zambesi. *Movunguvungu* is the tree; *mavunguvungu* is the fruit.

Identification: Medium-sized, deciduous **tree** with a rounded crown. **Bark** grey and smooth in younger specimens but rough with rounded flakes in larger trees. **Leaves** pale yellow-green on opening changing to dark green with age, roughly hairy becoming coarse and leathery with age, pinnately compound, 3–5 pairs of leaflets plus 1 terminal leaflet; side leaflets sometimes oblique at the base, margins entire. **Flowers** open as the leaves appear, strikingly large and dark red, corolla tube exterior often yellow-green; flowers hang in long open racemes, **pollinated** by bats and popular with birds as they contain large amounts of nectar. **Fruit** very distinctive sausage-shape, fibrous, hanging on long stems from the branches, up to c. 60 × 5 cm, some remain on the tree for most of the year.

Habitat: Widespread, but sparse, along treelines and on islands in the floodplain.

Flowering: Late in the cool dry period.

Uses and beliefs:
- Used as purgative.
- Because of the shape of the fruit Batswana, Wayeyi, Subiya and Hmbukushu men believe that if they chop the root and chew it, then drink the water, it will make their penis grow.
- The water from an infusion of the fruit is used to treat skin cancer.
- Wayeyi, Subiya and Hmbukushu use the wood to make *mekoro* (dug-out canoes).

FZ Vol.8 pt3, Blundell p.381, Ellery p.123, Palgrave updated p.1013, Tree Roodt p.137. PK, TK, CM, PN, SS, MT.

Campanulaceae (bluebell family)

Lobelia angolensis Engl. & Diels

Derivation: *Lobelia*, for the Flemish physician and botanist, Matthias de l'Obel (1538–1616); *angolensis*, of Angola.

Identification: Small erect (in some areas prostrate) **annual** or **short-lived perennial herb** with a **rhizomatous rooting system**. **Stems** often grooved with branches at the upper part of the plant. **Leaves** small, narrowly elliptical with smooth margins, narrowing to form a stalk at the base. **Flowers** borne singly in the upper leaf axils, pink/purple, c. 4 mm dia., 5 pointed lobes arranged asymmetrically; stamens dark blue-purple.

Habitat: Found rarely locally in the damp margins of the spillway and rivers.

Flowering: During the early rains and into the early part of the main rains.

FZ Vol.7 pt1, B Van Wyk Flowers New p.74, WFNSA p.420.

Commelinaceae (spiderwort family)

Floscopa glomerata (Schult. & Schult.f.) Hassk.

Derivation: *Flos*, flower [Latin], *scopa*, broom [Latin], alluding to the broom-like appearance of the inflorescence; *glomerata*, clustered into more or less rounded heads [Latin].

Identification: A sparsely branched **perennial herb**, up to c. 1 m tall growing supported in other vegetation. **Leaves** alternate, narrowly ovate, margins decurrent and forming a clasping sheath round the stem. Stems appear banded dark red and pale green. **Flowers** borne in a cluster of terminal spikes up to c. 6 cm tall, small, grey-purple, 3 petals; sepals densely covered in hairs; stamens bright yellow.

Habitat: Locally rare, growing in the margin of fast-flowing channels sheltered by *Cyperus papyrus*.

Flowering: During the rains.

B Van Wyk Flowers p.250, Ellery p.161, WFNSA p.34.

Convolvulaceae (morning glory family)

Ipomoea dichroa Choisy

Derivation: *Ips*, worm [Greek], *homoios*, like [Greek], referring to its trailing habit; *di*, two or twice [Greek], *chroa*, coloured [Greek].

Identification: Vigorous **annual climber**, up to c. 2.5 m. **Stems** heavily hairy. **Leaves** simple with palmate veining, 3 lobes, up to c. 30 cm (inc. petiole c. 15 cm) × 17 cm, base cordate, coarsely bristled on upper surface, a tangled mat of fine white hairs below. **Flowers** pink with a magenta centre, 5 lobes fused to form a funnel-shaped corolla tube, c. 2 cm dia. **Fruit** spherical.

Habitat: Locally widespread growing in light shade on islands and in the treeline along the floodplain.

Flowering: Late in the main rains.

FZ Vol.8 pt1 p.84

Convolvulaceae (morning glory family)

Ipomoea sp.
(similar to *Ipomoea dichroa* Choisy)

Derivation: *Ips*, worm [Greek], *homoios*, like [Greek], referring to its trailing habit.

Identification: Herbaceous climber, up to c. 2.5 m through the shade of its host; covered in a mass of white **hairs. Leaves** c. 12 cm wide, palmate with 5 deep lobes, margins smooth, underside silver grey. **Flowers** large c. 5 cm dia., bright pink, funnel-shaped, borne in clusters on c. 11 cm stalks at leaf axils; flowers close late in the afternoon. White **latex** present at breaks in the stem.

Habitat: Locally uncommon on sandy ridges in the mopane woodland.

Flowering: Flowers late in the main rains.

FZ vol.8 pt1.

Convolvulaceae (morning glory family)

Ipomoea eriocarpa R. Br.

Derivation: *Ips*, worm [Greek], *homoios*, like [Greek], referring to its trailing habit; *eriocarpa*, woolly-fruited [Latin].

Identification: Annual climbing **plant** up to c. 1 m tall depending on support. Young **leaves** sagittate, older ones cordate; margins entire, velvety hairs overall with longer hairs on the underside. **Flowers** pink, c. 1.3 cm dia., with linear sepals longer than the petals, almost sessile, growing in clusters in leaf axils. White **latex** present when the plant is cut.

Habitat: Widespread, in full sun, on grassland preferring the damper areas of the floodplain. Occasionally found growing in water.

Flowering: Late in the main rains.

FZ Vol.8 pt1.

Convolvulaceae (morning glory family)

Ipomoea pes-tigridis L.

Common names: English tiger-foot ipomoea.

Derivation: *Ips*, worm [Greek], *homoios*, like [Greek], referring to its trailing habit; *pes*, foot [Greek], *tigridis*, of the tiger [Latin].

Identification: A creeping and scrambling **plant**, up to c. 2 m tall (depending on support). **Stems** and **leaves** hairy. **Leaves** alternate, palmate with 7–9 deep lobes, margins entire. **Flowers** large, pale pink with a magenta throat, c. 5 cm dia., closing in the midday sun. White **latex** in the stems.

Habitat: Found occasionally locally either prostrate or climbing with the support of other vegetation in full sun; sometimes in shade in sandy areas of mixed woodland or scrub.

Flowering: Flowers late in the main rains.

FZ Vol.8 pt1.

Convolvulaceae (morning glory family)

Ipomoea shirambensis Baker

Common names: Subiya ungulu, womuzuka.

Derivation: *Ips*, worm [Greek], *homoios*, like [Greek], referring to its trailing habit; *shirambensis*, from Shiramba in the Lower Zambezi valley in Mozambique where the plant was collected for the first time.

Identification: Tall **deciduous woody climber** with pale grey **stems**. **Roots** grow tubers like sweet potatoes. **Leaves** cordate, slightly hairy, margins smooth. Flowers before the leaves are produced. **Flowers** funnel-shaped, pale pink with a darker centre, c. 6 cm dia.; flowers last only 1 day and close by mid-afternoon.

Habitat: Locally common in rocky riverine bush near the Chobe River.

Flowering: During the cool dry period.

Uses and beliefs:
- The roots are like sweet potatoes and can be eaten raw or cooked. To find the tubers, look for cracks in the soil among the rocks.

WFNSA p.336. GM.

Elatinaceae (waterwort family)

Bergia pentheriana Keissl.

Derivation: *Bergia* for Peter Jonas Bergius (1730–90), a Swedish botanist and physician, pupil of Linnaeus. *penta-*, five, *-anthera-*, stamens [Greek], with 5 stamens.

Identification: Small **herbaceous creeper** with bright red stems up to c. 75 cm long. **Leaves** stalkless, oval, opposite, decussate, margins hairy, **veins** showing white on both surfaces. **Flowers** pale pink, borne in clusters on short side stalks in the leaf axils, c. 4 mm dia., 5 petals and alternate sepals.

Habitat: Common and widespread in sandy areas of the floodplain.

Flowering: Throughout both early and main rains.

Euphorbiaceae (spurge family)

Euphorbia inaequilatera Sond.

Common names: English smooth creeping spurge.

Derivation: *Euphorbia*, in honour of Euphorbus, physician to King Juba II of Mauretania (24 BC); *inaequalis*, unequal, *latera*, side [Latin], referring to the asymmetrical leaves.

Identification: Small prostrate creeping **annual herb** c. 20 cm dia. with a single **taproot**. **Stems** dark pink. **Leaves** dark red/grey-green, asymmetrically oblong with a rounded tip, opposite, stalkless, smooth margined. **Male and female flowers** separate. **Female flowers** small c. 1 mm dia., dark red lobes with a white margin; **seed capsule** visible on a short stem hanging from the centre of the flower, gradually growing erect. **Male flowers** similar but without the capsule.

Habitat: Widespread in hard clay areas of the floodplain near seasonal pans and on lightly shaded islands.

Flowering: During the main rains.

Uses and beliefs:
• The whole plant can be used as a dusting powder for infants or smoked as a love potion.

FZ Vol.9 pt5, Hargreaves p.41.

Euphorbiaceae (spurge family)

Euphorbia polycnemoides Boiss.

Common names: English dove milk.

Derivations: *Euphorbia*, in honour of Euphorbus, physician to King Juba II of Mauretania (24 BC); *Polycnemum*, a genus of the *Chenopodiaceae*, *-oides*, similar to, looks like a *Polycnemum*. *Dove milk*, alluding to the latex in the stems.

Identification: Small blue-grey branching **herb** with pink **stems** up to c. 20 cm tall. **Leaves** c. 1.0 × 0.6 cm, grey-green, opposite, decussate, stalkless, oval with a rounded tip, margins serrated. **Male and female flowers** borne on the same plant at the apex and in clusters in the leaf axils, minute, white with red centres. **Fruit** small, trilobed, hanging from the centre of the flower. **Latex** present.

Habitat: Found occasionally locally on islands in the floodplain.

Flowering: During the main rains.

Uses and beliefs:
- Extensive research is being carried out into the possible use of this herb in the treatment of prostate cancer.

FZ Vol.9 pt5.

Fabaceae (pod-bearing family), Papilionoideae (pea sub-family)

Abrus precatorius L. subsp. *africanus* Verdc.

Common names: Setswana mophithi, baswabile, mophethe, mpitipiti; **San** xonequm; **Shiyeyi** chico, moXhwewe; **Kalanga** mabophe; **Herrero** vato-ohoni; **English** lucky bean climber, rosary pea, coral bead plant, crab's eyes, love bean.

Derivations: *Abrus*, origin uncertain but possibly from *abros*, delicate [Greek], referring to the delicate leaves; *precatorius*, relating to prayer [Latin], alluding to the use of the seeds to make rosaries; *africanus*, from Africa. *Baswabile* means 'Let them be ashamed'.

Identification: Scrambling, climbing **perennial herb** from a woody rooting stock. **Leaves** alternate, pinnately compound with at least 10 pairs of leaflets; leaflets oblong with smooth margins. **Flowers** dusky pink, borne in sparse clusters on a short spike. **Fruit** short flat pods that twist on ripening to reveal red **seeds** with a black spot, which are very poisonous.

Habitat: Found occasionally locally growing in light shade on islands and in mixed riverine woodland.

Flowering: During the early rains.

Uses and beliefs:
- The seeds are extremely poisonous; one is sufficient to kill an adult although effects are not immediately apparent and may take up to 2 days to appear.
- Smoke from burning the seed covered in petroleum jelly or fat is directed towards the eyes to treat eye infections; powdered seed can also be applied directly to the eyes at night.
- Wayeyi mix *Abrus precatorius* with other plants as an abortifacient.
- Batswana traditional doctors mix these beans with plants including *Neptunia oleracea* and *Glinus bainesii* to make good luck charms.
- Wayeyi also believe that to carry two seeds at all times will bring good luck.
- For a good result in a court case, San either wash themselves with an infusion of the plants and seed pods or apply a mixture of crushed leaves with fat or petroleum jelly to the body.
- Subiya and Wayeyi have similar traditions with regard to court cases. Subiya and Wayeyi also use the plant (particularly effective when mixed with *Neptunia oleracea* and *Glinus bainesii*) as a charm when they ask a girl to marry them: she will look shy and agree.
- Subiya girls should not step over this plant pre-puberty as their menstrual cycle will not begin.
- The seeds are widely used to make necklaces, often mixed with other seeds or with decorative beads woven from palm fronds.

Ellery p.152, Flowers Roodt p.83, Germishuizen p.37, WFNSA p.194. BB, BN, PN, KS.

Dolichos junodii (Harms) Verdc.

Derivation: *Dolichos*, long [Greek], also the Greek name for long-podded beans.

Identification: Climbing twining **perennial herb**, up to c. 2 m tall. **Leaves** alternate, trifoliate, upper leaflet is rhomboid, lower leaflet pair are asymmetrical with oblique bases. **Flowers** pea-like, the standard pink with light blue veining and a white throat, borne on long stems from the leaf axil. **Fruit** flat pods with few seeds, clearly defined restrictions between the seeds, style often retained.

Habitat: Locally rare in sandy areas of mopane woodland.

Flowering: During the main rains.

FZ Vol.3 pt5.

Fabaceae (pod-bearing family), Papilionoideae (pea sub-family)

Indigofera charlieriana Schinz var. *charlieriana*

Derivation: *Indigo*, the colour indigo, *fera*, bearing [Latin], referring to plants of this genus that are used to obtain indigo dye.

Identification: Small rather variable lax **annual herb**, up to c. 40 cm tall. **Leaves** smooth margined, alternate, compound with up to 3 pairs of leaflets plus a terminal leaflet; leaflets linear, in some plants quite broad in others very narrow. **Flowers** bright pink, borne in compact spikes from the leaf axils, which grow longer with age. **Fruit** small flat pods borne on a compact spike which forms a shape like a small Christmas tree.

Habitat: Widespread in sandy areas of woodland and along island and spillway margins. Found occasionally in lightly shaded areas of the riverine woodland.

Flowering: During the main rains.

Indigofera charlieriana Schinz var. *scaberrima* (Schinz) J.B.Gillett

Derivation: *Indigo*, the colour indigo, *fera*, bearing [Latin], referring to plants of this genus that are used to obtain indigo dye; *scaberrima*, very rough [Latin].

Identification: Erect **annual herb**, up to c. 45 cm tall. **Stems** and **leaves** covered in stiff hairs. **Leaves** grey-green, alternate, compound with up to 3 pairs of leaflets plus a terminal leaflet; leaflets linear, margins smooth. **Flowers** pale magenta pink, borne in rather sparse untidy racemes from the leaf axils; racemes covered in coarse brownish hairs. **Fruit** small flat pods borne on a compact spike which forms a shape like a small Christmas tree.

Habitat: Found occasionally locally growing in light shade on islands and in the treeline along the floodplain.

Flowering: During the main rains.

Fabaceae (pod-bearing family), Papilionoideae (pea sub-family)

Indigofera daleoides Benth. ex Harv. var. *daleoides*

Derivation: *Indigo*, the colour indigo, *fera*, bearing [Latin], referring to plants of this genus that are used to obtain indigo dye; *daleoides*, like *Dalea*, small herbs and shrubs of dry and desert places, named for Dr Samuel Dale (1659–1739), English botanist, apothecary and physician.

Identification: Sub-erect spreading **herb**, up to c. 30 cm tall, with a straight, strong, woody **taproot**. **Leaves** grey-green, compound with alternate leaflets plus a terminal leaflet, surfaces densely covered in velvety **hairs**. **Flowers** borne in a dense spike (c. 3 cm long) on a long stalk growing from the leaf axil; small (c. 3 mm dia.), dark pink with a darker red keel. Sepals as long as the petals, covered in brown hairs. **Fruits** small, cylindrical pods, c. 1.5 cm long.

Habitat: Locally uncommon in dry disturbed areas of riverine margin.

Flowering: During main rains.

B Van Wyk New p.208, Botweeds p.70, Germishuizen p.191, Turton p.89.

Indigofera flavicans Baker

Common names: Setswana tshikadithata, tshikadithate, mupidi.

Derivations: *Indigo*, the colour indigo, *fera*, bearing [Latin], referring to plants of this genus that are used to obtain indigo dye; *flavidus*, yellowish [Latin], referring to the colour of the young leaves. *Tshikadithata*, strong muscle, especially for men.

Identification: Prostrate creeping **annual herb**. **Stems** c. 30 cm long. **Leaves** alternate; younger leaves oval with a blunt bristled tip, margins smooth; mature leaves palmately compound with 3 oval leaflets, the upper leaflet larger, margins smooth, both leaf surfaces covered in thick yellow-white **hairs**; pairs of triangular **stipules** at the base of the leaf stalks. **Flowers** borne in an erect spike, strong rose pink with a darker keel.

Habitat: Locally common in full sun on sandy areas of the floodplain.

Flowering: During the main rains.

Uses and beliefs:
• Hmbukushu and Wayeyi use a decoction of the roots to clean the blood after intercourse with a pregnant woman.
• The roots are soaked and used as medicine for abdominal pain for both children and adults.
• Also used as a treatment for snake bites.

Ellery p.163, Hargreaves p.29. MMo, SS, MT.

Fabaceae (pod-bearing family), Papilionoideae (pea sub-family)

Lessertia benguellensis Baker f.

Common names: Setswana m- -fetola?

Derivation: *Lessertia*, for Jules Paul Benjamin de Lessert (1773–1847), a French industrialist, banker, amateur botanist and owner of an important private herbarium; *benguellensis*, from Benguella.

Identification: A tall **perennial herb** from a woody rooting stock, multi-stemmed but unbranching, up to c. 1.75 m tall. **Leaves** alternate, c. 7 pairs of leaflets plus a terminal leaflet; leaflets oblong and notched at the tip. **Flowers** pea-like, borne in many-flowered axillary racemes, standard cream with pink veining on both front and reverse, wing petals pink. **Fruit** oval and partially inflated with a distinct rim.

Habitat: Found occasionally locally in damp areas of the floodplain.

Flowering: During the main rains.

FZ Vol.3 pt7.

Tephrosia caerulea Baker f.

Derivation: *Tephros*, ashen [Greek], referring to the grey colour of the leaves of some species; *caerulea*, sky blue [Latin].

Identification: Erect branching **annual herb**, c. 0.75 m tall. **Leaves** grey-green, alternate, compound having 5 or 6 pairs of leaflets plus a terminal leaflet; leaflets linear with a blunt tip. **Flowers** borne on long (c. 25 cm) inflorescences above the plant, opening only after sunset, pea-like, bluey pink, wing petals are darker than the standard. **Fruit** harshly hairy flattened pods and the style is often retained.

Habitat: Locally uncommon in full sun on the spillway margin.

Flowering: During the early rains.

Fabaceae (pod-bearing family), Papilionoideae (pea sub-family)

Tephrosia lupinifolia DC.

Common names: Setswana luwira, namyati; **Afrikaans** vingerblaarertjie.

Derivation: *Tephros*, ashen [Greek], referring to the grey colour of the leaves of some species; *lupinus*, of the lupin family, *folia*, leaved [Latin], alluding to the lupin-like shape of the leaves.

Identification: Much-branching, occasionally creeping, **perennial herb**. **Stems** usually c. 30 cm long but may grow to c. 1.2 m. Stems and calyx covered in fine white **hairs**. **Leaves** compound with 3–5 palmately arranged linear leaflets; leaflets covered with hairs which become denser at the leaflet bases; small root-like bristles emerge from the centre of the compound leaf. **Flowers** dark pink, c. 9 mm long, in racemes at the apex of the branches. **Fruit** flattened pointed hairy pods, each seed delineated.

Habitat: Found occasionally locally in sandy areas in full sun.

Flowering: Middle to end of the main rains.

B Van Wyk Flowers p.214, Hargreaves p.31.

Fabaceae (pod-bearing family), Papilionoideae (pea sub-family)

Tephrosia purpurea (L.) Pers. subsp. *leptostachya* (DC.) Brummitt var. *pubescens* Baker

Common names: Setswana mamnyati; **English** ash vetch.

Derivation: *Tephros*, ashen [Greek], referring to the grey colour of the leaves of some species; *purpurea*, purple [Latin], in the colour of the flowers; *lepto*, thin or slender [Greek], *stachys*, spike [Greek], slender-spiked; *pubescens*, downy [Latin].

Identification: Much-branched bushy, **short-lived perennial** c. 50 cm tall. **Stems** zigzag, covered in fine hairs, often tinged with pink. **Leaves** c. 6 × 2 cm, grey-green, imparipinnate, up to 8 pairs of leaflets plus a terminal leaflet; leaflets with smooth margins, tips blunt with a bristle; pairs of linear **stipules** at the base of the leaf stalk. **Flowers** in long sparse spikes, dark pink, c. 9 mm long × 8 mm wide. **Fruit** flat, slightly curved pods, c. 3.5 × 0.3 cm, each seed clearly delineated.

Habitat: Common and widespread in full sun on sandy areas of the floodplain.

Flowering: From early in the rains through to the end of the cool dry period.

Uses and beliefs:
- San chew the plant but do not swallow their saliva as a cough remedy.
- Used medicinally against parasitic worms.
- The crushed plant is widely added to water as a fish poison.
- Used as a green mulch.

Ellery p.176, Hargreaves p.31. RM.

Fabaceae (pod-bearing family), Papilionoideae (pea sub-family)

Vigna unguiculata (L.) Walp.
subsp. ***unguiculata***
var. ***spontanea*** (Schweinf.) R.S.Pasquet

Vigna unguiculata (L.) Walp.
subsp. ***stenophylla*** (Harv.) Maréchal, Mascherpa & Stainier

Common names: Setswana nawa-yanaga, monawana, dinawa; **Afrikaans** catjang; **English** cowpea, black-eyed bean.

Derivations: *Vigna* in honour of the 17th century Italian botanist Domenico Vigna; *unguiculata*, bearing claws [Latin]; *spontanea*, spontaneous [Latin]. stenos, narrow [Latin], *phylon*, leaf [Greek]; narrow-leaved. *Nawa yanaga*, wild bean. *Monawana*, small bean.

Identification: Scrambling and climbing **annual herb**, up to c. 2 m tall. **Leaves** very variable, 3 leaflets, upper leaflet lobed at the base, lower leaflets oblique at the base with a single lobe; articulated with the central leaflet turning its underside towards the sun for part of the day whilst the 2 lower leaflets turn edge-on to the sun. **Flowers** pale magenta pink, c. 3 cm, borne in groups on long stalks; flower stalk grows longer as more flowers and pods develop; **stipules** in pairs at the leaf axils. **Fruit** long narrow cylindrical pods, up to c. 10 × 0.5 cm, with a coarsely hairy surface.

Vigna unguiculata subsp. *unguiculata* var *spontanea* (4 photos)

Subspecies differences: Subsp. *unguiculata* var. *spontanea* has more defined lobes (hastate) on the central leaflet. The leaves of subsp. *unguiculata* var. *spontanea* are slightly leathery, rough and **sandpapery** and have a pale green band across them, whereas the leaves of subsp. *stenophylla* are soft and a dull green. Subsp. *stenophylla* has dark purple markings on both the standard and the wing petals.

Habitat: A pan-tropical herb, indigenous to Africa, found occasionally locally in partial shade in sandy areas of the floodplain and woodlands, sometimes in full sun.

Flowering: During the main rains.

Uses and beliefs:
- The mature seeds, young pods and leaves may be eaten as vegetables either fresh, dried or cooked. Now harvested commercially in some areas.
- Some Batswana collect the leaves to cook as *morogo*, but these plants are generally considered not to taste very good.

FZ Vol.3 pt5, Turton p.50, Hargreaves p.32, Blundell p.121, WFNSA p.192, Wyk p.383. KS.

Vigna unguiculata subsp. *stenophylla* (3 photos)

Geraniaceae (geranium family)

Monsonia angustifolia A.Rich.

Common names: Setswana ramarungwana, phusana, tsatsalopane; **Afrikaans** angelbossie, teebossie, alsbos; **English** crane's bill.

Derivation: *Monsonia* for Lady Ann Monson (1714–76) who collected plants at the Cape of Good Hope; *angustifolius*, from *angusti*, narrow, *folius*, leaf [Latin] narrow-leaved.

Identification: Erect, bushy **annual herb**, c. 15 cm tall. **Stem** with long fine hairs. **Leaves** grey-green, opposite, decussate, linear, with lightly toothed and wavy margins. **Flowers** borne singly from the leaf axils on long stalks, c. 1.2 cm dia., palest pink or white, 5 almost translucent petals. Calyx with long fine hairs. **Fruit** seed pods that look like a cranes bill, c. 7 cm long, splitting open when dry to look like the arms of a tiny chandelier.

Habitat: Locally uncommon in sandy clay near seasonal pans.

Flowering: Early in the main rains.

FZ Vol.2 pt1, B Van Wyk Flowers p.216, Blundell p.46, Germishuizen p.79, Turton p.81, WFNSA p.204.

Loranthaceae (mistletoe family)

Plicosepalus kalachariensis (Schinz) Danser

Common names: Setswana palamêla; **Simbanderu** otjiraura; **Subiya** kasamu komoliro; **Shiyeyi** epalamela; **Afrikaans** voëlent; **English** acacia mistletoe.

Derivations: *Plicatus*, pleated [Latin], *sepalus*, sepal, the leafy outer part of the flower [Latin]; *kalachariensis*, of the Kalahari. *Palamêla*, the one that climbs, referring to the fact that the plant grows perched in trees. *Kasamu komoliro*, burning matches. *Voëlent*, grafted on by birds.

Identification: A shrubby **parasitic plant** growing on *Acacia* spp., especially *Acacia nigrescens* locally. **Leaves** hairless, linear to oblong with a rounded tip, smooth margins, 3 veins along their length, either opposite or alternate. **Flowers** borne in dense clusters, red and curved into a C-shape; lobes spring open and fling out the enclosed pollen at the slightest touch (an adaptation to pollination by birds or large insects). **Fruit** cylindrical, narrowing at either end, ripening to a pinky-red.

Habitat: Common and widespread growing on *Acacia* spp.

Flowering: Almost throughout the year but more abundantly in the cool dry period.

Uses and beliefs:
- Batswana traditional doctors mix this plant with others to make a charm that is used to obtain promotion (because it grows at the top so it is used to being at the top). The crushed plant mixed with vaseline is also applied to the body for good luck.
- During the rainy season, the plant is cut and placed around the houses to protect them from lightning.
- The fruit are boiled and mixed with groundnut oil to make a birdlime. When smeared on branches this catches small birds.
- The caterpillars of *Mylothris agathine*, *Stugeta bowkeri* and *Paraphnaeus hutchinsonii* feed on this plant.

Flowers Roodt p.109, Plowes & Drummond No 42, WFNSA p.126. BB, GM, PN, KS.

Nyctaginaceae (bougainvillea family)

Boerhavia coccinea Mill.
(syn. *Boerhavia diffusa*)

Common names: Setswana leralagori, mositi, masitis, mmadikokwana; **San** setlhabi.

Derivations: *Boerhavia*, for Herman Boerhaave (1668–1739), a Dutch physician and botanist, professor of botany and medicine at Leiden; *coccinea*, scarlet [Latin]. *Mmadikokwana*, the one that traps chicks; young birds may become tangled in the plants and die.

Identification: Variable low-growing much-branched **perennial herb**. **Stems** up to c. 35 cm long, hairy or hairless, often coloured pink. **Leaves** oval, alternate, stalks almost as long as the leaf blade, margins smooth or hairy and wavy, often with a bristle at the tip, **veins** clearly incised into the upper leaf surface making it look almost pleated. **Flowers** c. 4 mm dia., pale pink or cherry pink, almost red; flowers and **fruit** borne in long lax panicles, often in groups of 3. **Fruit** either ribbed and densely hairy or hairless and only lightly ribbed.

Habitat: Common and widespread in full sun and partial shade on sandy clay.

Flowering: Throughout the main rains.

Uses and beliefs:
• San (Bugakhwe) drink a decoction of boiled roots for the immediate relief of sharp pain.

FZ Vol.9 pt1.

Pedaliaceae (sesame family)

Ceratotheca sesamoides Endl.

Common names: Setswana kgotodua; **English** wild or false foxglove.

Derivation: *Keras*, horn [Greek], *thece*, case, chest or container [Greek], referring to the horned seed capsules; *sesamoides*, like the sesame plant [Greek].

Identification: A bushy, untidy **annual herb**, growing to c. 70 cm tall; bristly hairs overall. **Young stems** maroon. **Leaves** palmate, weakly trilobed with sharply serrated margins on lower lobes; glands in the leaf axils. **Flowers** tubular, 5-lobed, dark pink externally with white edges and internally pale pink with dark pink veins and a blotch of yellow. **Fruit** flattened and narrowly rectangular, with horns usually at right-angles to the capsule.

Habitat: Locally uncommon in deep sand areas of the floodplain and spillway and on the track side of sandy ridges through the mopane woodland, in full sun.

Flowering: During the main rains and into the beginning of the drier weather.

Uses and beliefs:
• In some areas, the seeds are collected like sesame seeds for their oil.

FZ Vol.8 pt3.

Pedaliaceae (sesame family)

Dicerocaryum eriocarpum (Decne.) Abels

Common names: Setswana legatapitse, lekanangwane, tshetlho-e-tonanyana, tshetlho-ya-dinku, samurai, malalamakatse, ibangu, lekatse, lematla, tallapoa ennyennyane; **Hmbukushu** maramata; **Subiya** lovangolo nzovu; **Afrikaans** elandsdoring, beesdubbeltjie, duiwelsdis; **English** large devil's thorn, studthorn, boot protector plant.

Derivations: *di*, two [Greek], *keros*, horn [Greek], *karyon*, nut [Greek], referring to its hard fruit with 2 conical spines on the top; *erion*, wool [Greek], *karpos*, fruit [Greek]. *Tshetlho*, devil's thorn, *e-tonanyana*, large devil's thorn. *Lekanangwane* means something that is upside down. *Lovangolo nzovu*, elephant thorn.

Identification: A mat-forming semi-perennial **creeping** herb; hairy overall. **Stems** up to c. 2.5 m in length. **Leaves** c. 7.5 cm (inc. c. 3 cm leaf stalk) × 4.5 cm, opposite, palmate with 3 lobes, deeply toothed margins. **Flowers** c. 2.3 cm wide × c. 2.5 cm. long, borne singly on c. 2.5 cm stalks in the leaf axils, pink with maroon blotching on the lower part of the throat. **Fruit** like a small flying saucer, c. 0.8 cm dia., with 2 sharp conical thorns from near the centre of the upper surface.

Habitat: Common in full sun in areas of clayey sand on the floodplain.

Flowering: Middle to end of the main rains.

Uses and beliefs:
• Hmbukushu men use a decoction of the root to treat bladder infections.
• Widely used in childbirth (also in animal births); in particular, a leaf infusion helps the expulsion of a retained placenta.
• Used as an aphrodisiac, as an enema for constipation, in the treatment of sexually transmitted diseases, and as a cleansing agent for couples after a miscarriage.
• Subiya men boil the roots and drink the decoction 2–3 times per day for a week to improve their sexual strength.
• It is also one of the ingredients of a good luck charm used when hunting.
• A soap substitute is made from crushed plant material soaked in water.

FZ Vol.8 pt3, B Van Wyk p.228, Ellery p.160, Hargreaves p.62, Turton p.113, Flowers Roodt p.129, WFNSA p.378. GM, MT.

Pedaliaceae (sesame family)

Harpagophytum procumbens (Burch.) DC. ex Meisn.
subsp. *procumbens*

Common names: Setswana and **Hmbukushu** sengaparile, legatapitse; **Setswana** makakare, lengakapitse, xwate; **Hmbukushu** senyaparele; **Ovambenderu** otjihangatene; **Subiya** ivangogu lye ingombe; **Afrikaans** beesdubbeltjie, beestedoorn, duiwelhaak; **English** devil's claw, grapple plant, grapple thorn.

Derivations: *Harpagos*, a hook [Greek], *phyton*, plant [Greek], referring to the hooks on the fruit; *procumbens*, prostrate, creeping [Latin]. *Legatapitsi*, hoof of the zebra. *Otjihangatene*, one medicine for six diseases. *Ivangogu lye ingombe*, cattle thorn, because it sticks in the cloven hooves of cattle.

Identification: Perennial herb with several prostrate annual stems, up to c. 60 cm long, from a succulent **taproot**; additional **tubers** on lateral roots. **Leaves** blue-green, opposite, oval with a rounded tip, margins deeply lobed. **Flowers** with a 5-lobed tubular corolla, pink with a yellow throat. **Fruit** flattened, oval with 4 rows of curved arms bearing recurved spines, length of the longest arm exceeds the width of the capsule proper, 2 spikes on the upper surface.

Before it fruits, *Harpagophytum procumbens* is easily mistaken for *H. zeyheri*. The fruit are more easily differentiated, those of *H. procumbens* having spines on the rim that are longer than the body of the fruits is wide whereas they are much shorter on *H. zeyheri*. The two species interbreed.

Habitat: Considered endangered but locally widespread in deeply sandy areas.

Flowering: During the main rains.

Uses and beliefs:
- Much used for medicinal purposes.
- Batswana, Hmbukushu, Subiya and Ovambenderu do not appear to differentiate between *Harpagophytum zeyheri* and *H. procumbens*, although the active ingredients are present in greater quantities in *H. procumbens*. Many use the roots of this plant for clearing the blood and the kidneys and for treating joints.
- Extracts of the roots, especially of *H. procumbens*, are widely used internationally in the treatment of joint pain, as either a remedial or a preventative treatment.
- The recurved spines on the fruit can cause animals to go lame or their jaws to lock, causing starvation.

FZ Vol.8 pt3, Flowers Roodt p.133, Hargreaves p.65, WFNSA p.380. BB, CM, GM.

Pedaliaceae (sesame family)

Harpagophytum zeyheri Decne. subsp. *subloblatum* (Engl.) Ihlenf. & Hartm.

Common names: Setswana tshero, makakarana, lengakare, sengaparile, legatapitsi; **Ovambenderu** otjihangatene; **English** grapple plant.

Derivation: *Harpagos*, hook [Greek], *phyton*, plant [Greek], referring to the hooks on the fruit; *zeyheri* for Carl Zeyher, a German naturalist who collected extensively in South Africa (1799–1858).

Identification: Creeping herb with a perennial tuberous root stock and annual stems up to c. 75 cm long, covered in short harsh **hairs** overall. **Leaves** blue-green, opposite, oval with a rounded tip, margins toothed especially on the upper half; undersides covered with dense hairs causing them to look silver-grey. **Flowers** are tubular ending in 5 lobes, usually pink fading to white at the throat which is yellow. **Fruit** flattened, oval with double spines on the rim and 2 spikes on the upper surface.

Before it fruits, *Harpagophytum zeyheri* is easily mistaken for *H. procumbens*. The fruit are more easily differentiated, those of *H. procumbens* having spines on the rim that are longer than body of the fruit is wide whereas those on *H. zeyheri* are much shorter. The two species interbreed.

Habitat: Widespread throughout the region in areas of deep sand.

Flowerifng: From the middle of the main rains into the cool dry period.

Uses and beliefs:
- Batswana, Hmbukushu, Subiya and Ovambenderu do not appear to differentiate between *Harpagophytum zeyheri* and *H. procumbens*, although the active ingredients are present in greater quantities in *H. procumbens*. Many use the roots of this plant for clearing the blood and the kidneys and for treating joints. Extracts of the roots, especially of *H. procumbens*, are widely used internationally in the treatment of joint pain, as either a remedial or a preventative treatment.
- The recurved spines on the fruit can cause animals to go lame or their jaws to lock, causing starvation.

FZ Vol.8 pt3, Germishuizen p.378, Turton p.47. BB, CM, GM.

Pedaliaceae (sesame family)

Sesamum angustifolium (Oliv.) Engl.

Common names: Ovambenderu onduri.

Derivation: *Sesamum* from the Greek name for sesame; *angusti*, narrow [Latin], *folius*, leaved [Latin].

Identification: Erect **annual herb**, up to c. 2 m tall. **Leaves** opposite, linear towards the top of the plant but lower down palmately compound with 3 or 5 leaflets, more usually 3. **Flowers** bright salmon pink with a darker maroon throat, 5 lobes fused to form a short tube, almost stalkless, borne at the leaf axils. **Seed pods** tubular, pointed at the tip, splitting to release the seed.

Easily confused with *Sesamum triphyllum*, which is generally larger and has soft grey-pink flowers.

Habitat: Found in groups, usually in full sun on the floodplain and occasionally in partial shade on the treeline margin, especially beside tracks.

Flowering: Middle to end of the main rains.

Uses and beliefs:
• Ovambenderu boil the leaves, flowers and stems and take the resulting decoction (approx. 125 ml twice a day) to treat fever.
• Used medicinally to treat smallpox and as a nasal and eye wash.
• The crushed leaves can be used to replace soap.

FZ Vol.8 pt3, Ellery p.175. BB.

Pedaliaceae (sesame family)

Sesamum triphyllum Asch.

Common names: Setswana motlhomaganyane, selaole, kgotodua, dhobi; **Hmbukushu** mongoma; **English** wild sesame, rain gauge.

Derivations: *Sesamum* from the Greek name for sesame; *tri*, three [Greek], *phyllon*, leaf [Greek], referring to the 3 leaflets of the upper leaves. *Motlhomaganyane*, to be in line, referring to the plant's single upright stem. Sometimes known as the *rain gauge* as its height is dependent on the amount of rain that falls.

Identification: Erect annual **herb**, up to c. 2.5 m tall. **Leaves** opposite, margins wavy, linear towards the top of the plant, compound with usually 3 or sometimes 5 leaflets lower down. **Flowers** greyish pink with a dark maroon-purple throat, 5 lobes fused to form a short tube, almost stalkless, at leaf axils. **Fruit** tubular, ridged, splitting in two to release the seed.

Easily confused with *Sesamum angustifolium*, which is generally smaller and has brighter salmon-pink flowers.

Habitat: Found occasionally, often in groups, in full sun on the floodplain, especially beside the tracks.

Flowering: Middle to end of the main rains.

Uses and beliefs:
- The seeds are edible and rich in oil.
- Wayeyi stir milk with a stalk of the plant until it thickens. The resulting concoction is drunk by men who wish to improve their fertility.
- The crushed leaves can be used in place of soap.
- Subiya plant the seeds mixed with the maize crop to encourage the maize to have more seed heads.
- Some Subiya believe that short children can be made to grow taller if this plant is rubbed into small cuts in the child's skin.
- San use the flexible tall stems to build shelters.

FZ Vol.8 pt3, B van Wyk p.228, BVW Photoguide ZA p.109, Flowers Roodt p.137, Turton p.105. TK, CM, GM, SS, MT.

Polygalaceae (milkwort family)

Polygala erioptera DC.

Derivation: *Poly*, much, *gala*, milk [Greek], because plants of this genus were believed in Europe to encourage milk production in stock; *erion*, wool [Greek], *pteron*, wing [Greek], referring to the upper parts of the flower.

Identification: A small erect, branching **annual herb**, c. 25 cm tall. **Leaves** linear, smooth margined, spiralling up the stem. **Flowers** minute, borne either singly or in short racemes spiralling up the stem, pale magenta pink, the keel a stronger shade of pink while the lighter wings are covered in a thick down; flowers only begin to open at 10.00 hrs. **Fruit** flattened, with a small wing along the edges.

Habitat: Widespread in the open grassland of the floodplain.

Flowering: During the main rains.

FZ Vol.1 pt1.

Polygonaceae (buckwheat family)

Persicaria glomerata (Dammer) S.Ortiz & Paiva

Derivation: *Persicaria*, the medieval name for a knotweed, from *persica*, peach [Latin], alluding to the shape of the leaves; *glomerata*, clustered into more or less rounded heads.

Identification: Slightly built, rather lax **aquatic plant**. **Stems** c. 80 cm long, thin, green. **Leaves** red-green, alternate, lanceolate with short stalks, margins smooth. **Flowers** pale pink, in small groups borne in spikes at the apex of the stems, becoming widely spaced further down the stalk, a narrow funnel shape with 5 lobes. **Calyx** red.

Habitat: Locally uncommon, growing in the margin of fast-flowing channels sheltered by *Cyperus papyrus* and various grasses.

Flowering: During the main rains.

Polygonaceae (buckwheat family)

Persicaria limbata (Meisn.) H.Hara

Common names: Setswana kubutona, letetemetso; **Shiyeyi** molemo wa segogwane; **English** knotweed.

Derivations: *Persicaria*, the medieval name for a knotweed, from persica, peach [Latin], alluding to the shape of the leaves; *limbata*, bordered [Latin], referring to the frilled leaf sheaths wrapping around the stems. *Kubutona*, the big hippo. *Molemo wa segogwane*, medicine of the frog.

Identification: A stout erect, freely branching, **perennial herb** of damp places, growing to a height of c. 1.2 m although often half that height, covered in harsh brownish **hairs**. **Stems** hollow and **root** at the lower nodes. **Leaves** stalkless, lanceolate with wavy margins. The major identifying feature is the large frilled, stipular, **leafy sheaths** fringed with hairs, wrapping around the stems. **Flowers** borne in paired terminal spikes, pale pink and white.

Habitat: Widespread in the region in damp areas.

Flowering: During the main rains.

Uses and beliefs:
* Wayeyi apply a paste made of pounded, dried leaves and roots, mixed with fat, externally to treat swollen necks and throats.

B Van Wyk New p.230, Flowers Roodt p.142, Germishuizen p.72, WFNSA p.130.

Polygonaceae (buckwheat family)

Polygonum decipiens (R.Br.) K.L.Wilson

Common names: Afrikaans slangwortel; **English** bistort, snake root.

Derivation: *Poly*, many [Greek], may be either *gonos*, seeds or *gonu*, knee or node [Greek], referring to either many seeds [Greek] or the many swollen nodes or sheaths of the plants; *decipiens*, deceptive or cheating [Latin].

Identification: Slim branching, sometimes lax, **annual, marginal herb**, sometimes deeply tinged red. **Stems** up to c. 60 cm long, red and green, long runs of slim stems rooting below the water. **Leaves** alternate, narrowly lanceolate, stalkless, smooth margined, red and green. Minute pink and white **flowers** borne in terminal spikes.

Habitat: A locally uncommon marginal plant, found in dried-out areas on the edge of the lagoon or on the margins of seasonal pans in damp conditions.

Flowering: Throughout the main rains.

B Van Wyk Flowers p.230, Blundell p.38.

Polygonaceae (buckwheat family)

Polygonum plebeium R.Br.

Derivation: *Poly*, many [Greek], may be either *gonos*, seeds [Greek] or *gonu*, knee or node [Greek], referring to either many seeds or the many swollen nodes or sheaths of the plants; *plebeium*, common [Latin].

Identification: Prostrate **annual herb**, up to c. 70 cm dia., making a radial pattern with short side branches. Main **stems** reddish-green, side stems green; stems and leaves **hairless**. **Leaves** alternate, linear, smooth margined; a dense ring of hairy **stipules** at the leaf axils. **Flowers** borne in almost stalkless clusters in the leaf axils, minute (c. 1 mm dia.), pink, 5 petals.

Habitat: Locally uncommon growing in dried-out seasonal pans.

Flowering: After the early rains when the pans dry out.

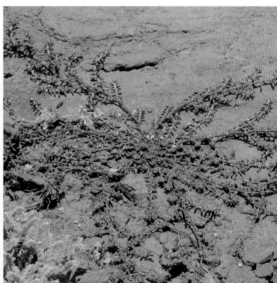

Portulacaceae (purslane family)

Portulaca hereroensis Schinz

Common names: Setswana serepe.

Derivation: *Portare*, to carry [Latin], *laca*, milk [Latin], i.e. milk carrying, originally referring to the milky latex in *Portulaca oleracea* and then applied to other plants of the genus; *hereroensis*, from the land of the Herero people.

Identification: Small **annual succulent**, much branched and finely stemmed; **plants** growing together to form loose mats. **Stems** red, rooting at the leaf nodes. **Leaves** fleshy, opposite, ovate, margins smooth, green. **Flowers** borne singly or in clusters of 2 or 3 at the apex of the plant, magenta-pink, 4 petals arranged alternately with the sepals; flowers surrounded by tight rings of **leaves** and bunches of long **hairs**. **Fruit** retain 4 sepals.

Habitat: Widespread growing in full sun in clearings in mopane woodland on clayey sand.

Flowering: During the main rains.

FZ Vol.1 pt2.

Potamogetonaceae (pondweed family)

Potamogeton nodosus Poir.

Derivation: *Potamos*, river [Greek], *geiton*, neighbour [Greek], referring to its preferred habitat; *nodosus*, having conspicuous joints or nodes.

Identification: A **perennial aquatic plant** with a **stoloniferous rooting** system. **Leaves** have 2 forms; upper floating leaves are bright green, alternate, narrowly oval, with parallel veins and entire margins; lower submerged leaves are dark brown-green, linear, margins entire. **Flowers** insignificant, growing in tight blunt spikes above the water from the leaf axils, beige-pink and green, with 4 petals.

Habitat: Common and widespread on the peaty margin of slow-flowing channels.

Flowering: During the main rains.

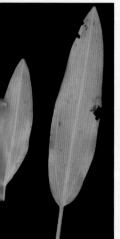

Scrophulariaceae (snapdragon or foxglove family)

Cycnium tubulosum (L.f.) Engl.

Common names: Setswana motshe-wa-badimo; **English** vlei ink-flower.

Derivations: *Cygnus*, swan [Latin], possibly as the tube of the flower resembles the neck of a swan; *tubulosum*, tubular [Latin]. *Motshe-wabadimo* has two meanings. To Wayeyi, it means a rainbow, and to Subiya it means a 'digging stick for the ancestors'. If the flowers are crushed they go blue, this and their *vlei* (damp meadow) habitat is the source of the English name.

Identification: An erect, delicate, sparsely branching, **hemi-parasitic herb**, c. 50 cm tall. **Leaves** opposite, linear, with a prominent mid-rib. **Flowers** borne in the axils of the upper leaves, a showy brilliant pink with a white throat, 5 lobes fused to form a corolla tube, the upper 2 joined higher up than the others.

Habitat: Widespread in damp open grassland of the floodplain.

Flowering: Throughout the main rains and into the cool dry period.

Uses and beliefs:
- The roots are boiled to make an ink-like decoction that is given to children, 3 times a day for a week, to make them strong and healthy.

FZ Vol.8 pt2, B Van Wyk Flowers p.86, Ellery p.159, Germishuizen p.366, Flowers Roodt p.145. CM, GM, PN.

Scrophulariaceae (snapdragon or foxglove family)

Sopubia mannii Skan
var. *tenuifolia* (Engl. & Gilg) Hepper

Common names: English common sopubia.

Derivation: *Sopubia* is possibly an anagram of *Bopusia*, another genus of this family; *mannii*, named after Gustav Mann, a Kew botanist based around Fernando Po and the Gulf of Guinea; *tenui*, slender, thin [Latin], *folia*, leaf [Latin], slender-leaved.

Identification: Branching erect **hemi-parasitic herb**, up to c. 60 cm tall, almost **hairless**. **Leaves** linear, almost needle-like, borne in groups of 3 around the stem. **Flowers** c. 2.2 cm dia., asymmetrical, pink and white, 5 petals, borne in loose terminal racemes.

Habitat: Locally common in grassy damp areas of the floodplain.

Flowering: Flowers throughout the rainy period.

FZ Vol.8 pt2.

Scrophulariaceae (snapdragon or foxglove family)

Striga asiatica (L.) Kuntze

Common names: Setswana matebelwe, molelwana, matabele; **Afrikaans** rooiblom, gewone-mielierooiblom; **English** witchweed, common mealie-witchweed.

Derivation: *Striga*, a straight rigid close-pressed rather short bristle-like hair [Latin]; *asiatica*, of Asia.

Identification: An attractive small erect **hemi-parasitic annual herb**, with rarely branching stems, c. 30 cm tall; covered in tiny harsh **hairs**. **Leaves** alternate, linear, smooth margined, stalkless, main vein strongly indented in the upper surface. **Flowers** borne in terminal spikes, it is rare to see more than 2 flowers open at one time on a spike; c. 7 mm dia., scarlet with yellow throats and yellow on the underside of the petals, asymmetrical with 5 lobes.

This species parasitises sorghum, maize and other grasses but has some chlorophyll and is partly self-sustaining.

Habitat: Locally rare in light shade in the mopane woodland.

Flowering: Flowers late in the main rains.

Uses and beliefs:
- A noxious weed.

FZ Vol.8 pt2, B Van Wyk Flowers p.236, Blundell p.378, Botweeds p.100, Germishuizen p.194, Turton p.65, WFNSA p.368.

Scrophulariaceae (snapdragon or foxglove family)

Striga gesnerioides (Willd.) Vatke

Common names: Afrikaans bloublom, rooiblommetjie; **English** purple witchweed, tobacco witchweed.

Derivation: *Striga*, a straight rigid close-pressed rather short bristle-like hair [Latin]; *gesnerioides*, like the genus *Gesneria* named in honour of Conrad von Gessner (1516–65) of Zurich, the most celebrated naturalist of his day.

Identification: Erect, occasionally branched, almost **hairless**, **hemi-parasitic herb**, up to c. 35 cm tall, growing in clumps. **Leaves** simple, linear, almost scale-like, spiralling up the stem. **Flowers** irregular, up to c. 8 mm dia., with 5 lobes, pink or white; pink variant flowers have white throats and magenta hairs on the upper petal. **Leaves** and **stems** range in colour from green, if the flowers are white or pink, through to deep magenta if the flowers are dark in colour. Plants **dry black**.

The photographed specimen is shown **parasitising** *Indigofera flavicans* but it parasitises a wide range of legumes, including *Tephrosia purpurea*, *Rhynchosia totta* and *Indigofera filipes*; it may also parasitise *Ipomoea coptica*.

Habitat: Widespread in years of heavy rainfall in deeply sandy open woodland and on the floodplain in full sun.

Flowering: Late in the main rains.

FZ Vol.8 pt2, B Van Wyk Flowers p.236, Germishuizen p.176, Blundell p.378, Hargreaves p.61.

Sterculiaceae (cacao or cola family)

Hermannia modesta (Ehrenb.) Planch.

Derivation: *Hermannia*, in honour of Paul Hermann (1646–95), a German botanist, professor of botany at Leiden in Holland, who visited southern Africa; *modesta*, modest [Latin]

Identification: Slight, erect **annual herb**, c. 85 cm tall; lightly covered in short coarse **hairs** overall. **Leaves** linear with smooth margins, spiralling up the stem. **Flowers** bright red, c. 0.5 cm dia., on c. 4 cm stalks from the leaf axils, 5 petals, sepals very short; stamens blue-grey and quite short. **Fruit** have 5 segments with a single stubby horn on the apex of each one.

Habitat: Found occasionally locally growing in mixed sandy open woodland and on sandy areas of the floodplain in full sun.

Flowering: During the main rains.

FZ Vol.1 pt2.

Verbenaceae (verbena family)

Lantana angolensis Moldenke

Common names: Setswana kgobe-tsa-badisana; **English** bird's brandy.

Derivation: *Lantana*, a late Latin name for *viburnum*, transferred to this genus; *angolensis*, from Angola.

Identification: An erect, branching aromatic **herb**, up to c. 1 m tall. **Leaves** opposite, decussate, oval, margins softly crenate; leaf surfaces velvety, veins sunken on the upper surface but prominent on the underside. **Flowers** borne in sessile clusters in the leaf axils, small, lavender pink, 5 lobes fused to form a tube, becoming yellow at the centre as they mature (and are pollinated). **Fruit** brilliant purple, spherical.

Habitat: Found occasionally locally growing in light shade on islands and in the treeline.

Flowering: Throughout the rainy period.

Uses and beliefs:
- Subiya boys at the cattle posts eat the fruit.

Ellery p.167. CM.

Yellow
or cream flowers

Acanthaceae (acanthus or spinyflower family)

Thunbergia reticulata Nees

Common names: English orb flower.

Derivations: *Thunbergia*, commemorates the Swedish botanist Carl Peter Thunberg, a pupil of Linnaeus often described as the 'father of Cape botany'; *reticulata*, netted, marked with a network [Latin], referring to veining of the leaves. *Orb* because of the shape of the fruit.

Identification: A shade-loving **annual climber** up to c. 1 m tall. **Stems** and **leaves** slightly hairy. **Leaves** sagittate, the margins forming wings down the leaf stalk to the plant stem. **Flowers** c. 1.5 cm dia., emerge in the early morning and last only a short time, apricot yellow, with 5 lobes fused to form a corolla tube, emerging from a pair of leafy bracts. Exudes a strong green plant smell when crushed.

Habitat: Widespread, growing in shade on islands and in the treeline along the floodplain.

Flowering: Early morning during the main rains.

Apocynaceae (oleander family)

Marsdenia sylvestris (Retz.) P.I.Forst.
(syn. *Gymnema sylvestre* (Retz.) Schult.)

Common names: Setswana lefswe.

Derivation: *Marsdenia* for the Irish-born traveller and plant collector William Marsden (1754–1836); *sylvestris*, growing in woods, forest-loving, wild [Latin].

Identification: Vigorous **perennial woody climber**, 5–6 m tall; slightly hairy overall. **Leaves** opposite, ovate with margins entire. **Flowers** minute, yellow, with 5 woolly lobes that are fused to form the corolla tube. **Latex** present.

Habitat: Locally rare growing near seasonal pans in the shelter of acacia thicket.

Flowering: During the main rains.

Uses and beliefs:
- Highly poisonous but there is a similar edible plant.

Apocynaceae (oleander family)

Orbea rogersii (L.Bolus) Bruyns

Derivation: *Orbis*, a circular shape or disc [Latin], referring to the disc in the centre of the flower; *rogersii*, for William Moyle Rogers (1835–1920) or his son Frederick Arundel Rogers (1876–1944).

Identification: Semi-recumbent leafless **perennial**. **Succulent** with grey-green branching stems and long fleshy teeth. **Flowers** borne singly or in pairs beside the fleshy teeth, pale yellow, with 5 long thin revolute lobes, the central part of the flower is covered with long, white, club-shaped hairs. Flowers have a strong cloying sweet vegetable **aroma**.

Habitat: Locally rare in a damp sandy clay area of mopane woodland with seasonal pans.

Flowering: During the main rains.

Arecaceae (palm family)

Phoenix reclinata Jacq.

Common names: Setswana tsaro, moxinxa-mokulane, nlala, thikerva; **Subiya** chibiringa; **Shiyeyi** oxhone; **Afrikaans** wildedadelboom; **English** wild date palm, dwarf date palm, feather palm.

Derivation: *Phoenix*, the Greek name for the date palm; *reclinata*, bent backward [Latin], alluding to the gracefully arching leaves.

Identification: Small **palm tree** sometimes with multiple trunks, c. 6 m tall. **Trunks** dark grey-brown, patterned with the scars of fallen leaves. **Leaves** feather-shaped, 3–4 m long. **Male and female flowers** grow on separate trees; **female flowers** creamy-white, borne in showy sprays; **male flowers** smaller, borne in small branching spikes amongst the young leaves, producing clouds of pollen. **Fruit** small, oval, orange, thinly fleshy.

Habitat: Common and widespread on islands in damp areas of the floodplain. Severely damaged by elephants and consequently most examples in the area exist as clumps of young plants. The only full-sized specimens found locally are male and less attractive to the elephants because they do not bear fruit.

Flowering: During the early rains.

Uses and beliefs:
- The fruit are edible. In the pan-handle, they are collected by local people to sell at the market in Maun.
- The fruit are pounded and mixed with milk to make a yoghurt-like drink.
- Both *Hyphaene petersiana* and *Phoenix reclinata* are the source of sap for making palm wine (*muchema* in Shiyeyi and *malovo* in Subiya) (see *H. petersiana* which is more normally used in this area).
- The leaves are used to make mats and simple hats.
- The roots produce a brown dye.

Ellery p.129, Palgrave updated p.97, Setshogo & Venter p.22, Tree Roodt p.19. GM, PN.

Asteraceae (daisy family)

Aspilia mossambicensis (Oliv.) Wild

Derivation: *Mossambicensis*, of Mozambique.

Identification: Erect much-branched **annual herb**, c. 70 cm tall. **Stems** dark red-brown and covered in long **hairs**. **Leaves** opposite and decussate, narrowly ovate, margins sandpapery with sparse small serrations. **Flowers** growing singly on long stalks from the leaf axils, composite, orange-yellow, c. 2.2 cm dia., with both disc and ray florets, ray florets with 2 deeply incised longitudinal veins in the petal.

Habitat: Locally uncommon, growing in partial shade in the margin of the riverine forest.

Flowering: During the main rains.

Blundell p.159.

Asteraceae (daisy family)

Bidens biternata (Lour.) Merr. & Sheriff

Common names: Setswana moonyana, mmonyana, setlhabakolobe, mokwelenyane; **English** black jack.

Derivation: *Bi*, two [Latin], *dens*, tooth [Latin], referring to the 2 bristles on each segment of the fruit; *biternata*, ternate leaflets are those borne in 3s, a plant is biternate when 3 leaf divisions each bear 3 leaflets. *Setlhabakolobe*, the thing that pricks the pig.

Identification: Erect **annual herb**, up to c. 1 m tall. **Leaves** compound, usually with 5–7 leaflets but locally rarely sufficiently mature or large enough to have more than 3 leaflets; the basal pair of leaflets sometimes divide further into 3 leaflets; margins serrated. **Flowers** composite borne terminally, yellow, including the sparse, irregularly spaced, ray florets.

Bidens biternata may easily be confused with *B. pilosa*, but the leaves of *B. pilosa* have 3 leaflets and ray-florets that, if present, are paler yellow. *B. pilosa* is less frequent in the area.

Habitat: A naturalised, widespread but uncommon weed found in partial shade on the margin of the riverine woodland.

Flowering: During the main rains.

Blundell p.160, Hargreaves p.66.

Asteraceae (daisy family)

Bidens pilosa L.

Common names: Setswana setlhabakolobe, moonyane; **Setswana** and **Kalanga** sina; **Ovambenderu** omupapaku; **Kgalgadi** nthoka-mosare; **San** xarexo, guexwe; **Subiya** monyana; **Afrikaans** knapsekerel; **English** blackjack, beggar sticks, bur marigold, Spanish needles, sweethearts, blanket-stabbers, cobbler's pegs.

Derivation: *Bi,* two [Latin], *dens,* tooth [Latin], referring to the 2 bristles on each segment of the fruit; *pilose,* hairy [Latin], with distinct long ascending hairs. Setlhabakolobe, the thing that pricks the pig. Guexwe, warthog.

Identification: Upright branching **herbaceous plant,** up to c. 1.3 m tall. **Stem** square, slightly buttressed. **Branches** almost at right angles to the stem. **Leaves** opposite and decussate, compound, trifoliolate, each leaflet palmate with 3 oval leaflets, margins serrated, each serration with a hair at its tip, more hairy on the lower surface than the upper. **Flowers** composite, mainly disk florets with a few irregularly spaced pale cream to yellow ray florets.

Bidens pilosa may easily be confused with *B. biternata*, but the leaves of *B. pilosa* have 3 leaflets and the flowers have ray-florets that, if present, are paler yellow. *B. pilosa* is less frequent in the area.

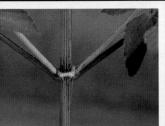 

Habitat: Locally widespread in shaded areas of mixed woodland. [Introduced.]

Flowering: From the middle of the main rains into the cool dry period.

Uses and beliefs:
- The leaves are used to make *morogo*, a vegetable relish.
- Bakgalgadi and San mix the seeds with animal fats or vegetable oil and burn them around sick family members.
- The plant is scientifically proven to have anti-bacterial and anti-fungal properties.

Roodt p.39, Turton p.57, B Van Wyk Flowers p.108. BB, CM.

Asteraceae (daisy family)

Bidens schimperi Sch. Bip. ex Walp.

Common names: Setswana monyana, setlhabakolobe, mosimama; **English** mnondo, bur marigold, yellow cosmos, black jack.

Derivations: *Bi*, two [Latin], *dens*, tooth [Latin], referring to the 2 bristles on each segment of the fruit; *schimperi* celebrates a German botanist, W.G. Schimper, who collected in Ethiopia in the mid-19th Century. *Setlhabakolobe*, the thing that pricks the pig.

Identification: A beautiful upright **annual daisy** up to c. 1.3 m tall. **Stems** ribbed, suffused maroon. Stems and leaves **pubescent**. **Leaves** opposite and decussate, compound, trifoliate, lower leaflets oval and deeply incised, the upper leaflet palmate with 3 lobes deeply incised. Yellow composite **flowers** with both disc and ray florets, c. 4.5 cm dia.

Habitat: Uncommon in dry years but widespread in years of heavy rainfall, in shaded areas of the treeline and riverine forest margin. [Introduced.]

Flowering: From the middle of the main rains into the cool dry period.

Blundell p.161, Flowers Roodt p.43.

Asteraceae (daisy family)

Calostephane divaricata Benth.

Common names: English wing stem daisy, a name used for several daisies that have wings on their stems.

Derivation: *Calo*, beautiful [Greek], *stephanos*, something that surrounds or encircles, a crown or wreath [Greek]; *divaricata*, spreading apart at a wide angle [Latin].

Identification: Erect **annual herb,** growing up to c. 1 m tall. **Leaves** bright green, alternate and ovate, c. 9 × 2 cm, margins serrated and decurrent, forming substantial wings on the stems, sparsely **hairy** especially along the margins and on the prominent veins on the lower surface. Yellow composite **flowers**, c. 3 cm dia., with well-spaced ray florets. Flower heads mature to a spherical ball of fruits with only very short pappi (the hairy calyx that aids wind dispersal of composite seeds).

Habitat: Widespread in years of heavy rainfall in light shade on islands and in woodland.

Flowering: From the middle of the main rains into the cool dry period.

Flowers Roodt p.43.

Asteraceae (daisy family)

Conyza aegyptiaca (L.) Aiton
(syn. *Conyza transvaalensis* Bremek.)

Derivation: *Conyza*, the ancient name for the genus *Inula* dating from the time of Pliny; *aegyptiaca*, Egyptian [Latin].

Identification: Erect **annual herb** branching at the base, up to c. 1.5 m tall. **Stems** ribbed. Stems and leaves covered with coarse silvery hairs. **Leaves** dark green, linear and alternate, spiralling up the stems in 3s, without stalks, margins deeply dentate. Yellow composite **flowers**, c. 7 mm dia., with only disc florets surrounded by silvery hairs, borne at the apex of the plant. The plant has a sweet lilac-like **perfume** when gently crushed.

Habitat: Locally uncommon in seasonally flooded areas of floodplain and river bank and occasionally in the shelter of acacia scrub.

Flowering: Throughout the rains and into the cool dry period.

Germishuizen p.129.

Asteraceae (daisy family)

Conyza bonariensis (L.) Cronquist

Common names: Setswana moromoswane; **Afrikaans** kleinskraalhans; **English** flax-leaf fleabane, horseweed.

Derivation: *Conyza* the ancient name for the genus *Inula*, dating from Pliny; *bonariensis*, of Buenos Aires, Argentina.

Identification: Erect, dark green **annual herb,** c. 1.2 m tall; covered in a velvety mass of **hairs**. Side branches around the same length as the main stem. **Stems** ribbed, with a pithy core. **Leaves** alternate and linear, c. 60 × 3 mm, smooth margined, without stalks. Racemes of yellowy-fawn composite **flowers**, c. 7 mm dia., without ray florets.

Habitat: Grows in damp areas of seasonal water courses. An international weed, native of South America. [Introduced.]

Flowering: Flowers during both early and main rains.

B Van Wyk Flowers p.42.

Asteraceae (daisy family)

Conyza stricta Willd.

Derivation: *Conyza* the ancient name for the genus *Inula*, dating from Pliny; *stricta*, upright or erect [Latin].

Identification: Strongly green coloured, slightly hairy **annual herb,** 0.3–1.2 m tall; covered with long soft **hairs. Stems** ribbed, with a pithy core. **Leaves** linear, smooth margined although some are slightly serrated, up to c. 5.5 × 0.5 cm, spiralling up the stem. **Flowers** in tight groups at apex of the stems, composite, without ray florets, silvery or fawn, c. 2 mm dia., turning into small feathery seed heads, c. 7 mm dia.

Habitat: Widespread in damp areas of seasonal water courses.

Flowering: During both early and main rains.

Blundell p.164.

Asteraceae (daisy family)

Crassocephalum picridifolium (DC.) S.Moore

Derivation: *Krasson*, thick [Greek], forceful, *kephale*, a head [Greek], may refer either to the thick heads of flowers or to the woolly heads of seeds prior to dispersal; *picros*, bitter, sharp, pungent [Greek], *folium*, leaf [Latin].

Identification: Erect, often single stemmed, **perennial herb**, up to c. 1 m tall. **Leaves** linear, stalkless, with toothed bristly edges, spiralling up the stem. **Flowers** borne at the apex of the plant and from upper leaf axils on long stalks, yellow, large, c. 1 cm dia., composite with only disc florets.

Habitat: Locally uncommon growing in the water margin of channels and lagoons.

Flowering: During the main rains.

Ellery p.158, WFNSA p.456.

Asteraceae (daisy family)

Flaveria bidentis (L.) Kuntze

Common names: Afrikaans smelterbossie; **English** smelter's bush.

Derivation: *Flavus*, yellow [Latin]; *bi*, two [Latin], *dentis*, toothed [Latin], double-toothed.

Identification: Erect, much-branched **annual herb**, up to c. 1.5 m tall. Older **stems** slightly pink and hairless. **Leaves** opposite and decussate, sessile, elliptical, hairless, with toothed margins. Apparent brilliant yellow **flowers** actually **bracts** fused to form a tube, arranged in terminal and axillary (secondary flowering) clusters. Actual flowers small and within the yellow bracts.

Habitat: Locally uncommon on seasonal pan margins and beside waterways (naturalised from tropical America). Becoming an invasive nuisance in Maun and the surrounding area.

Flowering: Throughout the rains.

B Van Wyk Flowers p.108, Turton p.21.

Asteraceae (daisy family)

Geigeria schinzii O.Hoffm.
subsp. *rhodesiana* (S.Moore) Merxm.

Common names: English known locally as the wing stem daisy, a name used for several daisies that have wings on their stems.

Derivation: *Geigeria* after the German pharmacist Dr L. Geiger; *schinzii* for the Swiss botanist Professor Schinz.

Identification: Short-lived bushy erect **perennial herb** with noticeable wings on its red-brown **stems**. Usually c. 50 cm tall but may reach c. 140 cm after good rainfall. **Leaves** narrowly oval, toothed margins decurrent producing distinct wings down the stem. **Flowers** deep yellow, composite, ray florets recurved and disc florets loosely spaced.

Habitat: Widespread and common on the floodplain.

Flowering: Throughout the year.

Uses and beliefs:
- Subiya believe that if this plant grows well in a particular year the harvest will be good.
- The boys take the individual florets of the flowers and suck them to get the nectar to give them energy.

CM, BN.

Asteraceae (daisy family)

Grangea anthemoides O.Hoffm.

Derivation: *Anthemoides*, resembling chamomile, from its Greek name *Anthemis*.

Identification: Low-growing silvery green multi-branched **annual** or possibly short-lived **perennial herb**, from a basal rosette up to c. 10 cm tall. Densely covered with fine white **hairs**. **Leaves** pinnatifid, deeply incised, c. 3 cm long. **Flowers** borne in dense clusters at the apex of the plant and from leaf axils, bright yellow, c. 4 mm dia., composite with only disc florets. A delicate **perfume** emanates from the plant but not from the flowers.

Habitat: Locally rare growing in the bottom of dried-up seasonal pans in hard clays and soil.

Flowering: During the early rains.

Asteraceae (daisy family)

Helichrysum argyrosphaerum DC.

Common names: Setswana masupegane; **Afrikaans** poprosie; **English** everlasting weed, wild everlasting.

Derivation: *Helios*, sun [Greek], *chrysos*, golden [Greek]; *argyros*, silvery [Greek], *sphaera*, a sphere [Greek], in both cases alluding to the flowers.

Identification: Prostrate, mat-forming, **annual herb**, c. 60 cm across, with a single slender **taproot**. Covered in soft long white **hairs**. **Leaves** blue-green, spathulate to oblanceolate, c. 1.5 × 0.5 cm, smooth margins. Spherical composite **flowers**, c. 5 mm dia., borne singly at the apex of the branches, surrounded by a mass of papery silver-white bracts. Flowers open pink but the disc florets at the centre are yellow, fading to white with age.

Habitat: An introduced plant spreading locally in flat dry sandy areas on islands and in mixed woodland. Also becoming widespread on the pavements and roadsides of towns and villages.

Flowering: During the main rains and into the cool dry period.

B Van Wyk New p.44, Botweeds p.20, Germishuizen p.64, Turton p.113, WFNSA p.438.

Asteraceae (daisy family)

Hirpicium gorterioides (Oliv. & Hiern) Roessler
subsp. *gorterioides*

Derivation: *Hirpex*, a harrow [Latin], possibly referring to the deeply serrated leaves of some specimens; *gorterioides* for the Dutch botanist David de Gorter (1717–83), physician, plant collector and professor of medicine.

Identification: Branching erect **herb**, c. 20 cm tall. **Stems** covered in hairs. **Leaves** vary from lanceolate with entire margins to oval with deeply incised margins, short hairs on the upper surface, a tangled mass of silver short hairs below. **Flowers** at the apex of the branches, showy, composite with yellow ray florets surrounding yellow disc florets, opening only in full sun. **Fruit** a mass of white fluff often more noticeable than the flowers.

Habitat: Found in sandy areas of the floodplain in full sun.

Flowering: During the main rains.

FZ Vol.6 pt1.

Asteraceae (daisy family)

Laggera decurrens (Vahl) Hepper & J.R.I.Wood

Common names: Setswana monamagadi, mokodi; **Hmbukushu** mangobombo; **Shiyeyi** mago; **Subiya** dixombombo; **English** dwarf sage, silky sage.

Derivation: *Laggera*, for the Swiss physician and botanist Dr. Franz Josef Lagger (1802–70); *decurrens*, running down [Latin], referring to the leaf margin that forms a flange or wing that merges with the stem below it. *Monamagadi*, means female wild sage.

Identification: Erect possibly **short-lived perennial** herb up to c. 1.5 m tall. Densely covered with velvety grey **hairs**. **Leaves** alternate, linear, c. 4 cm long, margins smooth running down to form 4 wings on the stems. Composite **flowers**, yellow, c. 2 mm dia., borne at the apex of the stems and from the leaf axils. They turn quickly into tufts of **seeds** with a fawny pappus. The plant is **aromatic** when crushed.

Habitat: Common, growing mainly along dry water courses and around seasonal pans.

Flowering: Throughout the year.

Uses and beliefs:
- Subiya burn this plant on a metal sheet with hot coals in chicken sheds, effectively eradicating fleas for weeks.
- Wayeyi and San boil the roots and drink the resulting decoction as a treatment for stomach pain.
- The stems are also used as a fly-whisk and mosquito repellent.

TK, MM, MMo.

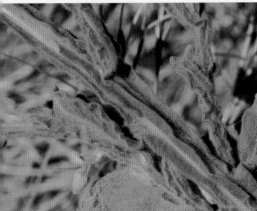

Asteraceae (daisy family)

Melanthera scandens (Schumach. & Thonn.) Roberty
subsp. *madagascariensis* (Baker) Wild

Common names: English yellow water-daisy.

Derivation: *Melas*, black [Greek], *anthera*, anthers [Greek], possibly referring to the black anthers; *scandens*, climbing [Latin], *madagascariensis*, of Madagascar

Identification: Rather lax, branching, scrambling, **annual herb**, growing to c. 1.8 m tall when supported by other vegetation. **Stem** square. Stems and leaves rough and sandpapery. **Leaves** hastate, opposite and decussate, margins toothed, with net veining. **Flowers** yellow, composite with both disc and ray florets, c. 25 cm dia.

Habitat: Locally uncommon on margins of flowing channels.

Flowering: During the main rains.

Uses and beliefs:
• Many medicinal uses ranging from cleansing external wounds and stopping bleeding to treating sore eyes, coughs and stomach disorders.

Ellery p.169.

Asteraceae (daisy family)

Melanthera triternata (Klatt.) Wild
(syn. *Melanthera marlothiana* O.Hoffm.)

Common names: English South American daisy.

Derivation: *Melas*, black [Greek], *anthera*, anthers [Greek], possibly referring to the black anthers of some plants of this genus; *triternata*, three times, for the three leaf sections in composite leaves.

Identification: Erect, branching, **annual herb**, up to c. 80 cm tall; short harsh hairs overall. **Leaves** opposite, oval, with toothed margins. **Flowers** yellow, composite with both disc and sparse ray florets, c. 1 cm dia.

Habitat: A common yellow daisy in the area, found on islands, near pans or in riverine forest, always in light shade.

Flowering: During the main rains and into the cool dry period.

Asteraceae (daisy family)

Nidorella resedifolia DC.

Common names: Setswana kgotodua; **English** nidorella.

Derivation: *Nitor* is the Latin for a strong smell. There are various possible derivations of *resedifolia*; it is suggested that it means 'leaves like mignonette', the garden flower, otherwise *resedo* meaning to heal and *folia* meaning leaf may refer to some healing property of the leaves.

Identification: Erect **herbaceous plant**, c. 80 cm tall. Often branched from the base or low on stem and again towards the flowering heads. Stem and leaves often harshly **hairy**. **Leaves** alternate, sessile, linear at top of plant, deeply incised (3–5 lobes) mainly on the lower part of the plant, c. 4 cm long. Bright yellow **flowers** arranged in groups at the apex of the stem, individual composite heads c. 3 mm dia., no ray florets.

Habitat: Widespread on the open floodplain, preferring areas near water.

Flowering: From the early rains into the cool dry period.

Ellery p.170, Turton p.111, Flowers Roodt p.47.

Asteraceae (daisy family)

Pseudognaphalium luteo-album (L.)
Hilliard & Burtt

Common names: Setswana motlalemetse, morethothobi, thitha; **Afrikaans** roerkuid; **English** cud weed, Jersey cudweed.

Derivation: *Pseudo*, false [Greek], *gnaphalion*, a downy plant with soft white leaves used to stuff cushions [Greek]; *luteo*, from *lutum*, dyer's greenweed, the source of a yellow dye [Latin], *album*, white [Latin].

Identification: A velvety, silvery green **annual herb**, up to c. 35 cm tall, with a whorl of leaves at the base. **Leaves** alternate, lanceolate, margins entire, c. 2 × 0.3 cm, covered in thick velvety white hairs. **Flowers** yellow, drying to orange, borne in small groups at the apex of the stem.

Habitat: Locally uncommon in dried-out areas on the edge of lagoons.

Flowering: During the main rains.

B Van Wyk Flowers p.114.

Asteraceae (daisy family)

Sclerocarpus africanus Jacq. ex Murray

Derivation: *Skleros*, hard [Greek], *-carpus*, fruited, hard-fruited [Greek]; *africanus*, African.

Identification: Erect branching **annual herb**, up to c. 1 m tall possibly taller later in season. **Leaves** opposite lower on the plant, alternate above, oval, slightly hairy, with dentate margins and 3 main veins. **Flowers** yellow, loosely formed composites with mainly disc florets and sparse rounded ray florets. A large receptacle holds the florets.

Habitat: Found only once in deep shade in riverine forest.

Flowering: During the main rains.

Asteraceae (daisy family)

Senecio strictifolius Hiern

Common names: Setswana mosimama; **San** gubagu; **English** ragwort.

Derivations: *Senex*, an old man [Latin], alluding to the white pappus of hairs on the seeds; *stricti*, straight, erect, upright [Latin], *-folius*, -leaved [Latin], with erect, possibly stiff, rather reduced leaves. *Mosimama* is the general term for all daisies.

Identification: Erect slim stemmed **annual** or **perennial herb**, c. 1 m tall, with its **roots** in slowly moving water. **Leaves** alternate, linear, c. 12 × 1 cm, recurved in their length, stalkless, margins slightly hooked dentate. **Inflorescence** a rather flat-topped terminal cluster of composite yellow **flowers**, c. 1.5 cm dia., with both ray florets and disc florets.

Habitat: Widespread but locally uncommon growing on the edge of the spillway.

Flowering: During the cool dry period.

Uses and beliefs:
- If a San woman is having birth problems, she bathes in an infusion of this plant. Afterwards, the water (with the plants still in it) is thrown on a path where people pass to take away the evil spirits.
- Caterpillars of *Dionychopus amasis* and *Trichoplusia exquisita* feed on this plant.

Ellery p.174, Flowers Roodt p.53. BB.

Burseraceae (myrrh family)

Commiphora edulis (Klotzsch) Engl. subsp. *edulis*

Common names: Setswana mokomoto, moroka; **Afrikaans** skurweblaarkanniedood; **English** rough-leaved commiphora, rough-leaved corkwood.

Derivation: *Kommi*, gum [Greek], *phoros*, bearing, carrying [Greek]; *edulis*, edible [Latin].

Identification: Small, many stemmed **deciduous tree** or **shrub**, c. 4 m tall. **Bark** pale grey peeling to an under-layer of green. **Stems** and leaves are covered in short **hairs**. Compound **leaves** borne in whorls at the apex of the branches, 3–4 pairs of leaflets with a slightly larger terminal leaflet, individual leaflets elliptical with net veining, margins smooth or occasionally scalloped and often with a hairy fringe. The falling leaves leave a strong **horseshoe scar** on the twig. **Flowers** small and inconspicuous, yellow, borne in long slender panicles. Fleshy **fruit** borne in long clusters, with 4 retained sepals, becoming apricot coloured on ripening.

Habitat: Found occasionally locally in riverine bush along the Chobe River.

Flowering: Before the main rains.

Uses and beliefs:
- The wood roots very easily from a cutting.
- Although the fruit is called 'edible', it is not eaten by humans, only by birds, baboons and rodents.

FZ Vol.2 pt1, Palgrave updated p.432, Setshogo & Venter p.38.

Burseraceae (myrrh family)

Commiphora merkeri Engl.

Common names: Afrikaans sebrabaskanniedood; **English** zebra-bark commiphora, zebra corkwood.

Derivation: *Kommi*, gum [Greek], *phoros*, bearing, carrying [Greek].

Identification: Small **deciduous tree** up to c. 5 m tall. **Bark** grey or greenish yellow and pitted in horizontal bands. **Leaves** borne in tight rosettes on short spine-tipped side branches, arranged alternately up the main branches. Leaves bluish-green with a greyish bloom, spathulate, stalkless, c. 9.5 × 4 cm, margin serrated around the upper 2/3 of the leaf. **Flowers** very small and yellowish, in inconspicuous clusters on the short side branches. **Fruit** ellipsoid, c. 1.3 × 1.0 cm, with a single seed, turning fleshy and pink as they ripen.

Habitat: Found occasionally locally in riverine bush along the Chobe River.

Flowering: At the beginning of the early rains.

FZ Vol.2 pt1, Palgrave p.365, Palgrave updated p.441, Setshogo & Venter p.39.

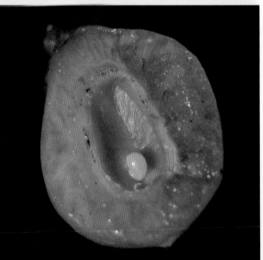

Burseraceae (myrrh family)

Commiphora mossambicensis (Oliv.) Engl.

Common names: Setswana moroka; **Afrikaans** peperblaarkanniedood; **English** pepper-leaved commiphora, pepper-leaved corkwood.

Derivation: *Kommi*, gum [Greek], *phoros*, bearing, carrying [Greek]; *mossambicensis*, of Mozambique.

Identification: Small **deciduous tree** c. 4 m tall. **Bark** smooth and pinkish grey with wrinkles around the branches making them look as if they have been set in toffee. **Leaves** compound with 3 leaflets, individual leaflets deltoid, shiny and leathery, with net veining, margins with a hairy fringe. Leaves have an aromatic peppery smell when crushed. **Flowers** borne on new growth in long racemes from the leaf axils, greenish cream, with 4 curled lobes, c. 4 mm dia. **Fruit** plum-like, borne in racemes on long pedicels, c. 3 cm, ripening to a blackish red. These trees readily exude a clear **resin** that has a pleasant spicy **aroma**.

Habitat: Found occasionally locally in riverine bush along the Chobe River.

Flowering: Before the main rains.

Uses and beliefs:
- The wood is used for kitchen utensils.
- Ashes from the bark are used to treat cuts on people and livestock.

FZ Vol.2 pt1, Hargreaves p.46, Palgrave p.366, Palgrave updated p.436, Setshogo & Venter p.39.

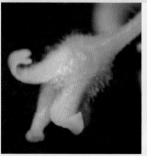

Capparaceae (caper family)

Cleome angustifolia Forssk. subsp. *petersiana* (Klotzsch) Kers

Common names: Afrikaans peultjiesbos, geelcleome; **English** yellow mouse whiskers.

Derivation: *Cleome* from Cleoma, a name used in the Dark Ages for a strong-tasting plant growing in damp places; *angustifolius*, from *angusti*, narrow [Latin], *folius*, leaf [Latin], narrow-leaved.

Identification: Erect annual **herb**, growing up to c. 1.2 m tall. **Leaves** palmately compound with fine needle-like linear leaflets. **Flowers** yellow, with 2 pairs of upright petals with dark brown-red markings on the throat, with 2 long recurved, fertile stamens and c. 10 short infertile stamens. **Fruit** long, narrow, tube-like pods.

Habitat: Common along partially shaded track sides in mopane and mixed woodland.

Flowering: Early in the main rains.

Blundell p.26, BVW Photoguide ZA p.37, Germishuizen p.99.

Guttiferae (mangosteen family)

Garcinia livingstonei T.Anderson

Common names: Setswana motsaodi, isika, mokonkono, mokononka, mokanonga, moralana, motsaudi; **Hmbukushu** moshika; **Subiya** insika, masika; **Shiyeyi** motsaudi; **Lozi** ghushika; **Afrikaans** Afrikageelmelkhout, laeveldse geelmelkhout; **English** African mangosteen, lowveld mangosteen, wild plum.

Derivations: *Garcinia* for Laurent Garcin (1683–1751), a French botanist; *livingstonei* for David Livingstone (1813–73), the missionary and explorer. *Insika* means one mangosteen, *masika* means many mangosteens.

Identification: Dark green **evergreen tree**, up to c. 10 m tall, with a dense crown of stiff straight branches rising at sharp angles. **Bark** irregularly fissured and dark grey. **Leaves** elliptical, with smooth margins and prominent white veining, upper surface leathery, dark and shiny, paler green below. Leaves produced in whorls and often in 3s up the branches. **Male and female flowers** borne on separate trees, male flowers the more visible. **Flowers** greenish cream, with 5 petals curving back to the stem and 2 sepals, like small pin-cushions, borne in clusters up the branches and with a cloying perfume and plentiful nectar. Ripe **fruit** are plum-shaped drupes, bright orange-red, c. 2 cm long, producing a sticky yellow latex when peeled.

Habitat: Locally common and widespread in damp areas along river banks and on islands, forming dense shade.

Flowering: Usually at the end of the cool dry period but also at the end of the main rains if conditions are right.

Uses and beliefs:
- The fruit are considered a great delicacy by humans, animals and birds. An alcoholic beverage is also made from the fruit. The fruit juice is mixed with milk to make a sort of yoghurt that is added to porridge.
- The roots are boiled and used as a mouthwash for toothache.
- Wayeyi and Subiya use a decoction of boiled roots to treat stomach pain and constipation.
- The wood is used for kraal fences and occasionally for making *mekoro* (dug-out canoes). The twigs are used to make stirring sticks for porridge.

FZ Vol.1 pt2, B&P van Wyk Trees p.360, Curtis & Mannheimer p.462, Ellery p.116, Hargreaves p.11, Palgrave updated p.738, P Van Wyk Kr Trees p.165, Setshogo & Venter p.45, Tree Roodt p.81. PK, BN, SS.

Combretaceae (bushwillow or combretum family)

Combretum apiculatum Sond. subsp. *apiculatum*

Common names: Setswana mohwidiri, mohudiri, mhudiri, mohulere, mogoliri, ntshingidtza, ntshingitsha; **Kalanga** ntshingitshi; **Lozi** mumpaumpa; **Ovambanderu** omumbuti; **Shiyeyi** mokgabi; **Afrikaans** rooibos(wilg), koedoebos; **English** (hairy) red bushwillow, kudu-bush.

Derivation: *Combretum*, believed to arise from the name of a climbing plant, used by Pliny; *apiculatum*, ending abruptly in a short point [Latin], referring to the leaves.

Identification: Shrubby **deciduous tree** c. 5 m tall. **Bark** dark grey, deeply fissured, scaly. **Leaves** hairless, opposite and decussate, elliptical, wavy margined, tip twisted, snapping off easily. Young leaves may be slightly glutinous. **Flowers** borne in short spikes from the leaf axils, greenish yellow. **Fruit** with 4 papery wings.

Habitat: Locally common in mixed sandy woodland.

Flowering: Middle to end of the main rains.

Uses and beliefs:
- Traditionally Batswana mix the roots with *Cucumis metuliferus* and other herbs to heal broken bones. A cross is made with the mixture at the place where the bone is broken in a ceremony called *thobega*.
- A decoction of the boiled roots is taken each time there is a motion until diarrhoea is cured.
- The bark is used for tanning leather.
- The wood is termite resistant and used for building and axe handles.
- The caterpillars of *Coeliades forestan* feed on the leaves.

FZ Vol.4 pt0, B Van Wyk Flowers p.292, B & P Wyk ZA Trees p.328, Curtis & Mannheimer p.468, Hargreaves p.32, P Van Wyk Kr Trees p.169, Palgrave p.663, Setshogo & Venter p.46. BN, RR, KS, SS.

Combretaceae (bushwillow or combretum family)

Combretum hereroense Schinz

Common names: Setswana mokabi, mokata, mungave, monwana, motsiara; **Hmbukushu** mofofo, mongave; **Kalanga** nswazwi, nthare; **San** mokabe; **Shiyeyi** ushuu; **Afrikaans** kierieklapper; **English** russet bush-willow, mouse-eared combretum.

Derivation: *Combretum*, believed to arise from the name of a climbing plant, used by Pliny; *hereroense*, from the region occupied by the Herero people.

Identification: Small **tree** or **shrub**, usually 3–5 m tall but occasionally up to c. 10 m. **Leaves** variable, may be alternate or opposite, hairy or smooth, elliptical to round with smooth margins. **Flowers** in dense terminal and axillary spikes, very small, greenish-cream, sweetly scented, with protuberant stamens. Four-winged **fruit** a bright russet colour, easily recognisable and borne profusely over the tree. This plant often has an untidy appearance as it is heavily browsed by elephants and other animals.

Habitat: Widespread in all kinds of woodland with both sandy and clay soils. Often found colonising drying areas of the floodplain.

Flowering: Before and during the early rains.

Uses and beliefs:
- San use this plant extensively. The fruit is used as an infusion (one fruit for one mug of tea). In other areas, both leaves and fruit are crushed to make an infusion.
- The roots are used to make a skin softening cream.
- Gargling with a decoction made from boiling 3 fruits is used to treat sore throats and loss of voice.
- A decoction of boiled roots is drunk for several (up to 5) days to enlarge the penis.
- The roots and bark are added to water and used in the curing of animal skins.
- Shiyeyi, Subiya and Hmbukushu use the wood for tool handles, axes, ploughs and for firewood.

FZ Vol.4 pt0, Ellery p.104, Hargreaves p.33, Palgrave p.802, Tree Roodt p.101, WFNSA p.274. RM, KN, SS, BT, MT.

Commelinaceae (spiderwort family)

Commelina africana L.
var. *barberae* (C.B.Clarke) C.B.Clarke
(This family is currently under revision and consequently these scientific names may change.)

Common names: Setswana tshoo-la-khudu; **English** yellow commelina, wandering jew, dayflower.

Derivation: *Commelina*, for the Dutch botanists Johan Commelijn (1629–92) and his nephew Caspar (1667–1731); *africana*, from Africa.

Identification: Self-supporting, erect, branching, bushy **perennial herb**, growing to c. 35 cm from a woody tuberous **rooting stock**. **Leaves** alternate, narrowly ovate, folding at the mid-vein and recurved along their length, margins decurrent and forming a stem sheath. **Flowers** bright lemon yellow, c. 2 cm across, with 2 showy petals and a third small, almost colourless one, emerging often in pairs from a folded leafy spathe that is c. twice as long as it is wide.

Habitat: Widespread but locally rare in light shade in mopane woodland.

Flowering: Late in the main rains.

B Van Wyk Flowers p.124, Botweeds p.114, WFNSA p.32.

Convolvulaceae (morning glory family)

Ipomoea obscura (L.) Ker-Gawl.

Common names: Setswana motangtanyane, kalake, mosokotsala, motsididi; **Afrikaans** wildepatat; **English** wild petunia, yellow ipomoea.

Derivation: *Ips*, worm [Greek], *homoios*, like [Greek], referring to its trailing creeping habit; *obscura*, indistinct, uncertain [Latin], possibly because its flowers are inconspicuous.

Identification: Scrambling, twining **perennial**, up to c. 2 m tall with suitable support. **Leaves** alternate, cordate, with smooth often hairy margins, covered in soft white hairs on both surfaces, especially along the veins, bright green above and lighter green below. Funnel-shaped, creamy-yellow **flowers**, c. 3 cm dia., with a 5-pointed star shape, borne in pairs on long stems from the leaf axils. **Fruit** hold 4 hairy seeds.

Habitat: Widespread but locally rare growing on a sandy tongue in mopane woodland.

Flowering: Late in the main rains.

Vol.8 pt1, B Van Wyk New p.126, Blundell p.372, Botweeds p.42, Germishuizen p.124, Pooley p.304, Turton p.69, WFNSA p.334.

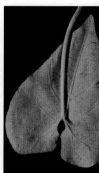

Convolvulaceae (morning glory family)

Ipomoea tuberculata Ker-Gawl.

Derivation: *Ips*, worm [Greek], *homoios*, like [Greek], referring to its trailing habit; *tuberculata*, tuberculate [Latin], covered with wart-like excrescences, referring to the stems, which may be tuberculate.

Identification: Vigorously growing **annual climber**, c. 2.5 m tall; hairless. **Leaves** palmate with deep oval lobes, the lower pair of lobes asymmetrically divided into 3 further lobes. **Flowers** showy, cone-shaped, creamy yellow with a magenta pink centre, c. 5 cm dia., borne in twos or threes on long stems. Flowers open at night and wilt as soon as the sun gets onto them. White **latex** present.

Habitat: Found occasionally locally on islands in the floodplain.

Flowering: During the main rains.

FZ Vol.8 pt1.

Convolvulaceae (morning glory family)

Merremia pinnata (Hochst. ex Choisy) Hallier f.

Derivation: *Merremia* for Blasius Merrem (1761–1824), professor of natural sciences at Marburg in Germany; *pinnata*, feather-like, having leaflets arranged on each side of a common stalk [Latin].

Identification: Creeping, climbing **annual herb**, stems c. 70 cm long. Whole plant, including the seed capsule, covered with sparse long **hairs**. **Leaves** alternate, spiralling up the stem, pinnatifid (shaped like deeply incised feathers), stalkless c. 3.5 cm long. **Flowers** yellow, funnel-shaped, c. 1 cm dia., borne usually singly on stems, c. 2.5 cm long. **Fruit** spherical, c. 5 mm dia., retains both the sepals and the stigma. **Latex** present.

Habitat: Locally uncommon in sandy areas of the spillway.

Flowering: Middle to end of the main rains.

Uses and beliefs:
- Palatable to animals and often grazed.

FZ Vol.8 pt1.

Convolvulaceae (morning glory family)

Xenostegia tridentata (L.) D.F.Austin & Staples
subsp. *angustifolia* (Jacq.) Lejoly & Lisowski
(syn. *Merremia tridentata* subsp. angustifolia (Jacq.) Ooststr.)

Common names: Setswana motantanyane; **English** miniature morning glory, merremia.

Derivations: *Tri*, three [Latin], *dens*, teeth [Latin], referring to the shape of the leaf; *angustifolia*, *angusti*, narrow [Latin], *folius*, leaf [Latin], narrow-leaved. *Motantanyane* is the generic name for climbing plants.

Identification: Herbaceous climber or **creeper**, most usually growing flat on the ground but occasionally erect, stems usually c. 50 cm long, occasionally climbing up to c. 2 m if suitable support is available. **Leaves** alternate, long and narrow, c. 3.5 cm (inc. c. 0.3 cm leaf stalk) × 1.5 cm, with a distinctly toothed base, number of teeth variable. **Flowers** yellow, c. 15 mm dia., funnel-shaped with acute tips to the petals, borne singly or in pairs on long flower stalks, c. 18 mm long.

Habitat: Widespread in full sun on the floodplain and amongst acacia scrub.

Flowering: From the middle of the main rains into the cool dry period. Seeming to produce only male flowers until adequate rain makes fruit viable.

Uses and beliefs:
- The trailing stems are used by the Kalanga to make a pad to protect the head when carrying heavy loads.
- San collect the leaves and cook them with water and salt to make *morogo* (*xhaa* in San), a vegetable relish.
- San also boil the roots and drink the resulting infusion to treat venereal diseases.
- In certain parts of Botswana, the plant is mixed with vaseline and smeared over the body as a medicine.

FZ Vol.8 pt1, Botweeds p.46, Pooley p.302. RM, RR.

Acanthosicyos naudinianus (Sond.)
C.Jeffrey

Common names: Setswana mokapane, mogapu, mokapana, magaga, lekawa; **G/wi** ka, ka jisa; **Hmbukushu** ropuiti; **Ju/'hoansi** n/uah; **Kgalgadi** kawa; **!Kung** cha; **Subiya** kazungula, umbwiti; **Shiyeyi** ntutuba; **Afrikaans** gemsbok komkommer; **English** gemsbok cucumber, herero cucumber, wild melon.

Derivation: *Acantha*, a spine or thorn [Greek], *sicyoideus*, gourd-shaped [Latin], i.e. swollen below with narrow neck above, referring to the spines on the fruit; *naudinianus* for the French botanist Charles Victor Naudin (1815–99).

Identification: Perennial **creeper** with stems up to c. 8 m long, from a large tuberous, carrot-like **root** that can be up to c. 1 m long. Heavily covered with rough **hairs** overall. **Leaves** palmate with 5 deeply toothed lobes, the central lobe much larger than the others. **Male and female flowers** on the same plant, **female flowers** about twice the size of the **male flowers**. Flowers yellow with 5 petals and 3 stamens, c. 2 cm dia. **Fruit** spherical, covered with spikes, becoming become pale yellow on ripening.

Habitat: Locally common and widespread on the floodplain, occasionally in sandy open areas in woodland.

Flowering: Throughout the year.

Uses and beliefs:
- Useful as a source of food and water for San. The fruit are popular with elephant and the foliage is heavily grazed by impala and kudu, especially in years of heavy rainfall when it is softer.
- The bitter roots are soaked in hot water for a few minutes and the resulting infusion is taken as a laxative.
- Both Wayeyi and Hmbukushu women rub the roots, which are very bitter, on their breasts to stop their babies suckling when it is time for them to be weaned.
- Wayeyi use the young stems to tie bundles of thatching grass or wood so that they can be carried back to the village.

FZ Vol.4 pt0, Botweeds p.48. CM, SS, MT.

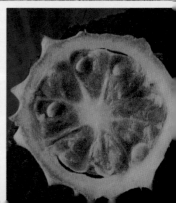

Cucurbitaceae (cucumber or gourd family)

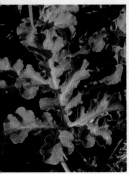

Citrullus lanatus (Thunb.) Matsum. & Nakai

Common names: Setswana kgengwe, legapu, katcama, mokate, legau, lekatane, tsama, mokaikai; **Subiya** kaniyangngombe, kaniyanzovu; **G/wi** n?a; **G//ana** n//an; **Afrikaans** tsamma, karkoer; **English** monkey apple, citron, citron melon, keme, tsamma, tsama melon, wild watermelon, bittermelon, bitter apple, edible seed melon.

Derivations: *Citrullus*, from *citrus* [Latin], from the appearance of the fruit and a slight citrus odour or flavour; *lanatus*, woolly [Latin], referring to the woolly hairs on the plant. *Kaniyangngombe*, cow dung. *Kaniyanzovu*, elephant dung.

Identification: Large climbing and creeping **annual herb**, up to c. 5 m. Harsh **hairs** overall except the flowers. **Leaves** with pointed tips, palmate, deeply 3–5-lobed with wavy and toothed margins. **Tendrils** at the leaf axils. **Flowers** usually with 5 petals, occasionally 6, c. 4 cm dia., **male and female flowers** on the same plant. **Fruit** large, green often with mottled bands of dark green running longitudinally around the fruit, c. 12 cm dia.

Habitat: Locally uncommon, occasionally on the trackside, on the floodplain and in the edge of the treeline.

Flowering: During the main rains.

Uses and beliefs:
- Subiya pound the dried and roasted seeds to make meal that is cooked with dried meat (*segwapa* in Setswana).
- *Legau* is eaten as a sweet melon.
- *Lekatane* and *tsama* are eaten by people and animals who need the water despite the bitter taste.
- San eat the fruit when they are happy and want to dance and sing, if they have hunted successfully or for some traditional occasions. They also fry the seeds to eat like peanuts.
- Traditionally, Subiya add chopped plants to the bath water of sick children, occasionally making the children drink a little. This treatment is believed effective for many illnesses because the plant is eaten by elephants, which eat 'medicinal' trees and plants. The fruit are the size and shape of elephant dung, and therefore thought to be a 'distillation' of all the benefits attainable from plants eaten by elephants.
- To win a court case, chew a small piece of root and swallow a little juice, then, holding the big toe of the left foot with the left hand, make a cross on the sole of the foot using the right hand and, while holding the plant, speak in the heart about wishing to win the case.
- *Mokate* is fed to cattle.
- The edible watermelon is a domesticated plant of this species.
- When elephants or cattle get into the fields and steal the melons, the seeds germinate in their faeces but they are no longer cultivated water melons but the wild ones.

FZ Vol.4 pt0, B Van Wyk Flowers p.128, Hargreaves p.17. CM, GM, RM.

Cucurbitaceae (cucumber or gourd family)

Coccinia adoensis (A.Rich.) Cogn.

Common names: Kalanga nyololo; **Subiya** malisaka; **English** wild spinach.

Derivation: *Coccineus*, scarlet [Latin], referring to the colour of the fruit; *adoensis* derived from a place name in Ethiopia.

Identification: Annual climber from a perennial **rootstock**, up to 2 m tall. **Leaves** alternate, almost hairless on upper surface, slightly hairy below, palmate, sometimes 3–5 lobes, margins with a bristle at the tip of each vein. **Tendrils** at the leaf axils. **Male and female flowers** separate, the male on longer stems, the female usually single on a short stem at the leaf axil, cream with 5 or 6 lobes, slightly furry, c. 1.5 cm dia.

Habitat: Locally uncommon on lightly shaded islands and treeline of the floodplain.

Flowering: During the main rains.

FZ Vol.4 pt0, B Van Wyk Flowers p.128, Germishuizen p.412, Blundell p.58.

Cucurbitaceae (cucumber or gourd family)

Cucumis africanus L.f.

Common names: Setswana mogabala?, kgatwe, legabala; **Kgalgadi** moshumo; **G/wi** n?o'nu; **English** bitter wild cucumber, horned cucumber, jelly melon.

Derivation: *Cucumis*, cucumber [Latin]; *africanus*, of Africa [Latin].

Identification: Annual scrambling creeper. Stems ribbed. Stems and leaves harshley **hairy**, giving the plant a silver-green colour. **Leaves** alternate, palmate with 5 lobes, margins toothed. **Tendrils** at the leaf axils. **Flowers** bright yellow with 5 petals, c. 2.5 cm dia., **male and female flowers** borne on the same plant.

Habitat: Locally rare growing in the clayey sand of the floodplain.

Flowering: Late in the main rains.

FZ Vol.4 pt0.

Cucurbitaceae (cucumber or gourd family)

Cucumis anguria L.

Common names: Setswana monyaku; **Kalanga** ntangabe; **Subiya** katamanwusea; **Afrikaans** rooi-agurkie, wilde komkommer, rooikomkommer; **English** wild cucumber, horned cucumber, jelly melon, gherkin, bur cucumber, gooseberry cucumber.

Derivation: *Cucumis*, cucumber [Latin]; *anguria* derived from the Greek for cucumber.

Identification: Scrambling, creeping and climbing **annual plant**, up to c. 1.5 m with a suitable support. **Tendrils** present. **Leaves** palmate, 5-lobed with occasional deep serrations on the margins, bristly on both surfaces, c. 7.5 mm (inc. c. 3.5 mm leaf stalk) × 4 cm. **Male and female flowers** on separate plants, the **male** in the leaf axil, the **female** on longer stalks, both c. 5 mm dia. and yellow. **Fruit** heavily softly spiny, green becoming yellow when ripe.

Cucumis anguria is easily distinguished from other *Cucumis* species by its prickly stems and leaf stalks.

Habitat: A pan-tropical herb originating from south-west Africa, found occasionally locally in light shade on islands and in farmed mopane woodland.

Flowering: From the middle of the main rains into the cool dry period.

Uses and beliefs:
- Batswana use the leaves for *morogo*, a vegetable relish. It is cooked and added to milk and a little mealie meal to thicken it, or else added to ground nuts or cooking oil.
- The young fruit are considered edible boiled or pickled, related to the gherkin.

FZ Vol.4 pt0, Fox & Young p.174, Hargreaves p.18, Wyk p.156.

Cucurbitaceae (cucumber or gourd family)

Cucumis metuliferus E.Mey. ex Naudin

Common names: Setswana thobega, mokapana, mokapa, mogabala; **San** gkaa; **Afrikaans** rooi-agurkie, wilde-komkommer, rooi-komkommer; **English** bitter wild cucumber, horned cucumber, jelly melon.

Derivation: *Cucumis*, cucumber [Latin].

Identification: Strong **annual climbing plant** up to 3 m; covered in coarse **hair**. **Leaves** are alternate, palmate, trilobed with cordate bases, margins irregularly serrated. **Flowers** yellow with 5 petals. **Fruit** up to 15 cm long, ovoid, at first blotched dark green but turning golden yellow when ripe, covered with solid, blunt conical **spikes**; **flesh** is emerald green.

Habitat: Widespread, in years of heavy rainfall, in shaded areas of riverine forest, treeline and islands.

Flowering: Middle to end of the main rains.

Uses and beliefs:
- San boil the fruit and eat them. The puréed fruit can be served as a chilled soup.
- The fruit may be used in salads or as dessert fruit. They are peeled, sliced and flavoured with lemon, salt and black pepper according to taste.
- It is also used as a treatment for worm under the skin. Squeeze the juice over the worm until it comes out.
- Traditionally, Batswana mix the roots with *Combretum apiculatum* and other herbs to heal broken bones. A cross is made with the mixture at the place where the bone is broken in a ceremony called *Thobega*.
- A decoction of boiled roots is taken each time there is a motion until diarrhoea is cured.

FZ Vol.4 pt0, WFNSA p.414, Wyk p.158. BB, RM, BN, SS.

Cucurbitaceae (cucumber or gourd family)

Kedrostis abdallai A.Zimm.

Common names: Setswana lechachalanoga, mampipinyane.

Derivation: *Kedrostis* is the classical Greek name for white bryony, a climbing plant. *Lechachalanoga*, 'the side of the snake', possibly referring to the patterning of the green fruit.

Identification: An **annual climbing herb**, c. 2 m tall; coarsely hairy overall. **Leaves** alternate, palmate with 5 lobes, margins serrated with a bristled point to each lobe. **Male flowers** green, c. 1.2 cm dia., with prominent orange anthers. **Female flowers** yellow c. 1.8 cm dia. **Tendrils** and flowers borne at the leaf axils. **Flowers** stalkless. **Fruit** carrot-shaped, green with white spots when under-ripe, ripening to brilliant red, splitting open on contact when ripe.

Habitat: Found occasionally in acacia scrub in full sun and on lightly shaded islands.

Flowering: During the main rains.

Uses and beliefs:
• This plant is generally believed to be poisonous. But some sources who may confuse it with the edible plant *Momordica balsamina*, say that the fruit are edible and can be rolled between the hands to allow the fruit to be sucked out.
• Some Batswana believe that snakes eat the fruit.

BN, KS.

Cucurbitaceae (cucumber or gourd family)

Kedrostis foetidissima (Jacq.) Cogn.

Common names: G//ana chunan; **Xade** karu, karo; **English** monkey pepper.

Derivation: *Kedrostis* is the classical Greek name for white bryony; *foetidissima*, very evil smelling [Latin].

Identification: Climbing plant, c. 2 m tall, with a tuberous **root**. Whole plant covered with club-shaped hairs, especially on leaf undersides. **Stem** ribbed. **Leaves** deeply cordate, margins entire. **Tendrils** from the leaf axils. **Flowers** small, yellow with 5 petals, c. 7 mm dia., in groups of both **male and female** in the leaf axils. **Fruit** red, hairy, spherical with a pointed tip, c. 7.5 mm dia. In dry weather, the plant has a strong unpleasant **smell**, also the open ripe fruit (like over-ripe camembert cheese).

Habitat: Locally widespread but not common growing in the shelter of thorn scrub and in light shade on islands.

Flowering: During the main rains.

FZ Vol.4 pt0, Germishuizen p.416.

Cucurbitaceae (cucumber or gourd family)

Momordica balsamina L.

Common names: Setswana mmampimpinyane, mantshegi, mmapupu, mmapuupuu, pingpingtshegatshega; **Subiya** malisako; **Shiyeyi** kgwakazenguro, maXatora; **Afrikaans** laloentjie; **English** balsam pear, balsam apple, African cucumber, balsamina.

Derivation: *Mordeo*, to bite [Latin], derived from the uneven surface of the fruit; *balsamina*, balsam, a sweet smelling resin. *Kgwakazenguro*, loosely translated means 'the curse' (see 'Uses and beliefs').

Identification: Strong **perennial climber** with a tuberous **root**. **Leaves** deeply palmate with 5 lobes, smooth with deeply dissected margins. **Tendrils** present. **Flowers** a soft creamy yellow with darker centre, 5 petals, c. 3 cm dia. **Male and female flowers** borne on the same plant. **Fruit** a sphere with pointed ends, ripening to a brilliant orange-red, spontaneously dehiscent.

Habitat: Widespread in full sunlight and partial shade on island margins and in the floodplain in clayey sand.

Flowering: Main rains through to the end of the cool dry period.

Uses and beliefs:
- The juicy seeds are very much sought by birds and animals and are edible when cooked. The leaves can also be used for cattle fodder.
- If there are step-children in a Wayeyi marriage and the parents quarrel, the step-mother may curse the child saying 'you will not eat my food'. The child will then be unable to eat until the traditional doctor gives the child a mixture of the (fresh or dried) fruit of *Momordica balsamina* and *Diospyros lycioides* with a little of the step-mother's porridge.
- Batswana believe that if you have an argument with someone and want to ill-wish them, you should clap the fruit between your hands (while speaking to your ancestors) and ask for something bad (e.g. a minor ailment or say a puncture in a tyre) to happen to that person.
- Subiya people living at the cattle posts use the plant as food.
- Older Subiya men sprinkle the water from boiling roots in the kraals to prevent hyaenas and lions entering. They beat children who misuse this plant.
- Doubts about the poisonous nature of this plant may arise from its being easily confused with *Kedrostis abdullai*, which is also found locally.

B Van Wyk Flowers p.296, Flowers Roodt p.73, Germishuizen p.60. BB, CM, PN, SS.

Cucurbitaceae (cucumber or gourd family)

Mukia maderaspatana (L.) M.Roem.

Common names: English rough bryony.

Derivation: *Mukia*, possibly from a Malayan name *mucca-piri* in which *mucca* means 'three-quarters' and *piri* means 'a spring', possibly referring to the coiled spring-shaped tendrils; *maderaspatana*, of Madras in southern India.

Identification: Coarsely hairy **climbing plant**, c. 1 m long. Tendrils present. **Leaves** alternate, broadly hastate with deeply cordate bases, margins are serrated to wavy. **Flowers** yellow, minute, separately **male and female**, with 5 petals, in clusters at the leaf axils. **Fruit** spherical, c. 1 cm dia., bright red.

Habitat: Found occasionally locally, especially in wet and dried out marshy areas, in peaty soils.

Flowering: During the main rains.

FZ Vol.4 pt0.

Ebenaceae (ebony family)

Diospyros lycioides Desf. subsp. *sericea* (Bernh.) de Winter

Common names: Setswana letlhajwa, motlhaje, motlhajwa, mothakeja; **San** tszini; **Subiya** musihantabwe; **Shiyeyi** rethajwa; **Afrikaans** bloubos; **English** (hairy) blue bush, red star apple, Kalahari star apple, toothbrush tree.

Derivations: *Dios*, divine [Greek], *pyros*, wheat [Greek], a name transferred to this genus with edible fruit; *lycioides* resembling the genus *Lycia*, a *Solanaceae* that comes from Lycia, Asia Minor; *sericeus*, silky [Latin], referring to the long straight silky hairs on the leaves. *Toothbrush tree*, see 'Uses and beliefs'.

Identification: Large, thick **shrubby bush**, c. 3 m tall. **Bark** dark grey, smooth. Silver-grey **leaves** spiralling up the new wood, obovate with smooth margins, with coarse **hairs** on both surfaces. **Flowers** borne singly at the leaf axils, pendulous, cream coloured, sweetly **scented**, with 5 recurved petals fused to form a short tube. A single **bract** below each flower. **Fruit** scarlet berries with a persistent recurved calyx, becoming brown with age.

Subsp. *sericea* is distinguished by the dense hairs on both surfaces of the leaves and by the **veins** on the leaves. These veins are indented on the upper surface, whereas they are prominent on the lower surface.

Habitat: Widespread and common as an understorey in woodlands on islands and in the treeline.

Flowering: During the cool dry period.

Uses and beliefs:
* Both Wayeyi and Subiya call the shrub the 'toothbrush tree' as the roots are used as toothbrushes. The roots are chewed and this dyes the tongue orange, making the teeth look white.
* The fruit are edible but rather bitter.
* If there are step-children in a Wayeyi marriage and the parents quarrel, the step-mother may curse the child saying 'you will not eat my food'. The child will then be unable to eat until the traditional doctor gives the child a mixture of the (fresh or dried) fruit of *Momordica balsamina* and *Diospyros lycioides* with a little of the step-mother's porridge.
* Wayeyi drink an infusion of boiled roots to clean the kidneys.
* In the past, teachers used sticks from this shrub to beat the children because it is whippy and does not break easily!
* The bark is used to make dark brown dye for baskets.

FZ Vol.7 pt1, B Van Wyk New p.130, Ellery p.108, Palgrave updated p.904, Setshogo & Venter p.51, Tree Roodt p.37, WFNSA p.292. TK, BN, PN, RR, KS, BT.

Eriospermaceae

Eriospermum bakerianum Schinz subsp. *bakerianum*

Common names: Setswana kgopokgolo.

Derivation: *Erion*, wool [Greek], *-spermum*, -seeded [Greek], woolly-seeded; *bakerianum*, for John Gilbert Baker (1834–1920), Keeper of the Kew Herbarium 1890–99.

Identification: Slender, erect, silvery-green **perennial herb**, c. 25 cm tall, from a **tuberous rooting** system. **Flowers** bright yellow, c. 1 cm dia., with 2 whorls of 3 tepals; borne in an open raceme on long stalks, appearing to flower before producing **leaves**.

Habitat: Locally rare in open, damp, clayey areas of mopane woodland.

Flowering: Late in the main rains.

Euphorbiaceae (spurge family)

Schinziophyton rautanenii (Schinz) Radcl.-Sm.
(syn. *Ricinodendron rautanenii* Schinz)

Common names: Setswana mongongo, monghongho, mokongwa, mungongo; **Hmbukushu** ngongo; **San** xumo; **Subiya** ingongo; **Lozi** mungongo; **Afrikaans** mankettiboom; **English** manketti tree, featherweight tree, mongongo nut.

Derivation: *Schinzio* for the Swiss botanist, Professor Schinz, *phyton*, plant [Greek].

Identification: Large, spreading **tree**, 15–20 m tall. **Bark** golden brown with grey patches. Young branches and twigs covered in rust-coloured **hairs**. **Leaves** palmately compound with usually c. 5 leaflets, leaflets obovate, margins entire, leaf stems and veining slightly reddish. **Male and female flowers** on separate trees. **Flowers** yellow, c. 1 cm dia., growing in long sprays. **Fruit** egg-shaped, c. 3 × 2 cm, hard and woody, light grey-green, covered in velvety hairs. Young wood exudes white **latex**.

Habitat: Locally rare in mixed woodland. (A protected species in both Namibia and South Africa because of its importance as a potential food.)

Flowering: During both the early and the main rains.

Uses and beliefs:
- Hmbukushu reckon this to be one of the best trees for *mekoro* (dug-out canoes).
- The nuts are chopped open, white kernels are peeled, pounded and boiled with water to extract a strongly flavoured bright yellow cooking oil. It has to be used very sparingly.
- The red aril on the kernel is also edible.

FZ Vol.9 pt4, B&P Wyk Trees p.474, Curtis & Mannheimer p.334, Hargreaves p.43, Palgrave p.432, Setshogo & Venter p.58. MM, RM.

Fabaceae (pod-bearing family), Caesalpinioideae (cassia sub-family)

Cassia abbreviata Oliv.
subsp. *beareana* (Holmes) Brenan

Common names: Setswana monepenepe, mokwankusha, ngaganyama, sifonkola; **Afrikaans** sjambokpeul; **English** sjambok pod, long tail cassia.

Derivation: *Cassia*, the Greek name for the genus of plants that produce senna pods and leaves; *abbreviata*, shortened, stunted [Latin]; *beareana*, for Dr O'Sullivan Beare who first collected this plant in Tanzania in the late 19th Century.

Identification: Slim **deciduous tree** with an open crown, c. 4 m tall. **Leaves** compound with up to 12 pairs of ovate leaflets, smooth margins. **Flowers** showy, brilliant yellow, short-lived, borne on long stems in short racemes at the apex of the twigs, sweetly **scented** and appearing before the leaves. **Fruit** long, cylindrical pods, up to c. 80 × 3 cm.

Habitat: Found occasionally in riverine bush near the Chobe River.

Flowering: During the cool dry period.

Uses and beliefs:
- The Batswana peel sections of the bark from the tree and boil them. The resulting bitter infusion is taken to clean the blood.

B&P van Wyk Trees p.422, Curtis & Mannheimer p.220, P Van Wyk Kr Trees p.88, Setshogo & Venter p.91, WFNSA p.166.

Chamaecrista falcinella (Oliv.) Lock

Derivation: *Chamae*, dwarf or low-growing [Greek], *crista*, crest or terminal tuft [Latin]; *falcinella*, resembling a small sickle [Latin], referring to the shape of the stigma.

Identification: Small, erect **herb**, branching at the base, c. 35 cm tall. **Leaves** alternate, sessile, pari-pinnately compound, the leaflets attached to the top side of the leaf stem. Leaves contact-sensitive and close up when touched. **Flowers** sparse, asymmetrical, yellow, c. 5 mm dia. Stigma sickle-shaped. **Seedpods** flat, c. 3 × 0.5 cm.

Chamaecrista falcinella is very easily confused with *C. mimosoides* and *C. stricta* but *C. falcinella* has shorter, slightly broader leaves, with shorter leaflets towards the apex and is a slightly smaller plant with smaller flowers.

Habitat: Widespread, found occasionally in deep sand and grassy areas near seasonal pans and channel margins.

Flowering: Late in the main rains and into the beginning of the cool dry period.

Fabaceae (pod-bearing family), Caesalpinioideae (cassia sub-family)

Chamaecrista mimosoides (L.) Greene
(syn. *Cassia mimosoides* L.)

Common names: Afrikaans visgraat-cassia, boesmanstee; **English** fish-bone cassia.

Derivation: *Chamae*, dwarf or low-growing [Greek]; *crista*, crest or terminal tuft [Latin]; *mimosoides*, like mimosa, [Latin].

Identification: Small, usually erect, bushy **annual herb** (occasionally becoming a **short-lived perennial** with woody stems), branching at the base, c. 50 cm tall. **Leaves** long, slender and straight, alternate, with a short stalk, pinnately compound, up to 65 pairs of leaflets attached to the top side of the leaf stem which has a distinct ridge along its upper surface, a distinct circular **gland** where the leaflets begin. Leaves contact-sensitive and close up when touched. **Flowers** sparse, asymmetrical, 5 petals, yellow, c. 8 mm dia., on stems whose length is c. twice the diameter of the flower, borne in groups of 1–3 at the leaf axil. Anthers and stigma distinctly hooked. **Seedpods** flat, c. 45 × 4 mm, turning black when mature.

Chamaecrista mimosoides is very easily confused with *C. falcinella* and *C. stricta*, but *C. mimosoides* has long narrow straight leaves with many pairs of leaflets and is a slightly larger plant with larger flowers.

Habitat: A widespread naturalised plant found occasionally in the open grassland of the floodplain margin.

Flowering: During the main rains.

B Van Wyk New p.122, Blundell p.93, Botweeds p.74, BVW Photoguide ZA p.36, Germishuizen p.101, WFNSA p.176.

Chamaecrista stricta E.Mey.

Derivation: *Chamae*, dwarf or low-growing [Greek], *crista*, crest or terminal tuft [Latin]; *stricta*, upright, erect [Latin].

Identification: Erect **annual herb** with hairy stems. **Leaves** long, slender, tapering towards the tip, stalkless, alternate, pinnately compound, up to 40 pairs of leaflets attached to the upper side of the leaf stem, almost overlapping. Leaflets covered in straggly **hairs**. Leaves contact-sensitive and close up when touched. **Flowers** irregular with 5 petals, borne singly at the leaf axils, petals paler yellow than the anthers. **Fruit** flat pods with rounded ends.

Chamaecrista stricta is very easily confused with *C. falcinella* and *C. mimosoides* but *C. stricta* has tapering leaves whilst the others are oblong.

Habitat: Found occasionally locally on sandy ridges in mopane woodland.

Flowering: During the main rains.

Blundell p.93.

Fabaceae (pod-bearing family), Caesalpinioideae (cassia sub-family)

Colophospermum mopane (Benth.) J.Léonard

Common names: Setswana mophane, (where the trees are growing in less favourable conditions looking as if they are 'farmed' by animals and quite small, they are known locally as *gumane*), mpani; **San** tlăba; **Subiya** muhane; **Yei** ogumane, opane; **Lozi** mupani; **Afrikaans** mopanie; **English** mopane, balsam tree, black ironwood, butterfly tree.

Derivation: *Kolla*, gum [Greek], *phoro*, to produce or bear [Greek], *sperma*, seed [Greek], referring to the gum produced by the fruit; *mopane* from the local name used by most local tribes.

Identification: Substantial deciduous **tree**, up to c. 20 m tall, or on poor alkaline soil, a woody shrub c. 1 m tall. Often damaged and heavily browsed initially by elephant then by antelope. **Bark** dark grey to black, deeply vertically fissured. **Leaf** distinctive bilobed, butterfly-shaped, easily recognisable and smelling strongly of turpentine when crushed. **Flowers** yellow, in pendular groups at leaf axils; anthers suspended on long filaments below the flower. **Fruit** flattened, kidney-shaped pods, covered in oil glands.

Habitat: Widespread, strongly growing woodland tree, usually found in one species woodland, possibly overtaking other species and colonising other areas of clayey, often alkaline, soils.

Flowering: During the main rains.

Uses and beliefs:
- Subiya bandage boiled bark onto cuts or wounds on humans and cattle.
- Wayeyi take an infusion of the roots to treat diarrhoea.
- Steam from an infusion of leaves is used to treat sore eyes.
- Batswana soak the bark in water for about an hour and drink the resulting infusion to cleanse themselves if they have had unprotected sex with a woman after miscarriage or within a month of fertilisation.
- Smoke from burning leaves is inhaled to drive away bad luck.
- The bark of old branches is burnt and the ash mixed with tobacco to make snuff.
- An excellent, virtually termite-resistant, wood for housing, carving, fences, pole bridges, etc. The bark is used as string for thatching and for tying bundles of firewood or grass. The wood burns well.
- Batswana believe that *mopane* trees attract lightning.
- The caterpillar of the emperor moth (*Imbrasia belina*), locally called *phane*, feeds on the leaves. These moths are a popular food source, eaten fried or dried, as they are high in protein. Dried, they can be stored for long periods.

B&P van Wyk Trees p.378, Curtis & Mannheimer p.202, Ellery p.103, Hargreaves p.28, Palgrave updated p.317, Setshogo & Venter p.13, Tree Roodt p.123. BB, BN, PN, RR, KS, BT.

Fabaceae (pod-bearing family), Caesalpinioideae (cassia sub-family)

Guibourtia coleosperma (Benth.) J.Léonard

Common names: Setswana motsaodi, motsaudi, nsibi, shi, tsaodi, tsaudi, ushi, gxwi; **Lozi** muzauli; **Afrikaans** bastermopanie; **English** (large) false mopane, African rosewood, (Rhodesian) copalwood, Rhodesian mahogany, Rhodesian teak.

Derivation: *Guibourtia*, for the French pharmacologist Nicholas Jean Baptiste Gaston Guibourt (1790–1861); *koleos*, a sheath [Greek], *sperma*, seed [Greek], referring to the seeds, which have a scarlet covering.

Identification: Large evergreen **tree** with a rounded, drooping crown, up to c. 20 m tall. **Trunk** reddish brown with dark brown and black markings. **Leaves** bi-lobed, butterfly-shaped, alternate, with a strong turpentine odour when crushed. **Flowers** small, cream coloured, with a sweet **scent**, borne in axillary and terminal panicles. **Fruit** circular, woody pods, becoming dark brown to black on ripening, 2 halves splitting to allow the single seed to emerge on a thread-like stalk. **Seed** has a conspicuous scarlet covering (aril) that attracts birds as a means of dispersal.

Habitat: Found occasionally locally in mixed riverine woodland. This tree is protected on state lands under Botswana's Forest Act.

Flowering: During the early rains.

Uses and beliefs:
- The red coverings of the fruit are removed with warm water and either eaten raw or made into a drink, which is said to be very nourishing. The seeds themselves are roasted or boiled, then pounded and eaten together with maize meal or the gravy from meat. It is an important staple food of the !Kung bushmen.
- The wood is known commercially as *machibi*.

B&P van Wyk Trees p.378, Curtis & Mannheimer p.204, Hargreaves p.29, Palgrave p.267, Setshogo & Venter p.92. KS.

Fabaceae (pod-bearing family), Caesalpinioideae (cassia sub-family)

Peltophorum africanum Sond.

Commxon names: Setswana mosetlha, setimamollo, mosiri, moyevu, moyethu, nzeze; **Afrikaans** dopperkiaat, huilboom; **English** weeping wattle, African wattle.

Derivations: *Pelthe*, a shield [Greek], *phoreo*, to bear [Greek], from the shape of the stigma; *africanum*, of Africa. *Mosetlha* can mean either the month of November or the colour yellow: the plant has yellow flowers in November.

Identification: Medium-sized **bush** or **tree**, often with more than 1 trunk, up to c. 6 m tall. **Leaves** alternate, bipinnately compound, leaflets small and oblong. **Flowers** yellow, in beautiful showy axillary racemes. **Fruit** flat pods, hanging in clusters, yellowish as they ripen becoming dark brown. Leaf stalks, flower stalks and sepals covered with dense, fine, brown **hairs**.

Habitat: Locally common growing in deep sand on the spillway and floodplain margins.

Flowering: From the early rains throughout the main rains.

Uses and beliefs:
• The tree has many medicinal uses from treating stomach complaints, to sore eyes.
• The heart wood is black and suitable for carving.
• Said to be one of the African 'rain trees' as, at the beginning of the rains, the larvae of a small insect, *Ptyleus grossus*, suck up the sweet sap and then eject almost pure water in large quantities, making the tree look as if it is raining.

Ellery p.128, Palgrave p.291, Setshogo & Venter p.93, Curtis & Mannheimer p.224.

Fabaceae (pod-bearing family), Caesalpinioideae (cassia sub-family)

Senna obtusifolia (L.) H.S.Irwin & Barneby
(syn. *Cassia obtusifolia* L.)

Derivation: *Senna* is a plant species found in Egypt that is used as a purgative; *obtusifolia*, blunt-leaved [Latin].

Identification: Erect **branching herb**, c. 80 cm tall. **Stem** ribbed, with pairs of long thin **stipules** at the leaf axils, c. 7 mm long. **Leaves** compound, c. 8 cm (inc. c. 2 cm leaf stalk) x 5.5 cm, with c. 3 pairs of oval leaflets, margins smooth with a small bristle at the tip. **Flowers** yellow, irregular, borne singly at leaf axils higher up the plant, c. 2 cm dia. **Fruit** straight pods with a square cross-section, c. 15 cm long.

Habitat: This naturalised or introduced herb is widespread on clayey sand in mopane woodland and near seasonal pans in the mopane.

Flowering: Middle to end of the main rains.

Uses and beliefs:
- Plants of the genus *Senna* are widely used as laxatives and purgatives.

Fabaceae (pod-bearing family), Mimosoideae (mimosa sub-family)

Acacia erioloba E.Mey.

Common names: Setswana mogotlho, mokala, omumbonde, mosu; **Lozi** muhoto; **Shiyeyi** ushuu; **!xõ:** //ah; **!kung** /ana; **Kûa** go; **Afrikaans** kameeldoring; **English** camel thorn, giraffe thorn, mimosa, thorn tree.

Derivations: *Acacia* from *akantha*, thorn [Greek]; *erioloba*, having woolly pods [Greek]. *Kameeldoring*, giraffe thorn. *Camel thorn*, a mistranslation of *Kameeldoring*. *Ushuu*, shaped like an umbrella.

Identification: Slow-growing **shrub** or **tree**, up to c. 10 m tall, with a spreading, rounded or umbrella-shaped crown. New growth reddish, slightly zig-zag. **Bark** dark grey to black, deeply longitudinally furrowed. **Thorns** white or pale, long (c. 5 cm) and straight, growing in pairs at the leaf axils. **Leaves** grey-green, alternate, often clustered on short side branches, bipinnately compound, with 2–5 pairs of primary leaflets each with 8–15 pairs of secondary leaflets. **Flowers** bright golden balls, c. 17 mm dia., growing on long stalks in clusters at the leaf axils. Flowers appear at the same time as the new leaves. **Fruit** thick curved flattened woody pods, covered in dense fine grey hairs, shaped rather like an ear-lobe. **Wood** dark red-brown.

Habitat: Common and widespread throughout the area, forming dense scrub with *Acacia hebeclada* on drying parts of the floodplain and large forest trees on the floodplain margins.

Flowering: During the cool dry period.

Uses and beliefs:
- The seeds may be roasted as a coffee substitute.
- The bark is burned and ground to make a remedy for headaches.
- Discharging and infected ears are treated with a powder of dried and crushed pods.
- A decoction of the roots may be used for coughs and nose bleeds.
- The gum is used as a remedy for colds and tuberculosis.
- The inner bark may be pounded and used as a perfume.
- The pods are a much-loved animal food.
- The wood is very strong and used for many purposes as it is termite resistant.
- These trees are said to attract lightning.
- Although there is no taboo against felling this tree, many tribes only allow the chief to cut its timber.
- The caterpillars of *Azanus jesous jesous* feed on the tree.

Ellery p.83, Hargreaves p.25, Palgrave updated p.278, Setshogo & Venter p.60, Tree Roodt p.157, WFNSA p.162. PN.

Fabaceae (pod-bearing family), Mimosoideae (mimosa sub-family)

Acacia nigrescens Oliv.

Common names: Setswana and **Shiyeyi** mokoba; **Setswana** more-o-mabele, mwanduchi, mughandutji, ugandu, nkogo, goshwe, zhinca, muwanduwehi, mokala; **San** ngawa; **Hmbukushu** moanduchi; **!kung** yi; **Shiyeyi** munga; **Afrikaans** knoppiesdoring; **English** knob-thorn.

Derivation: *Acacia* from *akantha*, thorn [Greek]; *nigrescens* becoming black [Latin], possibly because of the dark colour of the pods.

Identification: Large woodland **tree**, 10–20 m tall, easily recognised by the knobs on its branches and upper trunk (although some specimens do not have them). New **stems** dark pink, making the whole canopy appear pink before the flowers open. Younger branches with black, hooked **thorns** in pairs below the nodes. Remaining leafless for large parts of the year. **Leaves** blue-green, bipinnately compound. Leaflets distinctive because they are fewer, larger and rounder than those of most acacias. **Flowers** with a sweet lemony **perfume**, creamy-yellow, borne in spikes (c. 13 cm long) before the leaves appear. **Seed pods** flat, rather oblong, dark in colour. There are often many young seedlings to be found.

When young, *Acacia nigrescens* may be confused with *Acacia mellifera* but when mature, it usually has noticeable knobs on the trunks and its leaflets have net veining.

Habitat: A widespread and common tree throughout the islands and riverine mixed woodland.

Flowering: During the cool dry period.

Uses and beliefs:
- Wayeyi boil the ash of the bark with water, then mix it with petroleum jelly for use as a hair conditioner and styling gel.
- Wayeyi and Batswana infuse the chopped roots in warm water as a mouth wash for sore teeth.
- The knobs from the bark are pounded and applied to ulcers by San. If the ulcer reappears on another part of the body, the person bathes in water in which the bark has been soaked 1–3 times per day until the ulcer is cured.
- Hmbukushu and Wayeyi strip the white inner bark and use it as thatching twine and cord.
- Wayeyi dye their fish nets by boiling them with the red inner bark for about 3 hrs because fish are frightened by white nets. (The fishing nets are made either from the inner string of old car tyres or from the fibres of *Sanseveria deserti*).
- The heartwood is strong, close-grained and used for general purposes.
- Wayeyi regard the trees as 'male' if they have no knobs and 'female' if they have knobs.
- It is said that baboons prefer to sleep in the 'female' trees as leopards find them difficult to climb because of the knobs.
- The caterpillars of *Charaxes phaeus* feed on the leaves.

FZ Vol.3 pt1, Ellery p.91, Hargreaves p.26, Palgrave updated p.292, Tree Roodt p.171, WFNSA p.162. PK, TK, MM, RM, PN, BT.

Fabaceae (pod-bearing family), Mimosoideae (mimosa sub-family)

Dichrostachys cinerea (L.) Wight & Arn.

Common names: Setswana moselesele, mpangale; **Hmbukushu** mweye; **Ovambenderu** ojeete; **Afrikaans** sekelbos; **English** sickle bush, Kalahari Christmas tree.

Derivation: *Di*, two, *chroma*, colour, *stachys*, spike [Greek], *cinerea* [Latin], ash-like, referring to the colour of the bark.

Identification: Much-branched, **deciduous shrub** or **small tree** often forming a dense thicket. **Leaves** alternate, bipinnately compound with equal numbers of narrowly oval to linear leaflets on either side of the leaf stalks; small lateral branches terminating in hard **spikes**. [Can be confused with acacias which have paired thorns.] **Inflorescence** a pendulous axillary spike, upper part sterile pink flowers, lower part fertile yellow flowers. **Fruit** form a contorted spherical cluster of pods that remains on the tree for a long time, this often retains its shape after falling (until it rots under the tree) if it is not eaten.

Habitat: Common and widespread on almost all soil types. Rarely found locally because it is so very heavily browsed and then grazed.

Flowering: During and after the early rains.

Uses and beliefs:
- San use infusions of leaves to cure pains in the side; of 3 leaves, a spike and some bark for (phar)laryngitis; and of roots as a mouthwash for toothache (twice per day). They also use a rag soaked in a decoction of boiled leaves to treat wounds.
- The wood is used to make poles and tool handles. In the Tsodilo area, the wood is used to make spears.
- The roots of *Dichrostachys cinerea* are mixed in water with those of *Combretum mossambicense* and *Ximenia caffra* to make a preparation that is drunk each morning to make young boys strong when they start to be active sexually. To be more effective, the roots may be boiled with fresh milk from a cow that has had a male calf. Boys at the cattle post drink this concoction twice daily.
- The leaves are obviously highly palatable as they are much browsed.

FZ Vol.3 pt1, B van Wyk p.164 & p.224, Ellery p.107, Palgrave p.305, Tree Roodt p.181, Turton p.25, WFNSA p.170. BB, GM, RM, RR.

Fabaceae (pod-bearing family), Mimosoideae (mimosa sub-family)

Neptunia oleracea Lour.

Common names: Setswana baswabile; **English** yellow pan weed.

Derivations: *Neptunia* from *Neptunus*, god of the sea, rivers and fountains [Latin], referring to its watery habitat; *oleracea*, of the vegetable garden, applied to vegetables and pot herbs [Latin]. *Baswabile* means 'They failed' ('the shy plant'), 'Let them be ashamed'.

Identification: Creeping, floating, **annual herb** rooting from leaf nodes and growing in the mud of seasonal pans. **Stems** red. **Leaves** grey-green, alternate, bipinnately compound, very sensitive and fold up along their length when touched. **Flowers** a ball of yellow stamens. **Fruit** stand up above the floating stems and look like miniature palm trees. This **plant** can continue to flourish for some time even in the dried-out mud of a pan.

Habitat: Found occasionally locally, especially in years of high rainfall, rooted in the mud of seasonal pans.

Flowering: During the main rains.

Uses and beliefs:
- Wayeyi cook and eat the leaves of this plant as *morogo*, a vegetable relish.
- Traditional healers mix this plant with others with petroleum jelly, forming a concoction that is applied all over the body so that if a person is at fault in a court of law, they will still be declared innocent. For the charm to be particularly effective, it should be mixed with *Glinus bainesii* and *Abrus precatorius*.

FZ Vol.3 pt1, MM.

Fabaceae (pod-bearing family), Papilionoideae (pea sub-family)

Crotalaria laburnifolia L.

Common names: Afrikaans bosveld crotalaria; **English** rattle pod, bushveld crotalaria.

Derivation: *Krotalon*, rattle [Greek], as the seeds of many species rattle in the pod; *laburnifolia* [Latin] because the leaves resemble those of the laburnum tree.

Identification: Erect, branching **perennial herb**, up to c. 110 cm tall, growing from a woody root-stock. **Leaves** trifoliate with smooth margins, borne alternately up the stem, with very long stalks at least twice as long as the leaf blade. **Flowers** c. 4 cm long, borne in showy terminal racemes, standard and wing petals yellow, keel greeny-yellow, bent up at a right-angle. **Fruit** slightly inflated pods.

Habitat: Found occasionally locally in damp areas of degraded mopane woodland near seasonal pans.

Flowering: During the main rains.

Uses and beliefs:
- Although the toxicity of this particular species of *Crotalaria* is not known, many plants in this genus cause severe lung and/or liver damage in cattle.

FZ Vol.3 pt7, BVW Photoguide ZA p.40, Germishuizen p.105, WFNSA p.186.

Fabaceae (pod-bearing family), Papilionoideae (pea sub-family)

Crotalaria pisicarpa Welw. ex Baker

Derivation: *Krotalon*, rattle [Greek], as the seeds of many species rattle in the pod; *pisum*, pea [Latin], *carpus*, a fruit [Latin], with pea-like fruit.

Identification: Prostrate creeping **annual herb** with stems often buried in the sand, c. 50 cm dia.; **hairs** overall and particularly noticeable on new growth, calyx and fruit. **Leaves** alternate, trifoliate with smooth margins, leaflets elliptical, often with a rounded tip. **Flowers** borne in clusters of 2–3 at the apex of long stems from the leaf axils, pea-like, yellow with red veining on both sides of the standard. **Fruit** spherical, hairy pods, often found buried in the sand.

Habitat: Locally uncommon on sandy tracksides through mopane woodland.

Flowering: During the main rains.

Uses and beliefs:
• Although the toxicity of this particular species of *Crotalaria* is not known, many plants in this genus cause severe lung and/or liver damage in cattle.

FZ Vol.3 pt7.

Fabaceae (pod-bearing family), Papilionoideae (pea sub-family)

Crotalaria platysepala Harv.

Common names: English wild lucerne.

Derivation: *Krotalon*, rattle [Greek], as the seeds of many species rattle in the pod; *platy*, broad [Greek]; *sepalus*, sepalled [Latin and Greek], thus broad-sepalled.

Identification: Erect, branching, **annual herb**, up to c. 1 m tall. **Stems** often finely **hairy**, tinged dark red. **Leaves** alternate, palmately compound, (trifoliate) with 3 obovate leaflets with blunt tips, the terminal leaflet larger, long leaf stalks. **Flowers** in loose terminal racemes, yellow with fine reddish-brown lines on the standard, turning orange as they mature. **Fruit** almost spherical, silver green, hairy pods, sepals and stigma retained.

Habitat: Common and widespread in partial shade on islands and in woodland on sandy soils.

Flowering: Middle to end of the main rains.

Uses and beliefs:
• Although the toxicity of this particular species of *Crotalaria* is not known, many plants in this genus cause severe lung axnd/or liver damage in cattle.
• Although widely used as animal fodder in years of good rainfall (hence its English name), it appears to be avoided by game if there is an alternative.

FZ Vol.3 pt7, Ellery p.159.

Crotalaria sphaerocarpa Perr. ex DC. subsp. *sphaerocarpa*

Common names: Setswana malomaagorothwe, malomaagwerothwe, malomagorotwe; **Afrikaans** mielie-crotalaria; **English** wild lucerne, mealie crotalaria, crotalaria.

Derivation: *Krotalon*, rattle [Greek], as the seeds of many species rattle in the pod; *sphaericus*, globose, spherical [Latin], *-carpus*, fruited [Latin].

Identification: Rather variable, much branching, in some groups erect and in others straggling, **annual herb**, up to c. 110 cm tall. **Leaves** alternate, compound with 3 leaflets, leaflets linear, with smooth margins, folding at the mid-vein. **Stipules** at the leaf axils. **Flowers** borne in either tight or open racemes, yellow and slightly red veined on the reverse of the standard, up to c. 25 mm long × 6 mm wide. Flowers not very showy during the day, opening in the evening and at night. **Fruit** spherical, inflated like tiny balloons, calyx retained and either short (c. $1/3$ the length) or almost the length of the fruit.

Habitat: Common and widespread in full sun on the floodplain and in open scrub.

Flowering: Throughout the main rains.

Uses and beliefs:
- Widely grazed by game, particularly giraffe.
- The caterpillars of *Amphicallia bellatrix*, *Lampides boeticus* and *Euchrysops barkeri* feed on *Crotalaria*.
- Although the toxicity of this particular species of *Crotalaria* is not known, many plants in this genus cause severe lung and/or liver damage in cattle.

FZ Vol.3 pt7, B Van Wyk Flowers p.136, Botweeds p.66, Flowers Roodt p.87, Germishuizen p.105.

Fabaceae (pod-bearing family), Papilionoideae (pea sub-family)

Crotalaria steudneri Schweinf.

Derivation: *Krotalon*, rattle [Greek], as the seeds of many species rattle in the pod; *steudneri*, for Steudner who collected plants in Ethiopia and Eritrea in 1861–2.

Identification: Branching **annual herb**, c. 50 cm tall. **Leaves** alternate, compound, trifoliate, margins smooth. Lower leaf surface and stem sparsely covered with long **hairs**. Racemes of **flowers** emerge from the opposite side of the stem to the leaf axil. **Flowers** yellow, c. 7 mm long, heavily striped with brown or maroon over the whole flower. **Fruit** almost spherical, slightly hairy, blotched with brown or maroon on the upper surface, c. 8 mm dia.

Habitat: Found occasionally locally on sandy tracksides through the mopane woodland.

Flowering: During the main rains.

Uses and beliefs:
- Although the toxicity of this particular species of *Crotalaria* is not known, many plants in this genus cause severe lung and/or liver damage in cattle.

FZ Vol.3 pt7.

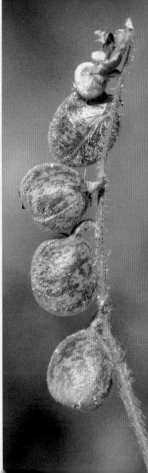

Rhynchosia minima (L.) DC.

Common names: Setswana lofse?, morupe.

Derivation: *Rhynchos*, a beak [Greek]; *minima*, smallest [Latin].

Identification: Small **perennial climbing plant** with twining stems, up to 2.5 m tall. **Leaves** alternate, compound, trifoliate, upper leaflet rhomboid, lower pair oblique, c. 7 cm (inc. leaf stem c. 2.5 cm) × 7 cm. Small yellow spots on reverse of leaves are glands. **Flowers** yellow, borne in racemes from leaf axils. **Fruit** flat pods, usually holding 2 seeds.

Habitat: Widespread, growing in partial shade on islands and in acacia scrub on the floodplain.

Flowering: Middle to end of the main rains.

FZ Vol.3 pt5, B van Wyk p.142, WFNSA p.180.

Fabaceae (pod-bearing family), Papilionoideae (pea sub-family)

Rhynchosia totta (Thunb.) DC. var. *fenchelii* Schinz

Common names: Setswana nawa-yanaga, tsebeatoje.

Derivation: *Rhynchos*, a beak [Greek], probably referring to the point on the seed pod; *totta*, precise meaning unclear, but may derive from the word *hottentot*, implying that it is common in South Africa.

Identification: Slender, scrambling **perennial herb** from a tuberous root-stock, up to c. 3 m tall. No tendrils. **Leaves** tri-foliate, up to c. 4 × 6 cm, spiralling up the stem, each leaflet articulated. In older leaves, the central leaflet has a tendency to turn in the opposite direction to the lower pair and away from the sun; lower leaflets oblique based. **Flowers** often in pairs on a short stalk, yellow standard and wings, green keel, sepals as long as the petals. A single bract behind the flower. **Fruit** broad, hairy pods, usually with 2 seeds, c. 3 × 0.7 cm.

Habitat: Found occasionally locally on sandy soil, lightly shaded in acacia scrub, mopane woodland and on islands.

Flowering: During the main rains.

FZ Vol.3 pt5, B Van Wyk Flowers p.142, Germishuizen p.178.

Sesbania microphylla Harms

Common names: Setswana mositanokana; **Ovambenderu** omunxumuhari.

Derivations: *Saisaban*, the common Arabic name for plants of this genus; *micro*, small [Greek], *phyllon*, leaf [Greek], small-leaved. *Mositanokana*, growing near water, a generic name for plants of this genus that thrive near water.

Identification: Erect, slender **herbaceous plant**, up to c. 2.5 m tall. **Leaves** alternate, long and slender, pinnately compound with c. 30 pairs of leaflets, leaflets linear and almost always held folded together to cut transpiration. **Flowers** borne in loose pendular racemes, c. 8 cm long, pea-like, yellow, the reverse of the standard flecked with dark blue or brown. **Fruit** long, slim, curving cylindrical pods.

Habitat: Widespread growing in damp marshy areas.

Flowering: During the main rains.

BB.

Fabaceae (pod-bearing family), Papilionoideae (pea sub-family)

Sesbania rostrata Bremek & Oberm.

Derivation: *Saisaban*, the common Arabic name for plants of this genus; *rostrata*, beaked [Latin].

Identification: Stocky **annual herb**, up to c. 1 m tall. **Leaves** hairless, compound with a variable number of pairs of leaflets and no terminal leaflet, leaflets oblong with a bristle at the tip. **Flowers** borne on short racemes from the leaf axils, yellow with brown spots on the standard, c. 2.3 cm long. **Fruit** a tubular pod, c. 14.5 cm long, with an etiolated point. They dry with each **seed** clearly delineated. Nitrogen nodules develop on both the roots and the stems.

Habitat: Common and widespread in seasonal pans.

Flowering: Late in the main rains and into the cool dry period.

Uses and beliefs:
- The nitrogen nodules could provide a useful supply of green fertiliser in *molapo* agriculture.

Stylosanthes fruticosa (Retz.) Alston

Common names: English wild lucerne.

Derivation: *Stylosus*, having a prominent or well-developed style [Latin], *anthos*, flower [Greek]; *fruticosa*, shrubby or bushy [Latin].

Identification: Erect, branching, short-lived **perennial herb,** c. 30 cm tall. **Leaves** alternate, trifoliately compound, leaflets ovate with sharply pointed tips, whitish veins conspicuous on the lower surface. **Flowers** yellow with a paler cream standard, c. 5 mm long, borne at the apex of the branches and partly hidden by leafy bracts. **Fruit** flattened pods with constrictions between the seeds.

Habitat: Locally rare in damp clayey areas of mopane woodland.

Flowering: Late in the main rains.

FZ Vol.3 pt6, B Van Wyk Flowers p.144, WFNSA p.188.

Fabaceae (pod-bearing family), Papilionoideae (pea sub-family)

Vigna luteola (Jacq.) Benth.

Derivation: *Vigna* in honour of Dominico Vigna (d. 1647), Professor of botany at Pisa, Italy; *luteola*, yellowish [Latin], referring to the flowers.

Identification: Scrambling, climbing **perennial herb**, up to 2.5 m tall. **Leaves** alternate, palmately compound, trifoliate, each leaflet jointed and lanceolate to narrowly ovate, margins smooth. A pair of triangular **stipules** at the base of the leaf stalk. **Flowers** yellow, borne in small terminal racemes on long stalks (c. 10 cm) from the leaf axils. **Fruit** long, slightly flattened pods with a slight constriction between each seed. The **plants** root in water. They emit a strong green odour when crushed.

Habitat: Widespread, growing in the water margin of channels and lagoons.

Flowering: Almost throughout the year.

Uses and beliefs:
• The leaves are used in the treatment of ulcers and syphilis.

FZ Vol.3 pt5, Ellery p.178.

Gentianaceae (gentian family)

Sebaea grandis (E.Mey.) Steud.

Common names: English primrose gentian, large-flowered sebaea.

Derivation: *Sebaea*, for Albert Seba (1665–1736), a Dutch apothecary, naturalist and writer; *grandis*, large [Latin], alluding to the flowers.

Identification: Erect, slender **annual herb**, c. 20 cm tall, with a **rhizomous rooting** system. **Leaves** ovate, scale-like, stalkless, opposite and decussate, margins smooth and decurrent forming a rib or small wing down the stem. **Flowers** borne at the apex of the plant on pairs of slender stalks from the leaf axils, creamy yellow, c. 2.5 cm dia., 5 overlapping lobes form the corolla tube, each lobe with a midvein and a point at the end of the vein. Sepals keeled.

Habitat: Locally uncommon, growing in open grassland near seasonal pans.

Flowering: Towards the end of the main rains.

FZ Vol.7 pt4, B Van Wyk Flowers p.146, Pooley p.298, WFNSA p.300.

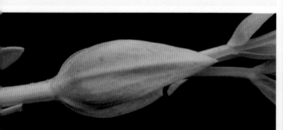

Hyacinthaceae (hyacinth family)

Albuca abyssinica Jacq.

Common names: Setswana ledutla?

Derivation: *Albus*, white [Latin], referring to the white flowers of many plants of this genus; *abyssinica*, from Abyssinia.

Identification: Prominent, erect, single-stemmed **perennial herb**, c. 1.8 m tall, arising from a bulbous **rooting** system. **Leaves** strap-like, c. 1.1 m long and c. 3 cm wide, margin hairy. **Flowers** yellow-green, borne in a straight raceme, c. 80 cm long. As the flowers grow, their stalks develop a distinctive loop. Below each flower bud is a long bract that dies back as the flower matures. **Bulb** c. 5 cm tall with a pale fibrous covering.

Habitat: Found occasionally locally in mopane woodland.

Flowering: Late in the main rains.

Blundell p.416.

Hydrocharitaceae (frog's-bit family)

Ottelia ulvifolia (Planch.) Walp.

Common names: English water lettuce.

Derivation: *Ottelia*, a Latinised form of the Malabar name *ottel-ambel*; *ulva*, the seaweed *Ulva*, *folius*, a leaf [Latin], alluding to the shape of the leaves which is like those of *Ulva*.

Identification: Perennial **aquatic herb** with submerged roots. **Leaves** red/green, c. 15 × 2 cm, forming a basal rosette, obovate, smooth margins, main veins running parallel to the leaf blade with cross-veins at right-angles. **Flowers** white or pale cream to yellow, c. 1.5 cm dia., solitary and growing up from the roots in a spathe, floating on the surface of the water. Flowers opening in the late morning.

Habitat: Locally uncommon but widespread in gently flowing waterways and channels.

Flowering: During the cool dry period.

Uses and beliefs:
• The ovary is inferior and easy to cut open. The Wayeyi use the water inside as eye drops.

Ellery p.147, WFNSA p.28. Kit, PK.

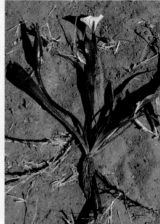

Lamiaceae (mint family)

Ocimum gratissimum L. var. *gratissimum*

Derivation: *Ocimum*, basil, from *okimon*, an aromatic herb [Greek]; *gratissimum*, very pleasing or agreeable [Latin].

Identification: Woody, short-lived **perennial herb**, c. 70 cm tall; covered overall in erect **hairs**. **Stems** square. **Leaves** oval with a pointed tip, opposite and decussate, margins serrated, leaf blade up to c. 10 × 1.5 cm. **Flowers** borne in terminal spikes, c. 10 cm tall, pale lemon yellow. **Fruits** 4 nutlets within the calyx.

Habitat: Locally rare in deeply shaded areas of islands and mopane woodland along the floodplain.

Flowering: During the main rains.

Lentibulariaceae

Utricularia gibba L.

Common names: English bladderwort.

Derivation: *Utriculus*, a small bottle [Latin], referring to the insect trapping bladders on the roots and leaves of these aquatic and marginal plants; *gibba*, swollen on one side [Latin].

Identification: Aquatic insectivorous herb forming a mat of fine-branching stems, c. 6–8 cm tall. **Roots** with minute ovoid transparent bladder-like **structures** that catch insects. **Leaves** growing on the surface, linear, like fine grass, c. 1 cm long. **Flowers** brilliant yellow with 2 lips and a spur growing parallel to the lower lip, borne singly or in pairs.

Habitat: Locally common growing in still or slow-flowing water on a, sometimes floating, mat of roots and grass or in suds in still or slow-flowing water.

Flowering: Almost throughout the year.

FZ Vol.8 pt3.

Malvaceae (hibiscus or mallow family)

Abutilon angulatum (Guill. & Perr.) Mast. var. *angulatum*

Common names: Setswana tsebe-yatlou, tshikadithata, dikurubede, dukurukane, damaqoq, masepaabanyana; **Ovambenderu** otjitjandoko; **Subiya** kothuinzovo; **English** elephant's ear.

Derivations: *Abutilon* is the Arabic name for the genus; *angulatum* refers to the buttressed angular stem [Latin]. *Tsebe-yatlou*, the ear of the elephant. *Tshik*, fibre and *thata*, strong, referring to the strong fibres produced from this plant, the name *tshikadithata* is used for several fibre-producing plants.

Identification: Erect, sparsely branched, **short-lived perennial herb**, growing to c. 2 m and occasionally to c. 4 m in years of heavy rainfall; covered in soft **velvety hairs** overall. **Leaves** broadly oval with a cordate base, leaf blade c. 45 cm (inc. c. 15 cm petiole) × 25 cm but much smaller towards the apex of the plant. **Flowers** apricot-yellow, c. 3 cm dia., borne singly or in clusters, on long stalks, in leaf axils and at apex of the plant. **Fruit** like a disc with a raised edge, c. 1.5 cm dia., with clearly defined seed segments.

Habitat: Locally common on inhabited islands in the floodplain.

Flowering: Late in the main rains and through the cool dry period.

Uses and beliefs:
- Leaves eaten raw or cooked are used as a remedy for hiccups.
- Batswana mix this plant with many other plants to reinforce their medical and magical uses.
- Batswana use fibres from the stems in rope making.
- The stems are used as 'fire sticks'.
- The leaves are considered to be a good replacement for toilet paper, even for babies, as they are soft and do not irritate the skin.
- Ovambenderu boys use the leaves as an apron for their traditional dress.

FZ Vol.1 pt2, Ellery p.152, Hargreaves p.14, Flowers Roodt p.113, Turton p.68. BB.

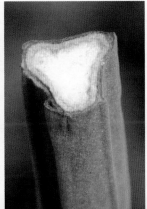

Malvaceae (hibiscus or mallow family)

Abutilon englerianum Ulbr.

Derivation: *Abutilon* is the Arabic name for the genus; *englerianum* for Professor H.G. Adolf Engler (1809–84), a German botanist.

Identification: Perennial, much-branched, **shrubby herb**, up to c. 2.5 m tall. **Stems** covered in 2 layers of highly irritant hairs, one straight and long and the other shorter and club-shaped. Pairs of **stipules** at the leaf axils. **Leaves** blue-green, soft and velvety, cordate with a dentate margin, spiralling up the stem. **Flowers** apricot yellow, 5 petals, c. 3 cm dia., on long stems. **Fruit** like a disc with a raised edge, c. 1.5 cm dia., with clearly defined seed segments.

Habitat: Found occasionally locally, growing in light shade on islands and in the treeline along the floodplain.

Flowering: Main rains into the cool dry period.

FZ Vol.1 pt2.

Malvaceae (hibiscus or mallow family)

Abutilon ramosum (Cav.) Guill. & Perr.

Common names: English branching abutilon.

Derivation: *Abutilon* is the Arabic name for the genus *ramosum*, branched [Latin].

Identification: Woody herb up to c. 1 m tall **branching** from low down; covered in stellate **hairs** with some longer soft hairs. **Leaves** spiralling up the stem, cordate, slightly 3-lobed with a pointed tip ending in a bristle, margins serrated with a tiny bristle at the tip of each serration. **Flowers** orangey yellow, c. 7 mm dia., with 5 overlapping petals. **Fruit** a capsule with segments each with a distinct point. Seed capsule shape is one of the major identifying features of this species.

Habitat: Locally rare in lightly shaded areas of islands and mixed woodland along the floodplain.

Flowering: During the main rains.

FZ Vol.1 pt2.

Malvaceae (hibiscus or mallow family)

Gossypium herbaceum L.
subsp. *africanum* (Watt) Vollesen

Common names: Setswana leloba, sesetlho, letseta, tlhale; **Afrikaans** katoenbossie, wildkatoen; **English** wild cotton.

Derivation: *Gossypium*, cotton [Latin]; *herbaceum*, herbaceous [Latin]; *africanum*, from Africa [Latin].

Identification: Small branching **shrublet**, c. 70 cm tall. Stems and leaves covered with dense short **hairs**. **Leaves** palmate, 5-lobed, each lobe bristled at the tip, base cordate, margins smooth. **Flowers** pendular, petals overlapping, yellow with red centres, 3 serrated **bracts** surround the flowers. **Fruit** almost spherical with a pointed tip. **Seeds** covered in cotton fluff.

Habitat: Locally uncommon, found in acacia scrub on the floodplain and occasionally on small islands.

Flowering: From the middle of the main rains into the cool dry period.

Uses and beliefs:
• A relative if not parent of cultivated cotton.

FZ Vol.1 pt2, BVW Photoguide ZA p.50, WFNSA p.256.

Malvaceae (hibiscus or mallow family)

Hibiscus caesius Garke

Derivation: *Hibiscus*, marsh mallow [Greek], possibly from the ibis which feeds on certain varieties of this genus; *caesius*, light blue [Latin].

Identification: Low **shrubby perennial** c. 1 m tall. **Leaves** alternate, palmately compound with 5 leaflets, leaflets elliptical with serrated margins, surfaces hairless. **Flowers** a delicate translucent yellow with a deep red centre, 5 petals, up to c. 10 cm dia. **Calyx** comprising 5 elliptical sepals, epicalyx of stiff linear projecting bracts.

Habitat: Found occasionally locally on islands and in the mixed woodland of the treeline.

Flowering: Middle to end of the main rains.

Uses and beliefs:
- Caterpillars of *Charaxes jasius* feed on hibiscus species.

FZ Vol.1 pt2

Malvaceae (hibiscus or mallow family)

Hibiscus cannabinus L.

Common names: Setswana moku, moelethaga; **Afrikaans** wilde-stokroos; **English** wild stockrose, kenaf, bastard jute.

Derivation: *Hibiscus*, marsh mallow [Greek], possibly from the ibis which feeds on certain varieties of this genus; *cannabinus*, hemp-like [Latin].

Identification: Erect, slightly spiny **annual herbaceous** plant. Often a single **stem** up to c. 1.5 m tall, covered in hooked bristles. **Leaves** c. 20 cm (inc. c. 10 cm leaf stalk) × 12.5 cm, alternate, palmate with c. 7 lobes, each lobe almost linear, margins serrated, bristly and often outlined in red, reverse of the leaf blade covered in short bristles. **Flowers** stalkless, pale mauve/cream with a dark red centre, not opening wide, c. 4 cm dia.

Habitat: Widespread in years of high rainfall in full sun on the floodplain.

Flowering: From the middle of the main rains into the cool dry period.

FZ Vol.1 pt2, WFNSA p.254, Blundell p.76.

Malvaceae (hibiscus or mallow family)

Hibiscus dongolensis Delile

Common names: Setswana letseta; **English** dongola hibiscus.

Derivation: *Hibiscus*, marsh mallow [Greek], possibly from the ibis which feeds on certain varieties of this genus; *dongola* is a place on the Nile in northern Sudan.

Identification: Bushy **perennial herb**, c. 1.4 m tall. **Leaves** variable from ovate to palmate, margin serrated, alternate, with a pair of stipules in the leaf axils, usually hairless but occasionally with lines of hairs running along the veins on the lower side of the leaves. **Flowers** pendant, c. 7 cm long, yellow with a red-brown centre, fully opening only briefly in full sun in the middle of the day. **Epicalyx** of 5 linear green bracts equal in length to the lobes of the calyx.

Habitat: Found occasionally locally, growing in well-drained soil on the margins of islands, termite mounds and the treeline.

Flowering: During the main rains.

FZ Vol.1 pt2, Plowes & Drummond 74.

Malvaceae (hibiscus or mallow family)

Hibiscus mastersianus Hiern

Derivation: *Hibiscus*, marsh mallow [Greek], possibly from the ibis which feeds on certain varieties of this genus.

Identification: Tall, single-stemmed **woody annual herb**, growing typically to c. 1.8 m; highly irritant hairs overall. **Stems** very prickly. **Leaves** alternate, palmate, either shallowly trilobed or deeply lobed with 5 lobes, margins serrated, a pair of c. 8 mm **stipules** at the leaf axils. Some plants with red edges to the leaves and red flecks on the stems. **Flowers** yellow with a red centre, some also with red edges on the underside of the petals, c. 7 cm dia., borne singly on short stems at the leaf axils.

Habitat: Locally common in wet years on sandy areas of the floodplain and on the sand ridges of mopane woodland.

Flowering: Late in the main rains.

FZ Vol.1 pt2.

Malvaceae (hibiscus or mallow family)

Hibiscus meeusei Exell

Common names: Setswana mmabashete, mmankgarwane, motswalakgoro, mmabasi; **English** wild stockrose.

Derivation: *Hibiscus*, marsh mallow [Greek], possibly from the ibis which feeds on certain varieties of this genus.

Identification: Erect, usually single-stemmed, **annual herb**, up to c. 1.5 m tall. **Stems** harshly bristled and irritant. **Leaves** alternate, palmate, mainly trilobed although more mature leaves sometimes have 2 additional smaller lobes, margins serrated. **Flowers** pale lemon yellow with a bright red centre, axillary, almost stalkless. **Fruit** woody, with sharp points when open.

Habitat: Widespread in years of heavy rainfall.

Flowering: Late in the main rains.

FZ Vol.1 pt2, Germishuizen p.114, Turton p.63.

Malvaceae (hibiscus or mallow family)

Hibiscus ovalifolius (Forssk.) Vahl

Common names: Hmbukushu moshawa; **English** (locally known as) wild stockrose.

Derivation: *Hibiscus*, marsh mallow [Greek], possibly from the ibis which feeds on certain varieties of this genus; *ovali*, oval [Latin], *folius*, leaf-like [Latin], referring to the shape of the sepals.

Identification: Low-growing, woody **shrublet**, up to c. 60 cm tall. **Stems** hairy with occasional star-shaped hairs. **Leaves** alternate, palmate with 3 lobes, c. 13 cm (inc. petiole c. 3.5 cm) × 9 cm, margins serrated, sparsely hairy. **Flowers** with 5 petals, lemon yellow with a dark red centre, c. 9 cm dia.

Often confused with *Hibiscus calyphyllus* but can be distinguished by the oval shape of the sepals.

Habitat: Common in the shade on islands and in the treeline along the floodplain.

Flowering: Early morning during the rains.

Uses and beliefs:
- Hmbukushu use the roots to sprinkle water over the belongings of a dead person.
- Hmbukushu soak the stems of *Hibiscus ovalifolius* until they are white then use them to weave mats.

MM, MT.

Malvaceae (hibiscus or mallow family)

Hibiscus schinzii Gürke

Common names: Setswana moshawa.

Derivation: *Hibiscus*, marsh mallow [Greek], possibly from the ibis which feeds on certain varieties of this genus; *schinzii* for the Swiss botanist, Professor Schinz.

Identification: Rather lax to almost prostrate **woody perennial**. **Hairs** irritate the skin and are strong enough to prick fingers. Main **stems** up to c. 80 cm long, pinkish with long hairs (up to c. 5 mm). **Leaves** palmate with 3–5 lobes, margins serrated, covered in stellate hairs. **Flowers** c. 5 cm dia., variable, with 5 petals and 5 sepals, pale lemon yellow with or without a red centre, borne on long stalks in the leaf axils. The epicalyx, the second row of the **calyx** made up of rather irregular linear bracts.

Habitat: Widespread, growing in full sun in sandy areas on the floodplain.

Flowering: From the early rains throughout the main rains and into the beginning of the cool dry period.

FZ Vol.1 pt2, B Van Wyk Flowers p.156, WFNSA p.254.

Malvaceae (hibiscus or mallow family)

Hibiscus sidiformis Baill.

Derivation: *Hibiscus*, marsh mallow [Greek], possibly from the ibis which feeds on certain varieties of this genus; *sidi*, the genus *Sida*, *-formis*, -shaped [Greek], meaning that it is similar in shape to *Malvaceae* of the genus *Sida*.

Identification: One of the smaller hibiscus, an erect **annual herb**, up to c. 55 cm tall; covered in short soft slightly sticky **hairs**; usually unbranched and supported by other vegetation. **Leaves** alternate, very variable, commonly trifoliate and smooth margined on the upper part of the plant, palmate sometimes with a cordate base on the lower part of the plant. Some plants in heavier shade with only palmate leaves. **Flowers** pale lemon yellow, 5 overlapping petals, c. 1.5 cm dia., on long stalks from the leaf axils, jointed c. 5 mm below flower; sepals c. half the length of the petals, linear.

Habitat: Common and widespread in sandy areas of the floodplain and mopane woodland in full sun.

Flowering: Middle to end of the main rains

FZ Vol.1 pt2.

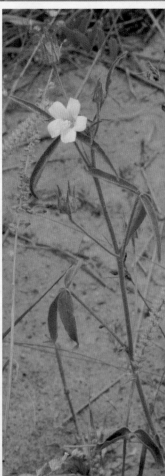

Malvaceae (hibiscus or mallow family)

Hibiscus trionum L.

Common names: Setswana delele-kwakwa; **Afrikaans** terblanbossie, uurblom; **English** black-eyed susan, bladder hibiscus, bladderweed, flower-of-an-hour.

Derivation: *Hibiscus*, marsh mallow [Greek], possibly from the ibis which feeds on certain varieties of this genus.

Identification: Low-growing **annual herb**, usually c. 20 cm tall but up to c. 45 cm in years of heavy rainfall; covered in long and short **hairs**, some stellate. **Stems** flushed with red. **Leaves** alternate and palmate with 3 deeply incised lobes, hairless above and slightly hairy below, c. 9 cm (inc. c. 2.5 cm leaf stem) × 5.5 cm. **Flowers** with 5 petals, translucent pale cream with a red centre, c. 4 cm dia. In bud, the **calyx** is often striped red; epicalyx fine linear bracts.

Habitat: Found occasionally locally, widespread in years of heavy rainfall, in full sun on the floodplain.

Flowering: Middle to end of the main rains.

B Van Wyk Flowers p.74, Germishuizen p.39, Blundell p.78.

Malvaceae (hibiscus or mallow family)

Hibiscus vitifolius L. subsp. *vulgaris* Brenan & Exell

Derivation: *Hibiscus*, marsh mallow [Greek], possibly from the ibis which feeds on certain varieties of this genus; *viti*, grape [Latin], *folius*, leaf [Latin], with leaves like the grapevine; *vulgaris*, common [Latin].

Identification: Small, branching **woody perennial**, c. 80 cm tall. **Leaves** alternate, palmate with 3–5 lobes, margins serrated, leaf underside with occasional harsh bifurcated or stellate **bristles** that are easily detached and highly irritant. **Flowers** pendular, pale creamy yellow with a dark red or purple centre, c. 4.5 cm dia., borne on long stalks from the leaf axils. The **epicalyx** is formed of c. 10 fine linear bracts.

Habitat: Locally uncommon growing in partial shade or full sun on islands and along the edge of the treeline.

Flowering: During the main rains.

FZ Vol.1 pt2.

Malvaceae (hibiscus or mallow family)

Pavonia burchellii (DC.) R.A.Dyer

Common names: Setswana matinose.

Derivation: *Pavonia* for José Antonio Pavon (1754–1840), a Spanish botanist; *burchellii* for William Burchell (1781–1863), an English naturalist who collected in Africa.

Identification: Erect, branching, **shrubby semi-perennial**, up to c. 1 m tall. **Leaves** alternate, palmate with 3–5 lobes, margins coarsely toothed, with stellate **hairs** on both surfaces. **Flowers** small, apricot-yellow, with 5 petals, borne singly on c. 1–4 cm long stalks with a swollen bend. A ring of leaf-like bracts below the flower.

Habitat: Rarely found in the shelter of bushes in areas of clayey sand near seasonal pans.

Flowering: During the main rains.

FZ Vol.1 pt2, B Van Wyk Flowers p.280, Germishuizen p.113, WFNSA p.254.

Malvaceae (hibiscus or mallow family)

Pavonia senegalensis Cav.

Common names: Setswana moshawa, marethe, mongalangala.

Derivation: *Pavonia* for José Antonio Pavon; *senegalensis*, from Senegal. *Moshawa* means "something that irritates the skin". This may come from the fact that both stems and leaves have groups of easily detached bristles.

Identification: Semi-prostrate **shrubby herb**, mainly growing in full sun but sometimes growing more upright in shade. **Stems** c. 2 m long with clusters of easily detached orange **bristles**. **Leaves** alternate, palmate with 5 lobes, hairy overall, up to c. 7 cm (inc. c. 2.5 cm leaf stem) × up to c. 7 cm, **smell** strongly when crushed. **Flowers** appear only in the morning, pink-tinged buds, open flowers cup-shaped, pale cream with dark red centre, c. 5 cm dia.

Habitat: Widespread on the floodplain in full sun also occasionally on islands and in the treeline of the floodplain.

Flowering: During the main rains.

FZ Vol.1 pt2

Malvaceae (hibiscus or mallow family)

Sida chrysantha Ulbr.

Common names: Setswana mosharashagana, kgotuduwa.

Derivation: *Sida*, a water plant [Greek]; *chrysos*, gold [Greek], *anthemon*, flower [Greek].

Identification: Erect **perennial** herb, up to c. 50 cm tall, woody at the base. **Leaves** alternate, oblong to narrowly ovate, c. 3.5 × 1.5 cm, margins serrated, darker blue-green above and paler silvery green below, hairy on both surfaces. **Flowers** golden yellow, c. 1.5 cm dia., in small groups on long stalks at leaf axils. **Seeds** black.

Habitat: Found occasionally locally on the spillway in full sun and in partial shade in the riverine forest.

Flowering: During the rains.

FZ Vol.1 pt2, Botweed p.84.

Sida cordifolia L.

Common names: Setswana motswalakgoro, bokunogu; **Kalanga** ntatatjiba; **Afrikaans** koekbossie; **English** flannel weed, heartleaf sida.

Derivation: *Sida*, a water plant [Greek]; *cordifolia* with heart-shaped leaves [Latin].

Identification: Semi-perennial **herbaceous plant**, up to c. 80 cm tall; covered in silver-grey velvety stellate **hairs**. **Leaves** with a velvety silver sheen, alternate and cordate, with serrated margins, c. 12 cm (inc. c. 4 cm leaf stem) × 6.5 cm. Frequently blotched yellow when growing in recently dried-out areas of spillway and lagoon, probably because of mineral imbalance. **Flowers** with 5 translucent dark yellow petals, c. 2 cm dia., in clusters at apex of the stems.

Habitat: Common and locally widespread in full sun on the floodplain and spillway.

Flowering: From the middle of the main rains into the cool dry period.

FZ Vol.1 pt2, B Van Wyk Flowers p.160.

Meliaceae (mahogany family)

Turraea zambesica Styles & F.White

Common names: Setswana motulu; **English** Zambezi honeysuckle tree.

Derivation: *Turraea*, for Giorgia della Torre (1607–88), professor of botany at Padua in Italy; *zambesica*, from the Zambezi area.

Identification: Arching slender **shrub**, up to c. 3 m tall. **Bark** a dark grey-brown splotched with a paler grey. Flowering and fruiting before the leaves emerge. **Leaves** elliptical, up to 8 × 4 cm. **Flowers** with a delicate citrus **perfume**, borne in clusters that spiral up the twigs, pale lemony yellow, with 5 long recurved petals; stamen filaments fused to form a tube from which emerges a stigma on a style twice as long as the stamen tube. **Fruit** slightly flattened spheres with striations on the surface, c. 1 cm dia.

Habitat: Locally rare in riverine bush near the Chobe River.

Flowering: During the cool dry period.

Curtis & Mannheimer p.307, Setshogo & Venter p.103.

Menispermaceae (monkey vine or curare family)

Cissampelos mucronata A.Rich.

Common names: Setswana mogatawapeba, mogatapeba, motantanyane; **Subiya** mokaikai.

Derivations: *Kissos*, a climber [Greek], *ampelos*, a vine [Greek]; *mucronata*, each leaf having a single short hair or bristle at its tip (a mucronate tip) [Latin]. *Mogatawapeba*, mouse tail. *Motantanyane*, 'the climbing one', is a generic name for plants that climb.

Identification: Shrubby climber with twining shoots, up to c. 4 m tall depending on support. **Leaves** oval with a cordate base, c. 8 × 7 cm, velvety and light green with a bristle at the apex; some patches of **hairs** appear golden. **Male flowers** clusters of greeny-yellow flowers with 4–5 petals, c. 2 mm dia., borne in the leaf axils. **Female flowers** long trailing spikes with each flower supported by a leaf-like bract. **Male and female flowers** borne on the same plant.

Habitat: Common and widespread, growing in light shade on islands and in the treeline along the floodplain.

Flowering: From early in the rains into the cool dry period.

Uses and beliefs:
• Cooked by Kalanga as *morogo*, a vegetable relish that is eaten with porridge.

FZ Vol.1 pt1. BB, K.

Menyanthaceae (bogbean family)

Nymphoides forbesiana (Griseb.) Kuntze

Common names: Setswana lesotho, tswii; **Subiya** isotho; **English** yellow water gentian.

Derivation: *Nymphoides* from *nymphaia*, a water nymph [Greek], resembling a water lily; *forbesiana* for John Forbes, an English plant collector who visited southern Africa and died in Mozambique in 1823.

Identification: Erect, rooted, **perennial aquatic herb**. **Stems** hollow and therefore floating. **Leaves** kidney-shaped, c. 10 cm dia., with a slight indentation at the outer edge, both surfaces green. **Flowers** feathery, yellow, with 5 lobes, c. 2 cm dia., borne in clusters on long stems on the leaf stalk, just below the leaf blade. These **plants** can withstand considerable changes in water level and even the drying out of pools.

Habitat: Widespread in shallow lagoons with little current.

Flowering: Early in the main rains.

Uses and beliefs:
- Wayeyi and Subiya consider this a delicious food plant. The roots are boiled in a mixture with meat, fish or bubble fish (cat fish).

FZ Vol.7 pt4. CM, GM.

Onagraceae (evening primrose family)

Ludwigia abyssinica A.Rich.

Derivation: *Ludwigia* in honour of the German botanist Professor Christian G. Ludwig of Leipzig; *abyssinica*, of Abyssinia.

Identification: Stout branching succulent **aquatic plant**, up to c. 1 m tall, supported in other vegetation; almost completely hairless and shiny. **Leaves** dark green with shiny red leaf stalks, lanceolate, with smooth margins, spiralling up the stems. **Flowers** yellow, plentiful, with 4 petals which alternate with the sepals, borne in clusters, sometimes on side shoots, at the leaf axils. **Fruit** tubular, bright red, c. 1 cm long, with the sepals retained at the apex.

Differentiated from the creeping *Ludwigia stolonifera* by being more or less erect and having smaller flowers, which also differentiate it from *L. leptocarpa*.

Habitat: Locally uncommon, growing in the margin of fast-flowing channels sheltered by *Cyperus papyrus*.

Flowering: During the main rains.

FZ Vol.4 pt1.

Onagraceae (evening primrose family)

Ludwigia leptocarpa (Nutt.) H.Hara

Common names: Setswana mutemo.

Derivation: *Ludwigia* in honour of the German botanist Professor Christian G. Ludwig of Leipzig; *lepto*, thin, slender [Greek], *carpus*, fruit [Greek], slender-fruited.

Identification: Stout, branching **perennial aquatic herb**, up to c. 1 m tall but more usually about 60 cm. Hairy, especially the young **stems** which are covered in brown hairs. **Leaves** alternate, elliptical narrowing to form the leaf stalk, margins wavy, covered in fine hairs. **Flowers** borne singly at the leaf axils, dark yellow, c. 2 cm in dia., with 5 petals that alternate with sepals. **Fruit** tubular, c. 3.5 cm long, hairy, with the sepals retained at the apex.

Differentiated from the creeping *Ludwigia stolonifera* by being more or less erect and from *L. abyssinica* by being hairy with larger flowers. The duller coloured *L. leptocarpa* does not have the brilliant shiny green and red leaves and stems of *L. stolonifera* and *L. abyssinica*.

Habitat: Found occasionally locally, but widespread growing in the water margin of channels and lagoons.

Flowering: During the main rains.

FZ Vol.4 pt1, Ellery p.168.

Onagraceae (evening primrose family)

Ludwigia stolonifera (Guill. & Perr.) P.H.Raven.

Common names: English willow herb.

Derivation: *Ludwigia* in honour of the German botanist Professor Christian G. Ludwig of Leipzig; *stolon*, a rooting runner [Latin], *fera*, bearing [Latin], having rooting runners.

Identification: Creeping branching **marginal aquatic herb**, c. 30 cm tall but mainly forming large spreading mats in the water margin; completely hairless. **Stems** bright shiny red. **Leaves** shiny, dark green, alternate, elliptical, margins smooth. **Flowers** bright yellow, shiny, c. 1 cm dia., on single stalks in leaf axils, petal number variable but usually 5; flowers usually disintegrate on picking. **Fruit** tubular, c. 2 cm long, with bumps where the seeds are visible. Long sepals are retained at the tip of the fruit.

Distinguished from *Ludwigia abyssinica* or *L. leptocarpa* because it is prostrate, creeping through the edge of the water and has brilliant shiny green and red leaves and stems. Its flowers are also large and shiny.

Habitat: Common and widespread along the shallow margins of permanent water whether still or flowing.

Flowering: Almost throughout the year but more abundantly during the rains.

FZ Vol.4 pto, Ellery p.169.

Orchidaceae (orchid family)

Ansellia africana Lindl. subsp. *africana*

Common names: Setswana palamêla; **English** leopard orchid, tree orchid, giant orchid.

Derivations: *Ansellia* for John Ansell, a mid-19th century English gardener and botanical collector; *africana*, from Africa [Latin]. *Palamêla*, 'it climbs'.

Identification: Robust **epiphyte**, growing from small, almost tubular, pseudo-bulbs. **Leaves** linear, bright green with some red-brown spots, c. 20 cm long, tending to fold at the mid-rib, margins smooth, hairless. **Flowers** yellow with scattered brownish-maroon spots, in branching racemes, c. 30 cm long.

Habitat: Rare but widespread growing on palm trees (*Hyphaene petersiana* in this area), sometimes found on *Acacia nigrescens*. Rapidly becoming less common as they are poached for garden decoration and traditional medicine. This plant is on the Botswanan red list of protected plants.

Flowering: During the early rains.

Uses and beliefs:
- Subiya believe planting this species in the garden around the house brings good luck.
- The pseudo-bulbs are boiled with milk and drunk by men to make them strong sexually.

FZ Vol.11 pt2, Blundell p.432, Germishuizen p.95, WFNSA p.118. GM.

Oxalidaceae (oxalis family)

Oxalis corniculata L.

Common names: Setswana tswaitswai?; **Afrikaans** tuinranksuring, ranksuring, steenboksuring; **English** oxalis, Jimson weed, wood sorrel, creeping lady's sorrel, creeping oxalis, yellow sorrel, creeping sorrel.

Derivation: *Oxys*, acid, sour [Greek], referring to the oxalic acid in the leaves of many species; *corniculata*, with short horns [Latin], possibly alluding to the shape of the fruit.

Identification: Rather variable small creeping **herb**, c. 10 cm tall. **Roots** rhizomous, running along under the soil. **Leaves** divided into 3 leaflets, indented at the tip, which droop or fold at night, often covered in sparse hairs. **Flowers** yellow, borne in small groups on long stalks. **Fruit** cylindrical but tapering at each end, covered in hairs.

Habitat: This cosmopolitan troublesome weed, which is poisonous to game, has recently been introduced into the area and favours short well-watered grassland.

Flowering: Throughout the rainy period.

FZ Vol.2 pt1, B Van Wyk New p.170, Blundell p.47, Germishuizen p.110.

Portulacaceae (purslane family)

Portulaca oleracea L. subsp. *oleracea*

Common names: Setswana serepe; **Afrikaans** varkkos; **English** (common) purslane, pigweed, purslain, pusky, wild purslane, pusley.

Derivation: *Portare*, to carry [Latin], *laca*, milk [Latin], i.e. milk-carrying, referring to the milky latex in this species; *oleracea*, pertaining to kitchen gardens, either as a pot herb or vegetable or as a weed [Latin].

Identification: Low-growing **succulent annual**, c. 30 cm tall. Long **taproot**. **Stems** red-brown when growing in damper areas. **Leaves** fleshy, hairless, broadly oval. **Flowers** yellow, minute, terminal and surrounded by a whorl of leaves.

Habitat: Found occasionally growing in heavily grazed short grass beside lagoons and channels. A cosmopolitan weed of disturbed soils. [Introduced.]

Flowering: Early in the main rains.

Uses and beliefs:
- Possibly introduced as a pot herb.

FZ Vol.1 pt2, B Van Wyk Flowers p.172, Blundell p.37.

Portulacaceae (purslane family)

Portulaca quadrifida L.

Common names: Afrikaans kanniedood, porselein; **English** purslane, pusley, wild purslane.

Derivation: *Portare*, to carry, *laca*, milk [Latin], i.e. milk-carrying, originally referring to the milky latex in *Portulaca oleracea* and then applied to other plants of the genus; *quadrifida*, cut into four [Latin].

Identification: Prostrate **annual succulent** with a somewhat swollen **taproot**. **Stems** red, rooting from the nodes where they touch the ground, c. 25 cm long. **Leaves** elliptical, with little or no stalk, arranged in pairs, c. 1 × 0.5 cm, smooth margins, grey-green on the upper surface and red on the underside, folding together against the heat. **Stipular hairs** at leaf axils. **Flowers** bright yellow (although in other regions they may be orange or red), shiny, c. 1.1 cm dia., opening after 14.00 hrs, 4 petals. Flowers are arranged in small groups at the apex of the stems, surrounded by 4 leaves and numerous hairs.

Habitat: A locally rare cosmopolitan weed found in sandy clay, damp areas of mopane woodland near seasonal pans.

Flowering: Late in the main rains.

FZ Vol.1 pt2, B Van Wyk New p.230, Blundell p.38, Germishuizen p.189, WFNSA p.134.

Portulacaceae (purslane family)

Talinum caffrum (Thunb.) Eckl. & Zeyh.

Common names: Setswana kgelegetla, thotamadi yo monamagadi.

Derivations: *Talinum*, the Senegalese name for certain species; *caffrum*, from Kaffraria, the old name for the Eastern Cape of South Africa. *Thotamadi yo monamagadi* is the female bloodsucker.

Identification: Rather lax **succulent** with shoots c. 35 cm long. Large **taproot** with small protruding fibrous roots, white when cut. **Leaves** grey-green, linear, spiralling up the stem, margins smooth. **Flowers** star-shaped, shiny yellow, 5 pointed petals, many stamens and only 2 sepals, c. 2 cm dia.; borne singly on long stalks from the leaf axils; opening only in the late afternooon. Flower stalks with a distinct joint and thickening. A pair of tiny **bracts** c. 1 cm below the flower head.

Habitat: Common and widespread in mopane woodland and occasionally on islands.

Flowering: Throughout the rains.

Uses and beliefs:
- Batswana use an infusion of the roots to clear the blood.

FZ vol.1 pt2, B Van Wyk Flowers p.172, WFNSA p.134. GM.

Portulacaceae (purslane family)

Talinum crispatulatum Dinter

Common names: Setswana kgelegetla, mojaphuti, thotamadi yomotunanyana.

Derivations: *Talinum*, the Senegalese name for certain species; *crispatus*, crisped, irregularly waved and twisted, kinky or curled [Latin], alluding to the margin of the leaves. *Thotamadi yo motunanyana* is the male blood sucker.

Identification: Rather prostrate **succulent** with stems up to c. 25 cm long. Large **taproot** dark orange-red when cut. **Stems** pinky red. **Leaves** grey-green, ovate, alternate and spiralling up the stem, margins smooth and irregularly waved. **Flowers** yellow, 5 shiny petals, with pointed tips and many stamens, only 2 pale green sepals. **Flowers** borne singly or in pairs on long stalks from the leaf axils, opening only in the late afternoon.

Habitat: Widespread in lightly shaded areas of islands and especially mopane woodland.

Flowering: Throughout the rains.

Uses and beliefs:
• Batswana use an infusion of the roots to clear the blood.

FZ Vol.1 pt2. GM.

Kohautia caespitosa Schnizl. subsp. ***brachyloba*** (Sond.) D.Mantell

Common names: Setswana mollo-wa-badimo.

Derivations: *Kohautia* in honour of Francis Kohaut who collected plants in Senegal in 1822; *caespitosa*, growing in dense clumps [Latin]; *brachyloba*, short-lobed [Latin]. *Mollo-wa-badimo*, fire from the gods.

Identification: Slim, erect, branching **annual** to **short-lived perennial herb,** c. 45 cm tall, arising from a **woody base**. Stems and leaf surfaces often densely **hairy**. **Leaves** borne in whorls, linear to lanceolate, stalkless, margins smooth. **Flowers** c. 3 mm dia., borne in stalkless pairs at the nodes, small, creamish to yellowish-green, with 4 lobes fused to form a long (c. 1.5 cm) corolla tube, which is slightly paler in colour, tube usually hairless. Scented at night. Calyx points retained on the **seed capsule**.

Easily confused with *Kohautia subverticillata* but the retained sepals on the seed capsule are smaller in *K. caespitosa* and its leaves tend to be narrower. The leaves of *K. caespitosa* are also more grey-green and covered in very short dense hairs. In this area, *K. caespitosa* is more common than *K. subverticillata*.

Habitat: Widespread, found occasionally locally growing in lightly shaded areas of the treeline and in mopane woodland.

Flowering: During the main rains.

FZ Vol.5 pt1, Blundell p.152, WFNSA p.404.

Scrophulariaceae (foxglove family)

Alectra orobanchoides Benth.

Derivation: *Alectron*, a cock [Greek], the flower resembles a cock's comb; *orobanchoides*, the plant resembles the genus *Orobanche*, another member of the same family.

Identification: Small, erect **hemi-parasite** with occasionally branching stems, c. 20 cm tall, covered with short coarse **hairs** overall. **Roots** bright orange, some attached to other plants. **Stems** maroon-purple. **Leaves** rudimentary, scale-like and spiralling up the stem, c. 8 × 5 mm. **Flowers** asymmetrical with 5 lobes fused to form a tube, yellow with maroon veining, c. 1 cm dia., borne singly at the leaf axils. These plants usually **dry black** when pressed.

Habitat: Locally uncommon but widespread in mopane woodland.

Flowering: During the main rains.

FZ Vol.8 pt2, B Van Wyk New p.176, WFNSA p.368.

Scrophulariaceae (foxglove family)

Alectra picta (Hiern) Hemsl.

Common names: Setswana matebelwe; **English** cowpea witchweed.

Derivation: *Alectron*, a cock [Greek], alluding to the resemblance of the flower to a cock's comb; *picta*, painted, brightly coloured [Latin].

Identification: Erect, occasionally branching, **perennial semi-parasitic herb**, c. 40 cm tall. **Roots** bright orange, some attached to other plants, mainly legumes. **Leaves**, or leaf-like bracts, small, oval, almost like scales, spiralling up the stem. **Flowers** borne singly, on short stalks, in the axils, bright lemon yellow with some red veining on the reverse; **stigma** curved to one side.

Habitat: Found rarely locally in heavy shade on islands in the floodplain.

Flowering: During the main rains.

Uses and beliefs:
- This plant is listed as a noxious weed. It parasitises mainly legumes and can kill crops of beans and peas.

FZ Vol.8 pt2, Blundell p.374, Botweeds p.98.

Scrophulariaceae (snapdragon or foxglove family)

Jamesbrittenia elegantissima
(Schinz) Hilliard

Derivation: *Jamesbrittenia*, for the British botanist, James Britten; e*legantissima*, very elegant [Latin].

Identification: Small aromatic, moisture-loving, **annual herbaceous plant,** up to c. 1 m tall at the end of the season but normally much smaller, branching from low down on the plant; **hairy** overall. **Leaves** pinnatifid and deeply serrated. **Flowers** asymmetrical, yellow, borne in the upper leaf axils and at the apex of the plant, c. 5 mm dia., with 5 rounded lobes and a corolla tube (c. 7 mm long).

Habitat: Common and widespread, found occasionally locally on the floodplain margin.

Flowering: From the early rains throughout the main rains.

FZ Vol.8 pt2.

Sterculiaceae (cacao or cola family)

Hermannia quartiniana A.Rich.

Common names: Setswana motantanyane.

Derivations: *Hermannia*, in honour of Paul Hermann (1646–95), a German botanist, professor of botany at Leiden in Holland who visited southern Africa; *quartiniana*, for Leon Richard Quartin-Dillon, a French physician and botanist. *Motantanyane*, the climbing one, a generic name for plants that climb.

Identification: Lax **perennial** creeping and scrambling **herb** from a woody **root stock**, c. 40 cm tall; covered in short coarse **hairs**. **Leaves** alternate, oblong, with serrated margins. **Flowers** cream-coloured, frequently produced in pairs from the leaf axils; bracts at the point where the flower stalk divides to carry the 2 flowers; petals long and rolled back. **Calyx** short.

Habitat: Locally uncommon in the open grassland of the floodplain margin and on islands.

Flowering: During the main rains.

FZ Vol.1 pt2, Germishuizen p.83, Turton p.69.

Sterculiaceae (cacao or cola family)

Melhania forbesii Mast.

Common names: Setswana muchima; **Hmbukushu** ndongo.

Derivation: *Melhania*, for Mount Melhan in Arabia; *forbesii*, for John Forbes an English plant collector who visited southern Africa and died in Mozambique in 1823.

Identification: Erect **woody shrublet**, 30–80 cm tall. **Stems** and **leaves** covered with a thick pelt of short matted grey **hairs** mixed with rusty-coloured longer hairs, hairs becoming more noticeable as the plant ages. **Leaves** alternate, oblong to ovate, margin serrated. **Flowers** c. 2 cm dia., with 5 shiny yellow petals opening only in the afternoon. Flowers borne in small groups at the apex of the plant or on stems, c. 5 cm long, from the leaf axils.

Habitat: Locally common in light shade on islands.

Flowering: Main rains through to the cool weather.

Uses and beliefs:
- This plant is edible.

FZ Vol.1 pt2, Hargreaves p.13. MT.

Sterculiaceae (cacao or cola family)

Waltheria indica L.

Common names: Setswana seretlwana, motswalakgoro, poo-khunung, sekoba; **Subiya** idarere; **Afrikaans** meidebossie.

Derivation: *Waltheria*, after the professor of medicine and keen gardener A.F. Walther; *indica*, from India.

Identification: Branching, erect **annual herb**, c. 50 cm tall; softly **hairy**. **Leaves** elliptical with a crenate margin, carried alternately on the stem. **Flowers** small, yellow, c. 3 mm dia., borne in dense clusters, on short stalks either at the apex of the plant or in the leaf axils. A prominent **bract** below the flower head. **Fruit** long, cylindrical pods.

Habitat: Locally uncommon. Found in open areas of the riverine forest and in light shade on islands in the floodplain. This plant seems to like areas that are heavily utilised by animals.

Flowering: Middle to end of the main rains.

Uses and beliefs:
• The leaves are used to make *morogo*, a vegetable relish.
• Used to treat barrenness, haemorrhaging, teething and coughs and as a purgative.

FZ Vol.1 pt2, B Van Wyk Flowers p.178, Ellery p.179, Turton p.75. GM.

Tiliaceae (jute family)

Corchorus tridens L.

Common names: Setswana ledelele; **Lozi** delele; **Hmbukushu** rwithe, dinyangombe; **Subiya** idarere.

Derivation: *Corchorus*, jute [Greek], another plant in the same genus; *tri*, three [Greek], *dens*, teeth [Latin], referring to the 3 spikes or bristles at the tip of the seed pods. *Rwithe* is *morogo* or vegetable relish.

Identification: Dark green **herb** up to c. 1.3 m tall. **Stems** tinged pink. **Leaves** are alternate, narrowly oval with a crenate margin, c. 15 cm (inc. c. 2.3 cm leaf stalk) × 3.8 cm. A **bristle** c. 1 cm long emerges from the last serration nearest the leaf stalk on each side. There are **stipules** in pairs at the leaf axils. Small yellow **flowers**, c. 0.8 cm dia., in groups of 2 or 3 in the leaf axils, open in the afternoon. **Fruit** are tubular **pods**, ribbed and striped red, c. 8 × 0.6 cm with 3 horns at the tip.

Habitat: Common and widespread throughout the area except in deeply shaded areas of the riverine forest. [Introduced.]

Flowering: During the main rains.

Uses and beliefs:
- The leaves of this plant are used to make *morogo*, a vegetable relish that is especially loved by Kalanga. Cook the leaves in water with a little bicarbonate of soda, salt and tomato. It tastes like okra and is eaten with mealie. The leaves can also be dried and stored.

FZ Vol.2 pt1. GM.

Tiliaceae (jute family)

Grewia bicolor Juss.

Common names: Setswana mogwana, ntewa, mambalane, monabo, kukuruthwe, motuu; **Subiya** motono; **Afrikaans** witblaarrosyntjie; **English** false brandy bush, bastard brandy bush, white-leaved raisin, white-leaved grewia.

Derivation: *Grewia*, for Dr Nehemiah Grew (1641–1712), an English doctor and botanist; *bi*, two [Latin], *color*, coloured [Latin], two-tone, possibly referring to the leaves, which are green above and almost white below, or to the sepals, which are yellow on the inside and green on the reverse.

Identification: Small **multi-stemmed shrub**, c. 2.5 m tall. **Bark** grey and smooth when young, becoming dark grey and deeply fissured with age. Young stems and leaves covered with fine white hairs. **Leaves** held horizontally or slightly drooping, alternate and obovate, with serrated margins and net veining towards the tip, base often obtuse, lower leaf surface often almost white with hairs. **Flowers** yellow, with 5 sepals that are yellow on the inside and green on the reverse; 5 petals very variable and sometimes very small, many stamens. **Fruit** either a single lobe or bilobed, turning brown on ripening.

Habitat: Widespread in mixed woodland.

Flowering: During the early rains.

Uses and beliefs:
- Fresh leaves are used to make tea.
- The fruit is edible and used for making *kgadi*, the traditional wine.
- Fibre from the bark is used to make rope and baskets. The cord may be plaited to make whips for herding cattle.
- The wood is used for the shafts of spears.
- Batswana boys cut the stems into sticks of about a finger's thickness and a length of 60 cm or more. They throw these sticks in a game called *nxai* or *xnavi* (nowadays 'javelin'). The sticks are cleaned so they will fly well.
- Twigs from the bush are used to make *lefetho*, a whisk for cooking.
- Caterpillars of the moths *Chasmina tibialis* and *Serrodes partite* feed on this species.

FZ Vol.2 pt1, Blundell p.68, Ellery p.118, Hargreaves p.12, Palgrave updated p.688, Setshogo & Venter p.20, Tree Roodt p.73. GM, KS.

Tiliaceae (jute family)

Grewia flavescens Juss.

Common names: Setswana mokgompatha, motsotsojane, mokankele, mpuzu, mangqore, gxoxe, mok-gomatha, mokidi; **Hmbukushu** moXhane; **Subiya** mompondo; **Shiyeyi** manXore; **Afrikaans** skurwerosyntjie; **English** rough-leaved raisin, sandpaper raisin, donkey berry.

Derivation: *Grewia* is named for Dr Nehemiah Grew, an English plant physiologist; *flavescens*, yellowish [Latin].

Identification: Branching, arching **shrub** or **small tree**, c. 3 m tall. **Trunks** heavily buttressed and square. **Leaves** large, held horizontally, alternate, obovate to lanceolate, with finely serrated margins on short leaf stalks, upper and lower surfaces are harshly hairy, with star-shaped hairs, giving them a sand-papery feel. **Stipules** in pairs at the leaf axils. **Flowers** borne in clusters at the leaf axils, with 5 yellow petals and 5 sepals that are arranged alternately with the petals, many stamens, c. 2.5 cm dia.

Habitat: Widespread on islands, in acacia scrub and in riverine woodland, usually on sand.

Flowering: During the main rains.

Uses and beliefs:
• The fruit are edible. They are usually eaten raw or used to make the traditional alcoholic beverage, *kgadi*.
• Batswana use the wood for spear shafts, fishing rods, bows and knobkerries.
• Lozi, Hmbukushu and Wayeyi use the young branches to make fishing baskets or traps (called *seXua* [Shiyeyi] and *thumba* [Hmbukushu]).

FZ Vol.2 pt1, Ellery p.119, Hargreaves p.12, WFNSA p.252. GM, KS, SS, MT.

Tiliaceae (jute family)

Triumfetta pentandra A.Rich.

Derivation: *Triumfetta* for G.B. Trionfetti, the professor of botany and director of the botanical garden in Rome; *penta*, five [Greek], *andros*, male [Greek], referring to the 5 stamens.

Identification: Annual branching **herb**, up to c. 90 cm tall, with a woody lower **stem** and a single **taproot**. **Stems** softly hairy and blotched red, occasionally with red bristles in more mature plants. Mature **leaves** trilobed, margins serrated, softly hairy, c. 10 cm (inc. c. 4 cm petiole) × 6 cm. **Flowers** yellow, c. 7 mm dia., borne in scorpioid spikes at the leaf axils. Flower buds tubular, red on the outside.

Habitat: Found occasionally locally in sandy areas in light shade on islands and in the scrub.

Flowering: Middle to end of the main rains.

FZ Vol.2 pt1.

Turneraceae

Tricliceras lobatum (Urb.) R.Fern.

Derivation: *Tricliceras*, 3-chambered [Greek], referring to the seed pods; *lobatus*, lobed [Greek], referring to the single lobes on the lower part of the leaves.

Identification: Erect, branching **annual herb**, c. 35 cm tall. **Leaves** broadly lanceolate with a lobe on each side at the base, stalkless, margins slightly dentate, spiralling up the stem. **Flowers** c. 5 mm dia., golden yellow, borne on long stalks from the leaf axils, with 5 pointed petals. **Seed pods** slim, tubular, c. 5.5 cm long, hanging from the stalks.

Habitat: Found only occasionally in full sun in deep sand and in mopane woodland.

Flowering: During the main rains.

FZ Vol.4 pto.

Vahliaceae

Vahlia capensis (L.f.) Thunb. subsp. *vulgaris* Bridson var. *vulgaris*

Common names: Setswana leetsane.

Derivation: *Vahlia*, for the Norwegian-born Danish botanist Martin Vahl (1749–1804), traveller, pupil of Linnaeus and professor of botany; *capensis*, of the Cape of Good Hope; *vulgaris*, common [Latin].

Identification: Multi-branched, erect **herb**, c. 30 cm tall; varying between being hairless and having short sticky hairs overall. **Leaves** fine and linear, growing in whorls up the stem. **Flowers** yellow, with 5 petals and 5 sepals alternately, c. 1 cm dia.; borne at the apex of the plant and on c. 5 mm stalks from leaf axils.

Habitat: A common and widespread moisture-loving plant that grows in open grassland on the floodplain.

Flowering: During the rains.

FZ Vol.4 pto, B Van Wyk Flowers p.182.

Vitaceae

Cyphostemma congestum (Baker) Wild & R.B.Drumm.

Common names: Setswana mohubuhubu?

Derivation: *Kyphos*, curved [Greek], *stemma*, a garland [Greek]; *congestus*, congested [Latin], arranged closely together.

Identification: Perennial **succulent climbing plant**, up to c. 2.5 m tall. **Leaves** arranged spirally up the stem, stalkless, compound, palmate with 5 leaflets, each leaflet oval with a blunt tip, a short stalk and toothed margins, surfaces downy. **Tendrils** on the opposite side of the stems to the leaf. Umbels of vestigial, creamy-yellow **flowers** borne at leaf axils. Flowers often mixed with **fruit**, which become bright red as they ripen.

Habitat: Locally uncommon but widespread and more plentiful in years of heavy rain. Grows with the support of shrubs and trees on islands and in the treeline.

Flowering: Throughout the main rains.

FZ Vol.2 pt2.

Xanthorrhoeaceae

Trachyandra arvensis (Schinz) Oberm.

Derivation: *Trachys*, rough, shaggy [Greek], *andros*, male parts [Greek]; *arvensis*, growing in or pertaining to cultivated fields [Latin].

Identification: Multi-branched **perennial herb**, c.˙80 cm tall, with a woody rhizomous **root**. Two types of **leaves**: each stem has c. 2 outer brown sheath-like leaves, the main leaves are fine, grass-like, hairless and slightly curved with no marked midrib. **Flowers** pale lemon yellow, small (c. 1 cm dia.), with recurved petals, pendulous, arranged in widely spaced racemes. **Fruit** erect, trilobed to globular.

Habitat: Found only occasionally in grassy, lightly shaded areas of the spillway margin and on island margins.

Flowering: Throughout the main rains.

FZ Vol.12 pt3.

Zygophyllaceae (calthrop or puncture vine family)

Tribulus terrestris L.

Common names: Setswana mosetlho, tshetlho, shosho, igogo-chitukunu; **Subiya** lovangu; **Shiyeyi** eshoshong; **Afrikaans** dubbeltjie; **English** devil's thorn, puncture vine, burnut, (land) calthrop.

Derivations: *Tribulus*, a 4-pronged iron implement, called a calthrop, used to impede the Roman cavalry, referring to the shape of the fruit; *terrestris*, of the earth, creeping [Latin]. *Mosetlho*, a plant bearing thorns

Identification: Vigorous **low-growing annual** with trailing stems, up to c. 1.5 m but more usually less than c. 50 cm. **Stems** orange-brown, hairy. **Leaves** pinnate, c. 8 pairs of leaflets, hairy on the underside, c. 5 × 2.2 cm. **Flowers** small (c. 1.7 cm dia.), yellow, with 5 petals, curling lengthways in the sun, borne at the leaf axils. **Fruit** unevenly 5-lobed breaking into segments, each bearing small hard spikes.

Habitat: A common and widespread nuisance in dry years in open sandy areas that are heavily trodden and grazed. This species is quickly invading other countries around the world.

Flowering: From the middle of the main rains into the cool dry period.

Uses and beliefs:
• Kalahari San boil the thorns and administer the decoction to women experiencing difficulty in childbirth.
• Eating this plant in a wilted condition may cause photosensitisation, blindness and even death.
• Subiya use this species as cattle fodder, it is collected during the rainy season, dried and bundled.

FZ Vol.2 pt1, Ellery p.177, Turton p.81, Flowers Roodt p.159, Germishuizen p.110, Blundell p.44, Van Wyk p.184. GM, PN.

White
or whitish flowers

Acanthaceae (*Acanthus* or spinyflower family)

Asystasia gangetica (L.) T.Anderson

Common name: English asystasia.

Derivation: *Asystasia*, inconsistency [Latin], as the corolla tube is almost regular or symmetrical, which is unusual in this family; *gangetica* probably means that it occurs along the River Ganges.

Identification: A scrambling **perennial herb**, c. 80 cm tall. **Stems** ribbed and hairy. **Leaves** opposite, decussate, oval, margins entire, covered with short fine hairs on both surfaces, c. 8.5 cm (inc. c. 2.5 cm leaf stem) × 3.5 cm. **Flowers** in a 1-sided raceme, white tubular corolla with 5 lobes, the lower one enlarged and with purple-blotched markings in the throat. **Seed pods** bristly and fiddle-shaped. Some plants are more hairy than others.

Habitat: Found occasionally locally in deeply shaded areas of the riverine forest, on islands and in the treeline along the spillway. Also found in sandy areas of acacia scrub in lighter shade.

Flowering: Middle to end of the main rains.

Uses and beliefs:
- The leaf may be used as an antidote for snake bite.
- The plant is used to ease pain in childbirth.

Ellery p.156, Pooley p.200.

Acanthaceae (*Acanthus* or spinyflower family)

Barleria lugardii C.B.Clarke

Common name: Setswana magogodi-a-noka.

Derivation: *Barleria* for Jacques Barrelier, a Dominican monk of Paris; *lugardii*, Edward James and Charlotte Eleanor Lugard visited Ngamiland in the late 1890s, many of the plants that they collected were named in their honour.

Identification: A low-growing compact **woody shrub**, c. 40 cm tall. **Leaves** narrowly oval, opposite and decussate, margins entire. **Flowers** white with 4 asymmetrical lobes, emerging from a pair of coarse leafy sepals; 2 stamens with indigo anthers that become white as they are charged with pollen.

Habitat: Locally uncommon in lightly shaded mopane woodland.

Flowering: During the main rains.

Uses and beliefs:
- Often grazed by game.

Acanthaceae (*Acanthus* or spinyflower family)

Blepharis maderaspatensis (L.)
B.Heyne ex Roth

Derivation: *Blepharon*, eyelid [Greek], referring to the bracts; *maderaspatensis*, of Madras in southern India.

Identification: Shrubby **herbaceous** plant. **Stems** up to c. 80 cm long, slightly hairy, branching at each leaf axil. **Leaves** sessile, narrowly oval, margins entire with coarse bristles, undersides a paler green, borne in whorls up the stems, c. 5 × 2 cm. **Flowers** in clusters at leaf axils, white, c. 3 mm dia. **Bracts** around flower heads have long bristles (c. 3 mm long) along margins.

Habitat: Locally uncommon on sandy soil in light shade, on islands and in the treeline along the floodplain.

Flowering: Middle to end of the main rains.

Blundell p.389.

Acanthaceae (*Acanthus* or spinyflower family)

Dicliptera paniculata (Forssk.) I.Darbysh.
(syn. *Peristrophe paniculata* (Forssk.) Brummitt)

Common names: English lady flower.

Derivation: *Diklis*, double-folding [Greek], *pteron*, a wing [Greek], alluding to the 2 wing-like parts of the seed capsule; *paniculata* refers to the type of inflorescence, having flowers arranged in panicles [Latin].

Identification: A branching, erect **semi-perennial herb**, up to c. 1 m tall. **Leaves** few in number, oval with an elongated point and narrowing to form a leaf stalk, margins entire, opposite and decussate on main stem, c. 12 cm (inc. c. 2 cm leaf stem) × 5.5 cm, often shed as the weather becomes drier. **Flowers** c. 8 mm wide, white with lavender or mauve markings on the enlarged lower petal of the tubular corolla; borne in panicles from the leaf axils and at the apex of the plant on fine stiff stalks. **Fruit** fiddle-shaped pods, c. 1.5 cm long, splitting at mid-line to release c. 6 seeds.

Habitat: Widespread, found occasionally locally growing in light shade on islands and in the treeline along the floodplain.

Flowering: Late in the main rains through to the end of the cool dry period.

Turton p.83.

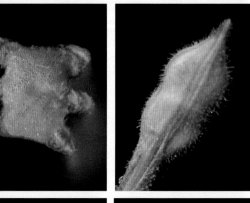

Acanthaceae (*Acanthus* or spinyflower family)

Duosperma crenatum (Lindau) P.G.Mey.

Derivation: *Duo*, two [Latin], *sperma*, fruit [Latin], twin-fruited; *crenatum*, scalloped, crenate [Latin].

Identification: Woody **shrubby perennial**, c. 30 cm tall. **Bark** peels in papery sections as it ages. **Leaves** are opposite and decussate, oval with serrated margins, tending to fold along the central vein and slightly decurved. **Flowers** borne in pairs of small clusters at the axils of the leaves, c. 5 mm dia., white with magenta markings on the lower central lobe; 2 lips fused to form a tube, the lower lip with 3 lobes, the upper with 2 lobes.

Habitat: Locally common and widespread in mopane woodland.

Flowering: During the main rains.

Acanthaceae (*Acanthus* or spinyflower family)

Hypoestes forskaolii (Vahl) R.Br.
(syn. *Justicia forskaolii* Vahl)

Common names: English white ribbon bush.

Derivation: *Hypo*, under [Greek], *estia*, house [Greek], referring to the way the calyx is covered by bracts; *forskaolii*, in honour of Pehr Forsskål (1732–1804), a Swedish botanist, traveller and student of Linnaeus.

Identification: Lax, branching **herb**, c. 60 cm tall; covered in short fine **hairs**. Main **stems** slightly squared; **nodes** enlarged, bent and rooting where they touch the ground. **Leaves** ovate with a narrow tip, opposite, decussate, margins entire. **Flowers** held diagonally across the stem; lower lip white, upper lip white with purple spots; anthers bright red.

Habitat: Locally widespread in lightly shaded areas of islands and in mixed woodland.

Flowering: Almost throughout the year.

Blundell p.392, Pooley p.200.

Acanthaceae (*Acanthus* or spinyflower family)

Justicia betonica L.

Common names: English leafy bract justicia, paper plume.

Derivation: *Justicia* for James Justice, a Scottish gardener; *betonica*, may be a variation of *vettonica*, a similar plant that grows in Spain, or an anagram of *nicoteba*, a name used for this plant in the past.

Identification: A branching, weakly erect, **annual** to **short-lived perennial herb**, c. 30 cm tall. **Leaves** opposite and decussate, almost stalkless, oval, hairless, tending to fold along the mid-vein and to be slightly decurved along their length. Erect **flower spikes** make the plants easily recognisable with their sharply pointed **leafy bracts**, white with green veining. **Flowers** emerging from the bracts; 2-lipped, the upper lip vestigial; white with pinkish-maroon marking in the throat. **Fruit** fiddle-shaped and dehiscent.

Habitat: Common, often in light shade, on islands, in woodland and in scrub.

Flowering: During the main rains.

Uses and beliefs:
- The Batswana boil the roots and drink the decoction to treat kidney problems and the blood system.
- The whole plant may be collected and burned when a member of the family dies. The ashes are spread in the yard as a means of communication with the ancestors, to bring peace within the family.

B van Wyk Flowers p.34, Blundell p.393, Pooley p.200, Flowers Roodt p.11, WFNSA p.396. BB.

Acanthaceae (*Acanthus* or spinyflower family)

Justicia exigua S.Moore

Derivation: *Justicia* for James Justice, a Scottish gardener; *exigua*, weak, feeble, little [Latin].

Identification: A small, rather lax, **annual herb**, c. 20 cm tall. Both stems and leaves with fairly long bristles. **Leaves** opposite, ovate, margins entire. **Flowers** c. 2 mm dia., borne in short single-sided racemes on long fine stalks from leaf axils, white with magenta blotches symmetrically in the throat. **Fruit** with the calyx remaining, fiddle-shaped as is typical of the *Justicia* genus.

Habitat: Found occasionally in damp areas in partial shade at the edge of seasonal pans and on islands, often in the shelter of fallen trees

Flowering: During the main rains.

Acanthaceae (*Acanthus* or spinyflower family)

Ruellia patula Jacq.

Derivation: *Ruellia* for Jean Ruel, herbalist to François I of France; *patula*, spread or outspread [Latin].

Identification: A semi-prostrate **perennial herb**, branching at the leaf axils, c. 30 cm tall, slightly hairy overall. **Leaves** opposite and decussate, oval with a rounded tip and slightly wavy margins, c. 4.0 cm (inc. c. 0.5 cm leaf stalk) × 1.5 cm. **Flowers** white, 5 lobes fused to form a corolla tube, c. 2.5 cm long.

Habitat: Widespread growing in deep shade in the riverine forest.

Flowering: From the beginning of the early rains.

Turton p.22.

Alismataceae (water plantain family)

Caldesia parnassifolia (L.) Parl.
(syn. *Caldesia reniformis* (D.Don) Makino)

Derivation: *Caldesia*, possibly in honour of Francisco José de Caldas (1771–1816), a Colombian botanist and patriot; *parnassus*, Mount Parnassus in Greece; *-folia*, -leaved [Latin], referring to the shape of the leaves of Parnassia, a small herb that grows on Mount Parnassus.

Identification: Perennial aquatic herb with a submerged **rooting system**, up to c. 80 cm tall. **Leaves** forming a basal rosette, floating on the surface, kidney-shaped with smooth margins, c. 7.5 × 6.5 cm. **Flowers** borne in an erect panicle above the water, white with 3 petals; 6 stamens with yellow anthers. **Fruit** a receptacle with a cluster of pointed seeds.

Habitat: A widespread naturalised herb found occasionally locally, rooting in the peaty soil of flowing waterways and channels.

Flowering: Early in the main rains.

Amaranthaceae (pigweed or cockscomb family)

Aerva leucura Moq.

Common names: Setswana tlhogotshweu; **San** kaguhe; **Afrikaans** aambeibossie; **English** foam bush, haemorrhoid bush, white head, aerva.

Derivations: *Aerva* is the Arabic name for this genus; *leuco*, white [Greek], referring to the flowers. *Tlhogotshweu*, white head, refers to an older person with white hair. *Kaguhe*, 'come and see'.

Identification: A woody, branching, **perennial herb**, up to c. 1.5 m tall. **Stems** lightly ribbed. **Leaves** simple, alternate, narrowly oval to linear, margins wavy, up to c. 17 cm long and narrowing to form a leaf stalk, dull green, covered with dense white hairs. **Flowers** tiny, green but appearing white due to the mass of white hairs surrounding them, borne in pendular spikes up to c. 10 cm long.

Habitat: Widespread, found occasionally locally on the open floodplain in full sun but also on lightly shaded islands on the floodplain.

Flowering: From the middle of the main rains into the cool dry period.

Uses and beliefs:
- The leaves are cooked as *morogo*, a spinach-like vegetable relish eaten with porridge.
- The plant is also used to treat haemorrhoids.
- Hmbukushu and Wayeyi use the plant, sometimes mixed with *Enicostema axillare*, in two ways. First, if a child is unlucky (e.g. he buys cattle and they die), his mother or father cut the leaves and flowers and soak them overnight. At sunrise, the child must face the sun while the parent takes a mouthful of the liquid and sprays it over the child towards the sun. It brings good luck. Second, if babies are sleeping restlessly, it is believed that there are *tokolosi* (bad spirits) in the yard. The leaves, flowers and roots are burned to drive away the evil spirits.
- Batswana use this plant to treat *dikgaba* (evil spirits). They sleep with *A. leucura* under their pillow and/or bathe night and morning in water soaked in this plant.
- It is also used in the Okavango Delta to cure children who do not look after their parents properly.
- Wayeyi burn the leaves and flowers during a good luck ceremony to give them good hunting, as a means of communication with their ancestors.
- Wayeyi and the Ovambenderu use the flowers to stuff pillows and mattresses.
- San from the central Kalahari collect the flowers to use as a filter when smoking a pipe.

FZ Vol.9 pt1, B Van Wyk Flowers p.36, Ellery p.154, Flowers Roodt p.17. BB, MM, PN, KS.

Amaranthaceae (pigweed or cockscomb family)

Alternanthera sessilis (L.) DC.

Common name: English amaranthus weed.

Derivation: *Alternus*, alternate, *anthera*, the anther (part of the stamen) [Latin], referring to alternating sterile and fertile stamens; *sessilis*, stalkless, sessile [Latin].

Identification: A straggling **perennial herb**, up to c. 50 cm tall, growing in the water margin; generally **hairless**. **Stems** square, may be pinkish and may **root** at the nodes. **Leaves** linear to spathulate, smooth margined, opposite and decussate. **Flowers** small silvery-green balls, usually borne singly or in clusters on short stalks at the leaf axil; 3 yellow stamens.

Habitat: Widespread but locally uncommon in damp areas of the floodplain and on channel margins.

Flowering: During the main rains and into the cool dry period.

Uses and beliefs:
• The plants must be highly palatable as they are heavily grazed and the upper shoots often have severe insect infestations.

FZ Vol.9 pt1, Ellery p.156.

Amaranthaceae (pigweed or cockscomb family)

Celosia trigyna L.

Derivation: *Keleos*, burning [Greek], alluding to the brilliant colours of some flowers of the genus; *tri*, three [Greek], *gyne*, female [Greek], alluding to the tripartite female organs of the flower.

Identification: An erect **annual herb** branching near the base, c. 60 cm tall; **hairless** overall. **Stems** square, tinged pink. Leafy **stipules** in the leaf axils curve around the stems. **Leaves** ovate, with smooth margins that are slightly decurrent, spiralling up the stem. Leaves tend to fall before the plant fruits. **Flowers** borne in spikes at the apex of the plant and in the axils, white to pinkish. There is an unpleasant strong green vegetable **odour** when the plant is crushed.

Habitat: Locally uncommon in the margin of the riverine forest.

Flowering: Flowers late in the main rains.

Uses and beliefs:
- Attractive to animals and frequently grazed.

FZ Vol.9 pt1.

Amaranthaceae (pigweed or cockscomb family)

Gomphrena celosioides Mart.

Common names: Setswana tsamai?, mositanoka; **Hmbukushu** diyanga; **Afrikaans** mierbossie; **English** batchelor's button, globe amaranth, prostrate globe amaranth.

Derivation: *Gomphrena*, the Latin name for a kind of amaranth; *celosia*, keleos, burning [Greek], may refer to the shape of the flowers which are like tiny candles.

Identification: A small **perennial herb** growing in thick tufts up to c. 15 cm tall. **Leaves** almost stalkless, oval with smooth margins, usually slightly hairy on both surfaces. **Flowers** very small with yellow anthers protruding from each floret, densely crowded in a tight silvery-white ball, c. 1.2 cm dia.

Habitat: An early introduction from South America now common throughout the region. Usually growing in clayey grassland areas in full sun near water.

Flowering: From the early rains into the cool dry period.

Uses and beliefs:
• Hmbukushu make a paste of crushed leaves and roots that is smeared around the umbilical cord of a newborn baby to aid healing.

FZ Vol.9 pt1, B van Wyk Flowers p.36, Blundell p.42. MT.

Amaranthaceae (pigweed or cockscomb family)

Guilleminea densa (Willd.) Moq.

Common names: Setswana mohulapitse; **English** carrot weed.

Derivation: *Guilleminea*, named for the French botanist Jean Baptiste Antoine Guillemin (1796–1842); *densa*, compact, dense [Latin].

Identification: A creeping **annual herb** forming a dense mat c. 50 cm across, with a long thick **taproot**. **Leaves** arranged in pairs, ovate with an oblique base, asymmetrical, smooth margins. **Flowers** borne in clusters at the leaf axils, whitish-green, small (c. 2 mm dia.); 5 yellow anthers. **Calyx** stiff, papery, often sticks into bare feet or catches fur or hair.

Habitat: This introduced, pan-tropical weed is found occasionally locally in areas of mixed riverine woodland. It may also be found growing along the pavements in town.

Flowering: During the main rains.

FZ Vol.9 pt1.

Amaranthaceae (pigweed or cockscomb family)

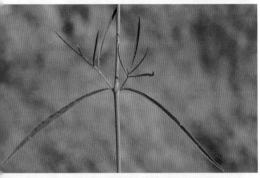

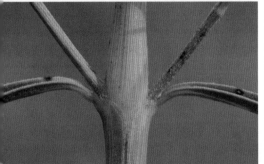

Kyphocarpa angustifolia (Moq.) Lopr.

Common names: Setswana mogatlawammutla, nnyoyammutla, mosonowamutha, ntshê; **Shiyeyi** mokera o shoro; **Herrero** okapi; **English** hare's tail bush, silky burweed.

Derivations: *Kyphocarpa*, from *cyphos*, bent or hunchbacked [Greek], *carpos*, fruit [Greek]; *angustifolius*, from *angusti*, narrow [Latin], *folius*, leaf [Latin]. *Mosonowamutla* means 'backside of the scrub hare'. *Nnyo-yammutla* is apparently a fairly rude reference to the female genitalia of the scrub hare. *Mokera o shoro*, also means tail of the scrub hare.

Identification: An erect grass-like **herbaceous annual**, up to c. 1 m tall, branching at the base. **Stem** lightly ribbed. **Leaves** opposite and decussate, stalkless (sessile), linear with smooth margins, c. 12 × 0.7 cm; tending to fall early as the plant ages. **Flowers** in terminal spikes, curving as they grow longer, c. 13 cm tall, 5 petals, white and green, densely hairy, with occasional spiky bracts.

Habitat: Widespread occasional plants in full sun on the open floodplain, in areas of deep sand, often heavily trampled.

Flowering: Main rains and well into the cool dry period.

Uses and beliefs:
• Ovambenderu make an infusion of the roots for stomach ache.
• Because they believe the hare to be the cleverest animal, some Subiya believe that if you rub this plant into a small cut on the forehead of a young child, he will become very intelligent.
• San use the roots as an aphrodisiac.

FZ Vol.9 pt1, Germishuizen p.73, Flowers Roodt p.25. BB, CM, PN.

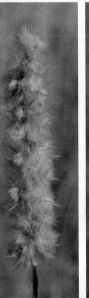

Amaryllidaceae (amaryllis or vlei-lily family)

Crinum harmsii Baker

Common names: Setswana mogaga.

Derivation: *Crinum*, a lily [Greek].

Identification: Large bulbous **lily** up to c. 85 cm tall. **Leaves** in a basal rosette, erect and linear with tufts of bristles along the margins, up to c. 125 × 4.5 cm. **Flower** head of c. 20 flowers is c. 28 cm dia. Flowers c. 15 cm dia., greenish-white, underside of petals pink blotched with green; stamens and stigma maroon. **Flower buds** hang downwards but turn upwards on opening allowing the petals to hang down limply, as if the flower is already wilting.

Habitat: Rare, found on open floodplain and amongst acacia scrub, in full sun.

Flowering: During the main rains.

Uses and beliefs:
• When a Batswana or San woman's husband or child dies, she will lie in the house until the funeral is over. When she leaves the house, she drops layers of a *Crinum* bulb as she walks around the village, especially in the goat and cattle kraals. This is because Batswana believe that female elephants, if they have a miscarriage or lose a baby, dig this bulb and throw it on the pathways. A decoction of this with other plants is also drunk by a widow to cleanse herself.

RM, KS.

Anthericaceae

Chlorophytum sphacelatum (Baker)
Kativu
subsp. *sphacelatum*
(syn. *Anthericum whytei* Baker)

Common names: Setswana tshuga.

Derivation: *Chloros*, green [Greek], *phyton*, plant [Greek]; *sphacelatum*, withered as if dead [Latin].

Identification: Tall rhizomous-rooted **perennial herb**, c. 80 cm tall. Grey-green, erect **leaves** mainly in a basal rosette but also at nodes up flower stems, strap-like, folded at the mid-rib. **Flowers** in single spikes, white with protruding yellow anthers. Flowers and fruit mixed all the way up the inflorescence. **Fruit** 3-lobed.

Habitat: Found only occasionally in grassy areas of light shade on islands and the spillway margin.

Flowering: In the middle of the main rains.

Turton p.139.

Anthericaceae

Chlorophytum sphacelatum (Baker) Kativu
subsp. *milanjianum* (Rendle) Kativu
(syn. *Anthericum milanjianum* Rendle)

Derivation: *Chloros*, green [Greek], *phyton*, plant [Greek]; *sphacelatum*, withered as if dead [Latin]; *milanjiana*, probably from Mt Mulanje in Malawi.

Identification: Perennial **herb** with a **rhizomous rooting** system, c. 70 cm tall. **Leaves** strap-like, folding slightly at the mid-rib, c. 30 × 1.2 cm. **Flowers** pure white, 3 petals and 3 tepals (sepals that look like petals); slightly pendant and mixed with the fruit along the stem; anthers yellow.

The yellow anthers of this plant, unlike those of *Chlorophytum sphacelatum* subsp. *sphacelatum*, do not protrude beyond the petals.

Habitat: Found occasionally locally in deeply shaded areas of mopane woodland.

Flowering: During the main rains.

Apocynaceae (oleander family)

Marsdenia macrantha (Klotzsch) Schltr.

Common names: Hmbukushu mayungu anyambi.

Derivations: *Marsdenia* for William Marsden, secretary to the British Admiralty, orientalist and traveller; *macrantha*, large-flowered [Greek]. *Mayungu anyambi*, pumpkin, because the fruit looks like a small pumpkin.

Identification: A **shrubby climber** from a **woody rootstock**, up to c. 2 m tall depending on support. **Bark** covered in splits that appear warty in texture. **Leaves** opposite, mainly on side shoots, elliptical, hairless, wavy margined, folding at the mid-rib and slightly recurved. **New growth** is covered in matted brown **hairs**. **Flowers** borne in clusters at the leaf axils, greenish white, c. 2.5 cm dia., strongly perfumed; 5 revolute, twisted and hairy petals and 5 alternate sepals make a double flat star. White **latex** in the cut stems.

Habitat: Found only occasionally in both full sun and light shade on islands, in the mopane woodland, and in the floodplain.

Flowering: From the early rains into the cool dry period.

MT.

Apocynaceae (oleander family)

Orthanthera jasminiflora (Decne.) Schinz

Common names: Setswana rapologwane; **Hmbukushu** kamongamba; **Subiya** mosamo womosana; **Ovambenderu** oruzenga; **Shiyeyi** manXatura; **Afrikaans** sandmelktou; **English** star jasmine.

Derivation: *Orthos*, straight [Greek], *anthera*, anthers [Greek]; *jasminiflora*, resembling jasmine flowers.

Identification: A mat-forming **creeping perennial** with stems up to c. 8 m long; covered in **harsh hairs**. **Leaves** narrowly cordate to oval, arranged in opposite pairs, with net veining, margins wavy and harshly hairy. **Flowers** borne in clusters on a short stalk at the leaf axils, star-shaped, white and green on the reverse becoming yellow with age, 5 narrow pointed lobes fuse to form a tube. Flowers open at sunset and are highly **perfumed**. **Fruit** green woody cylindrical pods narrowing to a point.

Habitat: Common and widespread in floodplain grassland.

Flowering: From the early rains into the cool dry period.

Uses and beliefs:
- In certain areas, Hmbukushu cook the young leaves and fruit and use them as *morogo*, a vegetable relish for pap or mealie porridge. In other areas, they peel the young fruit and eat the seeds raw.
- Ovambenderu eat the young pods.
- *Orthanthera jasminiflora* has a large root which Batswana boil in order to drink the resulting decoction (3 times per day) as a treatment for backache.
- Wayeyi women rub the sliced roots on their breasts to discourage children from suckling when they should be weaned. They also believe that eating the fruit will enhance their bust.
- If dogs are not keen to hunt, Wayeyi rub the roots on their teeth and they become aggressive and courageous.
- Hmbukushu use the young stems as string to tie bundles.
- Young Ovambenderu boys blow the seeds in the air as a toy.
- The roots are also much enjoyed by elephant and ostrich.

B Van Wyk Flowers p.38. BB, GM, MM, RM, SS, MT.

Arecaceae (palm family)

Hyphaene petersiana Klotzsch ex Mart.

Common names: Setswana mokolwane, mokolane, mokulane (the fruit is mokonkolwane); **Subiya** kakoma; **Shiyeyi** ongawu; **Afrikaans** nordelike lalapalm; **English** real fan palm, lola palm, ilala palm, northern ilala palm, northern lala palm, vegetable-ivory palm.

Derivations: *Hyphaino*, to entwine [Greek], alluding to the fibres encasing the fruit; *petersiana* in honour of the 19th century German botanist, Professor W.C.H. Peters. *Ilala*, is the Zulu name for the tree. It means 'sleeping' and is probably the result of over-indulgence with palm wine.

Identification: A medium-sized **palm tree** with a single trunk, c. 15 m tall. **Trunk** patterned with the scars of the fallen leaves. **Leaves** fan-shaped having long stalks with recurved spines. **Male and female flowers** grow on separate trees; **female flowers** cream coloured, hanging down in large branching sprays; **male flowers** a strong yellowy green, also borne in long sprays that are retained on the tree. **Fruit** spherical, dark reddish-brown.

Habitat: Common and widespread on islands in the floodplain. This species is said to be an indicator of high salinity.

Flowering: Before the early rains.

Uses and beliefs:
- The bases of the young leaf shoots are used as a vegetable known locally as *gau*.
- The milk present in the young fruit is similar to coconut milk and is relished by local people.
- The pulp below the hard outer casing of the fruit is said to resemble gingerbread.
- Both *Hyphaene petersiana* and *Phoenix reclinata* (the wild date palm) are sources of palm wine (*muchema* in Shiyeyi and *malovo* in Subiya), but the extraction of the sap kills the tree. Up to 60 litres of sap may be produced by an individual tree. It is very refreshing, slightly alcoholic and tastes like ginger beer. If fermented for up to 36 hrs it becomes very intoxicating. Palm wine is distilled to make a highly potent spirit.
- The roots are boiled and the resulting decoction is used to treat coughs and by men (only) to clean the kidneys. It is very bitter and just one dose is sufficient.
- The major species used in the basket industry in Botswana. The young unopened leaves are stripped to their individual fibres and used to weave baskets. Wayeyi and Subiya women boil the leaves with maize meal to bleach them white, with *Berchemia discolor* (*motsentsela*) to colour them brown, with *Euclea divinorum* (*motlhakola*) for dark brown-black and with *Sansevieria aethiopica* and *Aloe sp.* for yellow.
- The inner part of the ripe fruit resembles vegetable ivory and although quite small is carved to make trinkets.
- The leaf stems are used as fire wood.
- The palm leaf hearts and fruit are very popular with elephants and baboons.

Ellery p.122, Hargreaves p.76, Palgrave updated p.99, Setshogo & Venter p.4, Tree Roodt p.15. CM, GM, BN, PN, KS.

Asparagaceae (asparagus family)

Asparagus africanus Lam.
(syn. *Protasparagus africanus* (Lam.) Oberm.)

Common names: Setswana tshobatshobane, mositwasitwane, mositwane, mhalatsamaru; **Hmbukushu** chovachova; **English** catbush, wild asparagus, sparrowgrass.

Derivation: *Aspharagos*, edible asparagus [Greek], originally from *sparasso*, to tear or cut, referring to the sharp thorns on the stems; *africanus*, of Africa.

Identification: A rambling, climbing **perennial herb** up to c. 2 m tall, from a fibrous root-stock. **Stems** have whorls of what look like needle-like leaves, c. 1 cm long. These are actually **cladophylls**, sections of stem that perform the function of leaves. There are sharp hooked **spines** at the nodes. **Flowers** small (c. 3 mm dia.), white with green lining, 5 petals. **Fruit** spherical, bright red.

Habitat: Locally common on islands and in the treeline, usually growing in full sun.

Flowering: Towards the end of the early rains and at the beginning of the main rains.

Uses and beliefs:
- The young shoots are boiled and eaten as a vegetable.
- This plant has many medicinal uses as it has strong antibacterial and anti-microbial properties.
- Batswana mix *Cyathula orthacantha* and *Asparagus africanus* to neutralise snake, scorpion and spider bites. They also drink an infusion of shoots to treat stomach problems.
- Wayeyi make an infusion of the shoots to treat bladder and kidney problems.
- It is also said to be an aphrodisiac.
- Hmbukushu take the tuberous roots of asparagus, pound them and leave them to soak overnight to provide soap for washing.

Blundell p.420, Turton p.123, Flowers Roodt p.35. BB, MM.

Asparagaceae (asparagus family)

Asparagus nelsii Schinz

Common names: Setswana mhalatsamaru, choba-choba, diyanambo; **English** sandveld asparagus, wild asparagus.

Derivation: *Aspharagos*, edible asparagus [Greek], originally from *sparasso*, to tear or cut, referring to the sharp spines on the nodes.

Identification: A shrubby, climbing **perennial herb** from a fibrous root-stock, up to c. 1.5 m tall. The stems have dense whorls of what look like **needle-like leaves**, c. 2 cm long, that make the plant look rather shaggy overall. These are actually **cladophylls**, sections of stems that perform the function of leaves. There are sharp short **spines** at the nodes. **Flowers** small, white, with 5 petals; stamens long and protruding with yellow anthers. **Fruit** spherical, bright red.

Habitat: Widespread in damp areas in acacia scrub with seasonal pans.

Flowering: Flowers early in the main rains.

Uses and beliefs:
• The people of Botswana do not necessarily differentiate between the different species of *Asparagus* and may use them all as much as they use *Asparagus africanus*.
• The roots are mixed with other herbs and boiled to treat kidney problems.
• If you burn *Asparagus nelsii* when it is cloudy, the rain will go away (used for ploughing or parties, especially the Christmas choir gathering).

KS.

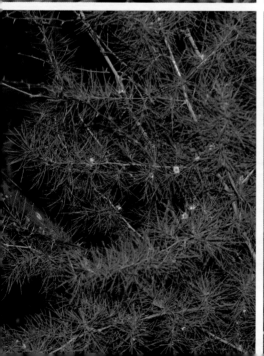

Asparagaceae (asparagus family)

Sansevieria aethiopica Thunb.

Common names: Setswana mokgôtshe-ô-monamagadi, mokgotshe, mogotse, mosokelatsebing, mosokelatsebeng, mosokalatshebe, makgolela-a-dinaka-tsa-pudi; **Kalanga** anaka; **Shiyeyi** ompopusa; **Afrikaans** aambeiwortel; **English** mother-in-law's tongue, bowstring hemp.

Derivations: *Sansevieria* for Pietro Sanseverino, Prince of Bisignano, in whose garden near Naples a plant of this genus was found growing; *aethiopica*, African, usually southern African. *Mokgôtshe-ô-monamagadi*, the female Mokgôtshe. *Mosokelatsebeng*, to twist into the ear, alluding to the medicinal use of the roots in the Serowe area where pieces of root are twisted into the ear to treat earache. *Tsa-pudi*, for goats. *Aambeiwortel*, haemorrhoid bush.

Identification: A **succulent perennial**, up to c. 50 cm tall, with a **rhizomous rooting** system. **Leaves** linear, hard and erect, smooth margined, grey-green with grey-brown horizontal blotches, 'U'-shaped in cross-section. **Flowers** in spikes growing from a basal rosette of spiky leaves, individual flowers in groups spiralling up the stem, creamy white, c. 3 cm dia., corolla tubes c. 2.5 cm long. **Fruit** spherical, red berries.

Habitat: Widespread in deep shade in well-drained areas of mixed woodland, preferring termite mounds.

Flowering: During the early rains.

Uses and beliefs:
- San use the roots as a source of water in dry areas, although they have a rather bitter taste.
- Wayeyi cook the leaf slowly in the fire and leave it to cool. They then squeeze the juice into the ear to treat earache. San use the juice of the leaves without cooking them for the same purpose.
- Widely used to make string and rope. The leaves are crushed and soaked in water for a month to clean the fibres. The resultant string and rope is used to make snares for animals, bow-strings, mats and baskets.
- The roots of *Sansevieria aethiopica* are boiled with the leaves of *Hyphaene petersiana* to make a yellow dye for baskets.

B Van Wyk Flowers p.70, Ellery p.174, Flowers Roodt p.79. RM, MMo, PN, KS.

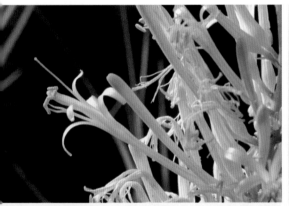

Asparagaceae (asparagus family)

Sansevieria pearsonii N.E.Br.

Common names: Setswana mokgotshi, mokgotshe ômontonanyana; **San** gwi; **Afrikaans** bobbejaan-sedood; **English** spiky mother-in-law's tongue, gemsbok horn, stiff bowstring.

Derivations: *Sansevieria* for Pietro Sanseverino, Prince of Bisignano, in whose garden near Naples a plant of the genus was found growing; *pearsonii*, for the English botanist Professor Henry Harold Welch Pearson (1870–1916), founder of the South African National Botanic Gardens, Kirstenbosch. *Mokgôtshe ômontonanyana*, the male mokgôtshe. *Gemsbok horn*, because the shape of the plant resembles the gemsbok's horns.

Identification: Perennial succulent, up to c. 1 m tall. **Leaves** erect, pointed, fibrous, with a curved cross-section, growing from a tuberous **rooting** system. **Flowers** on an erect basal raceme, turning from white or green to fawny pink with age; 5 strongly recurved petals fused to form a tube; stamens long and protuberant.

Habitat: Widespread in shaded riverine woodland, often associated with termite mounds as it favours well-drained soils.

Flowering: Early in the main rains.

Uses and beliefs:
• San squeeze the juice from the stems as a treatment for earache.
• The fibrous leaves are widely used for making string (e.g. cords for bows and fishing nets).

Flowers Roodt p.79, Pooley p.100, WFNSA p.42. BB, PN.

Asteraceae (daisy family)

Blainvillea acmella (L.) Philipson

Common names: English paniculated spot flower, para cress, Paraguay cress, spot flower.

Derivation: *Blainvillea*, for the French biologist Henri Marie Ducrotay de Blainville (1777–1850); *acme*, the highest point [Greek].

Identification: Erect branching **annual herb**, up to c. 1.6 m but more usually c. 0.8 m tall; covered in rather sparse **hairs** overall. **Stems** blotched brown and green. **Leaves** opposite and decussate, ovate with a tapering point, hairy on both surfaces, margins are sparsely toothed, with 3 main veins and net veining. **Flower** heads borne in groups at the apex of the plant on long individual stalks from the leaf axils, compound of insignificant, white disc florets.

Habitat: Common and widespread in riverine woodland, especially, mopane woodland.

Flowering: During the main rains.

Asteraceae (daisy family)

Eclipta prostrata (L.) L.

Common names: English eclipta.

Derivation: *Ekleipo*, to cease, to stop, to be deficient [Greek], alluding to the absence of a pappus, the fluffy part of the fruit; *prostrata*, flat on the ground, prostrate [Latin].

Identification: An erect much-branched **herb**, c. 50 cm tall; covered in harsh short hairs. **Stems** pink. **Leaves** opposite and decussate, narrowly elliptical, stalkless, margins are slightly toothed. **Flowers** borne in small clusters on stalks at leaf axils and from the apex of the plant, white, composite with both ray and disc florets; phyllaries (scale-like sepals of *Asteraceae*) unusual being almost egg-shaped and longer than the ray florets. The flower heads become hard green discs of **seeds** without a pappus.

Habitat: Uncommon on the margin of seasonal pans. [Naturalised.]

Flowering: During the main rains.

Bombacaceae (baobab family)

Adansonia digitata L.

Common names: Setswana mowana, muwane, moana, mobuyu, mbuyu, ibozu; **San** nium, emm; **Hmbukushu** divuyu; **Subiya** ivozu; **Afrikaans** kremetartboom; **English** baobab, upside-down tree.

Derivations: *Adansonia*, for the 18th century French botanist Michel Adanson; *digitatus*, digitate, shaped like an open hand [Latin]. *Kremetartboom*, cream of tartar tree, because of its presence in the fruit. *Baobab*, 'bu hobab', the name given to the tree by Egyptian merchants in the 16th century.

Identification: Probably the best known African **tree**, short but rather fat, up to c. 15 m tall and c. 28 m in circumference but the few specimens remaining in the area are smaller. **Leaves** compound, palmate with up to 5 oval leaflets, with smooth margins and net veining; borne in whorls at the apex of the twigs. **Flowers** pendulous, large, 5 overlapping crinkled white petals, 5 greenish-white hairy sepals; many stamens on a fused central column. Probably pollinated by bats.

Habitat: Uncommon but widespread on islands and along the forest margin of the spillway.

Flowering: Before and during the rains.

Uses and beliefs:
- Fruit and seeds are eaten and a rich source of vitamin C.
- Hmbukushu and San eat the fruit either straight from the tree or chopped with fresh milk. They also collect the young leaves to cook as *morogo*, a vegetable relish.
- The seeds can be roasted and used as a substitute for coffee.
- Subiya women chew the bark and use the resulting decoction to wash babies for the first 3 months of life. It is believed that these children will grow strong and healthy. If children get very thin, the Batswana cut the bark and tie it round the child's waist so that they will grow strong, like the tree.
- "Before cutting the tree, you should pray to your ancestors because they live in it. If you do not, they will become angry and you will go mad."
- The fibrous bark is pounded for use in making rope and mats.
- Hugely attractive to elephants, which strip the bark and eat branches (elephant damage shown in bottom left image).

FZ Vol.1 pt2, Hargreaves p.13, Palgrave p.587, Setshogo & Venter p.35, Tree Roodt p.74. CM, GM, RM, BN, RR.

Boraginaceae (forget-me-not family)

Heliotropium ciliatum Kaplan

Derivation: *Helios*, the sun, *trope*, to turn [Greek], from an old disproved idea that the flowers turned with the sun; *ciliatum*, fringed with hairs, like eyelashes [Latin].

Identification: Small dark, green, erect branching **perennial herb**, c. 25 cm tall, growing from a woody base. **Leaves** alternate, ovate with a rounded tip and wavy margins. **Flowers** borne in 1-sided terminal spikes that curl at the tip like a scorpion's tail, white with dark green or brown centre; lobes star-shaped with elongated tips fused at the base to form a tube.

Habitat: Found occasionally locally growing in mixed sandy open woodland.

Flowering: During the main rains.

FZ Vol.7 pt4, B Van Wyk Flowers p.46.

Boraginaceae (forget-me-not family)

Heliotropium ovalifolium Forssk.

Common names: Setswana motlhatswapelo; **English** forget-me-not, grey-leaf heliotrope.

Derivations: *Helios*, the sun [Greek], *trope*, to turn [Greek], from an old disproved idea that the flowers turned with the sun; *ovali*, oval [Latin], *folium*, leaf [Latin]. *Motlhatswapelo* means to wash the heart.

Identification: Lax plant sometimes upright, if supported in other vegetation up to c. 45 cm tall; silver-green because of a covering of white **hairs** of variable length. **Leaves** alternate, narrowly oval, margins entire, c. 4 cm (inc. c. 1 cm petiole) × 1 cm, slightly recurved with net veining. **Inflorescence** coils to look like a scorpion's tail of buds at apex of branches. **Flowers** white, often with yellow centres, c. 3 mm diameter in pairs on one side of the stalk.

Habitat: Widespread in years of heavy rain throughout the floodplains. If the weather is dry, then found only occasionally near water.

Flowering: From the early rains into the cool dry period.

FZ Vol.7 pt4. BN.

Boraginaceae (forget-me-not family)

Heliotropium strigosum Willd.

Derivation: *Helios*, the sun [Greek], *trope*, to turn [Greek], from an old disproved idea that the flowers turned with the sun; *strigosum*, with stiff bristles [Latin].

Identification: Small, bright green, erect, branching **annual herb**, up to c. 20 cm tall; covered in silvery harsh **bristles**. **Leaves** alternate, linear with entire margins. **Flowers** borne alternately on 1-sided terminal spikes that curl slightly at the tip like a scorpion's tail; white with a yellow centre; lobes star-shaped and fused to form a tube.

Habitat: Found occasionally locally in grassy damp areas with seasonal pans, not actually on the pan margin.

Flowering: Early in the main rains.

FZ Vol.7 pt4.

Boraginaceae (forget-me-not family)

Heliotropium supinum L.

Derivation: *Helios*, the sun [Greek], *trope*, to turn [Greek], from an old disproved idea that the flowers turned with the sun; *supinum*, prostrate [Latin].

Identification: Prostrate, silver-grey **annual plant**, up to c. 1 m dia. **Leaves** alternate, ovate, base oblique, margins entire; veins deeply incised into upper surface of the leaf and appearing almost laid onto the surface below; upper surface covered in a silvery down. Leaf underside, stems and flower spikes covered in longer hairs. **Flowers** arranged in tightly curled double rows c. 7 cm long, emerging from a hairy calyx, minute (c. 1 mm dia.), white, 5 lobes.

Habitat: Widespread in the clay of dried-out seasonal pans.

Flowering: Throughout the cool dry period and into the early rains until the seasonal pans are flooded.

FZ Vol.7 pt4.

Capparaceae (caper family)

Capparis tomentosa Lam.

Common names: Setswana motawana, modyangwe; **San** motawana; **Subiya** tchuvongololo; **English** woolly caper bush.

Derivations: *Capparis*, the ancient Greek name for these shrubs; *tomentosa* densely woolly, with matted hairs [Latin]. *Motawana* means 'young lion' (the thorns look like lion's claws).

Identification: Very variable **perennial climbing shrub**, up to c. 10 m tall depending on support. Second- and third-year **wood** remaining quite green. **Leaves** oval, thick and woolly, with net veining, tending to fold at the mid-vein, variable size (c. 3–6 cm long). Some plants with vicious **hooks** in pairs below leaves (even on new wood) but most develop thorns only on second- and third-year wood. **Flowers** produced in groups at the leaf axils, white, asymmetrical; with 2 sepals and 2 whorls of green petals, the outer whorl with 2 petals and the inner whorl with 4 petals (2 petals of which are fused); a mass of pink-tinged stamens; ovary green, c. 5 cm long, emerging on a long stalk from among the stamens. The flowers have a spicy **aromatic perfume** and last only 1 day. **Fruit** lemon-shaped, mainly with a shiny surface but some rough, becoming dark orange as they ripen.

Habitat: Common and widespread growing up trees, often on termite mounds along the treeline and on islands. This species may also form patches of scrub in seriously over-grazed areas.

Flowering: During the cool dry period.

Uses and beliefs:
- The plant is strongly antiseptic and has many medicinal uses.
- If serious problems arise between two people, e.g. one has stolen the other's cows or has sent *tokolosi* ('evil spirits') to cause trouble, then the offended person pounds the roots with rock salt. He then uses the mixture to make a line in front of his door and a cross on his bed. He ties another piece of root to the roof tree of his house. Then, he adds ash from a tree that has been struck by lightning to the mixture to send thunder and lightning to kill his enemy.
- San place branches of the plant around the door of their house to keep away *tokolosi* and to help peaceful sleep.

FZ Vol.1 pt1, Ellery p.100, Flowers Roodt p.57, WFNSA p.146. GM, MM, RM, BN.

Capparaceae (caper family)

Cleome gynandra L.
(syn. *Gynandropsis gynandra* (L.) Briq.)

Common names: Setswana rothwe, leketa, rotho; **Kalanga** nyevi; **San** seshungwa; **Subiya** sinshungwa; **Afrikaans** snotterbelletjie; **English** spider wisp, cat's whiskers, African herbage, bastard mustard.

Derivation: *Cleome* from Cleoma, a name used in the Dark Ages for a strong-tasting plant growing in a damp place; *gynae*, female [Greek], *andros*, male [Greek], referring to the union of the male and female parts of the flower.

Identification: Erect, bushy, annual **herb**, up to c. 70 cm tall. **Leaves** palmately compound with 5 oval leaflets. **Flowers** with white petals each elongated to form a stalk; brown-red reproductive organs borne prominently on a long stalk. **Fruit** long, narrow tubular pods. The plant has a distinctive mustard **odour** when crushed.

Habitat: Found occasionally in lightly shaded areas of the riverine woodland.

Flowering: During the main rains.

Uses and beliefs:
- The leaves are used to make *morogo* (vegetable relish). It has to be cooked a long time and tastes like mustard spinach but is always rather sandy. If it is cooked and then dried, it can be stored for some years. "Always tear the leaves from the plant, do not pinch them as this will make the *morogo* hot, like chilli."

FZ Vol.1 pt1, B Van Wyk Flowers p.48, Blundell p.27. RM, BN, KS.

Caryophyllaceae (carnation family)

Polycarpaea eriantha Hochst. ex A.Rich. var. *effusa* (Oliv.) Turrill
(syn. *Polycarpaea corymbosa* (L.) Lam. var. effusa Oliv.)

Derivation: *Poly*, many [Latin], *carpus*, fruit [Latin], with many fruit; *corymbosa*, with flowers arranged in a corymb.

Identification: Erect, silvery-green, **annual herb** with sparse branches, up to c. 45 cm tall. **Stems** covered in fine white **hairs**. **Leaves** linear, borne in well-spaced whorls up the stem. **Flower buds** silvery. **Flowers** borne in rather flat clusters at the apex of the plant, minute (c. 2 mm dia.), sepals white and papery, petals orange and stamens yellow.

Habitat: Found occasionally locally on the floodplain in areas of clayey sand. [Introduced.]

Flowering: During the main rains.

FZ Vol.1 pt2, B Van Wyk Flowers p.272.

Combretaceae (bushwillow or combretum family)

Combretum mossambicense
(Klotzsch) Engl.

Common names: Setswana motsheketsane, motsweketsane; **Lozi** mubesuba, chiromiundi; **San** drui; **Subiya** motombolo; **Afrikaans** knoppiesklimop; **English** shaving brush combretum, knobbly combretum, Mozambique combretum, knobbly creeping bushwillow.

Derivations: *Combretum*, the Latin name for a climbing plant but not of this genus; *mossambicense*, of Mozambique. 'Knobbly', refers to the rigid knobs or spines on the stems.

Identification: Scrambling shrub, usually c. 2.5 m tall but may climb to c. 10 m with good support. Downward-facing **blunt pegs** or **spines** develop from the stalks of fallen leaves which remain on the plant, usually in pairs but sometimes threes. **Leaves** opposite, obovate, oval or oblong, smooth margined, hairy on both surfaces, with net veining, veins deeply incised on the upper surface and prominent on the lower surface. **Flowers** in dense clusters, borne on bare branches at the leaf nodes, white with long stamens and red or pink anthers, sweetly scented. **Fruit** different from those of all other Combretaceae in that they are 5-winged, although they can be 4-winged; wings papery, dark golden brown to straw-coloured.

Habitat: Locally common in sandy soils on islands and along the treeline of the floodplain.

Flowering: This species and *Acacia nigrescens* are the first plants to flower at the end of the cool dry period, well before the early rains.

Uses and beliefs:
- The roots are used to treat back-ache.
- *Combretum mossambicense* is mixed with *Dichrostachys cinerea* and *Ximenia caffra* to make a preparation that will make young boys strong when they start to be active sexually. A mixture of the roots is placed in a container with water and the preparation drunk each morning. To increase the effect, the roots are boiled with fresh milk from a cow that has had a male calf. Boys at the cattle post would drink this concoction twice daily.
- Kalanga use the wood as pestles for stamping mealie.
- San use the dry wood as fire sticks.
- Wayeyi attach a metal hook to the end of long stems and use this tool to catch spring hares in their burrows.

FZ Vol.4 pt0, Ellery p.105, Hargreaves p.34, Palgrave p.672, Setshogo & Venter p.48, Tree Roodt p.89, Curtis & Mannheimer p.482. GM, RM, KS, SS.

Combretaceae (bushwillow or combretum family)

Pteleopsis myrtifolia (M.A.Lawson) Engl. & Diels

Common names: Setswana moanzabalo, mofungi, motindi; **Afrikaans** stinkboswilg; **English** pteleopsis, (two-winged) stink bushwillow, two-winged pteleopsis, myrtle bushwillow.

Derivation: *Ptelea*, elm tree [Greek], *opsis*, aspect, appearance [Greek], hence resembling an elm tree; *myrti*, myrtle [Latin], *folia*, leaf [Latin].

Identification: A branching, many trunked, densely leaved **shrub**, c. 4 m tall. **Trunk** with large pale grey lichen blotches on darker grey. Very variable in the appearance of both leaves and fruit as this species hybridises freely with *Pteleopsis anisoptera*. **Leaves** c. 7 × 2.5 cm, dark green, leathery, either opposite or alternate, turning to lie in the same orientation, oval, margins smooth or occasionally with a fine hair fringe. Young leaves may be **hairy** and these hairs are sometimes retained on the underside of the leaf. **Flowers** in balls (c. 2 cm dia.) borne at the leaf axils, small (c. 0.6 cm dia.), cream coloured, with a strong unpleasant **scent**. **Fruit** 2 or occasionally 3-winged, greenish-yellow drying to light brown. **Wood** red and very hard.

Habitat: Locally common in evergreen riverine bush along the Chobe River.

Flowering: Early in the main rains.

Uses and beliefs:
- The wood is good for furniture making.

FZ Vol.4 pt0, Palgrave p.678, Palgrave updated p.814, Setshogo & Venter p.48.

Combretaceae (bushwillow or combretum family)

Terminalia sericea DC.

Common names: Setswana mogonono, mokuba, moshosho, nsuru-ntukunu, dzau, mobonona; **Kalanga** nsusu; **Ovambenderu** omuseasetu; **Lozi** muhorono; **Hmbukushu** ghushosho; **Shiyeyi** uXhouwa; **Afrikaans** geelhout, vaalboom; **English** silver terminalia, silver cluster-leaf.

Derivation: *Terminus*, end [Latin], referring to the leaves being clustered at the end of the stems; *sericeus*, silky [Latin], referring to the long straight silky hairs on the leaves.

Identification: Usually a small **shrub** but can grow to be a substantial **tree**, c. 15 m tall, with a spreading crown. **Bark** either dark or light grey and deeply ridged longitudinally; freshly torn bark is pale rusty-red. **Leaves** blue-green with a silvery sheen from their covering of silky hairs, narrowly obovate to elliptical; borne spirally around and at twig terminals. **Flowers** silky, beige or white, borne in axillary spikes; an unpleasant nauseous odour. **Fruit** flattened, oval, pinky-red, with a papery edge, single-seeded. Newly cut **wood** is bright yellow.

Habitat: Found on deep sand throughout the region in either bush or tree form.

Flowering: From just before the early rains to well into the main rains.

Uses and beliefs:
- Batswana use the bark as an antidote to arrow poison.
- Batswana apply a powder made from dried bark or the roots to infected wounds. The roots are also pounded and soaked in water to make an antiseptic infusion used for bathing infected skin or as an eye-wash.
- Pounded and mixed with water, the plant is also used to cure animal skins.
- Subiya use the wood as handles for ploughs. It is also used for *mokoro* poles.
- San use the flexible young branches to make snares.
- In the Tsodillo area, the wood is used to make spears.
- Both Lozi and the Subiya use the bark to make cord.
- The caterpillars of *Hamanumida daedalus* butterflies feed on this species.

FZ Vol.4 pt0, B&P van Wyk Trees p.174, Curtis & Mannheimer p.498, Ellery p.135, Germishuizen p.276, Hargreaves p.35, P Van Wyk Kr Trees p.181, Palgrave p.684, Setshogo & Venter p.49, Tree Roodt p.93. Kit, PK, GM, PN, KS.

Convolvulaceae (morning glory family)

Astripomoea lachnosperma (Choisy) A.Meeuse

Derivation: *Astri*, star-like [Greek], *ips*, worm [Greek], *homoios*, like [Greek], referring to the trailing habit of other ipomoea; *lachni*, woolly or downy [Greek], *sperma*, seeded [Greek].

Identification: Downy, erect, **annual herb**, up to c. 50 cm tall; covered in velvety hairs overall. **Leaves** alternate, broadly oval, with wavy margins on the upper part and entire below; densely covered in stellate hairs and paler in colour below. **Flowers** borne in stalked clusters at leaf axils and at the apex of the plant, white with a magenta centre, c. 2 cm dia., with 5 pointed lobes fused to form a flattened funnel shape.

Habitat: Uncommon on partially shaded track sides in dry mixed woodland.

Flowering: During the main rains.

FZ Vol.8 pt1, Blundell p.192.

Convolvulaceae (morning glory family)

Convolvulus sagittatus Thunb.

Common names: English wild bindweed.

Derivation: *Convolvo*, to twine around [Latin]; *sagittatus*, arrow-shaped [Latin].

Identification: A **creeping plant** with stems c. 50 cm long. **Leaves** linear with 2 small lobes at the base, which may be divided into 2 or 3 points. **Flowers** cone-shaped, with 5 pointed lobes, white sometimes with centre tinged pink or maroon.

Habitat: Found occasionally locally in full sun on the floodplain and on islands.

Flowering: Late in the early rains and throughout the main rains.

Uses and beliefs:
• San boil the roots and drink one dose of the resulting decoction to treat venereal diseases.

FZ Vol.8 pt1, B van Wyk Flowers p.52. RM.

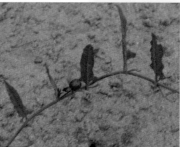

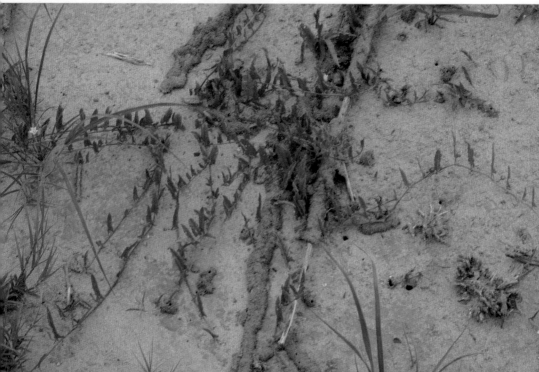

Convolvulaceae (morning glory family)

Falkia oblonga C.Krauss.

Derivation: *Falkia*, for Johan Peter Falk (1733–74), botanist, traveller and pupil of Linnaeus; *oblonga*, oblong [Latin], referring to the shape of the leaves.

Identification: A dwarf mat-forming **perennial herb**, c. 5 cm tall. **Leaves** stalked, growing closely together almost as a basal rosette, covered in fine harsh **hairs**, base and apex rounded. **Flowers** borne singly at the leaf axils, white with a faint pink line on the reverse of the lobes, 5-lobed. The flowers open during the morning when warmed by the sun and close as the sun goes off them.

Habitat: Widespread on the floodplain in full sun on areas of hard clay soil with little other vegetation.

Flowering: Throughout the early and the main rains. It is often the first flower to appear after the first showers.

FZ Vol.8 pt1, B Van Wyk Flowers p.52.

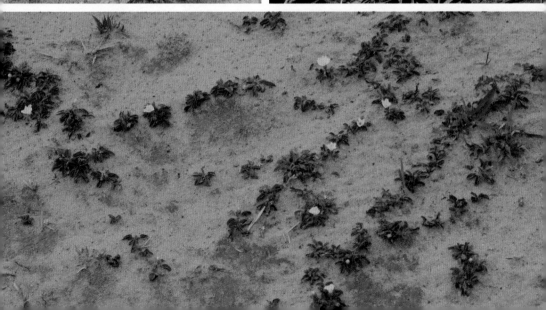

Convolvulaceae (morning glory family)

Ipomoea chloroneura Hallier f.

Derivation: *Ips*, worm [Greek], *homoios*, like [Greek], referring to its trailing habit; *chloros*, green [Greek], *neuron*, vein or nerve [Greek], green-veined.

Identification: Scrambling **annual herb**, branching from the base of the stem, some branches raised, others prostrate, c. 1 m dia.; covered with golden brown **hairs** apart from the leaves. **Leaves** narrowly obovate, margins entire, spiralling up stem, underside silver, upper side darker green. **Flowers** in small groups on long stalks, surrounded by leafy bracts, funnel-shaped, white, some with a magenta throat, corolla lobes hairy along the mid-line. **Latex** from the broken plant is sticky.

Habitat: Widespread growing in full sun in clearings in mopane woodland.

Flowering: During the main rains.

Uses and beliefs:
- San use this plant as medicine for the fontanelles (*phogwana*) of babies. If the baby vomits and has a distended stomach and constipation, they burn the green plant over coals on a crock and let the child inhale the smoke. Then mix the ashes with petroleum jelly and make a cross on the top of the child's head, on the stomach and on the back.

FZ Vol.8 pt1. KS.

Convolvulaceae (morning glory family)

Ipomoea coptica (L.) Roem. & Schult. var. *coptica*

Common names: Setswana kgane; **English** hand-palm ipomoea.

Derivation: *Ips*, worm [Greek], *homoios*, like [Greek], referring to its trailing habit; *coptica* possibly from the Greek *koptos*, chopped small, referring to the leaves.

Identification: Trailing, creeping **annual herb**. **Stems** c. 60 cm long, ribbed and twisted. **Leaves** palmate, 5-lobed, margins deeply serrated to deeply incised. Pairs of **stipules** of similar form to leaves at leaf axils. **Flowers** white sometimes with a magenta centre, lobes fused to form a funnel, c. 2 cm dia., on a c. 5 cm (sometimes up to c. 9 cm) flower stalk. White **latex** present in cut stems.

Habitat: Common in deeply sandy areas in full sun throughout the region.

Flowering: Middle to end of the main rains.

FZ Vo.8 pt1, Flowers Roodt p.65.

Convolvulaceae (morning glory family)

Ipomoea leucanthemum (Klotzsch) Hallier f.

Derivation: *Ips*, worm [Greek], *homoios*, like [Greek], referring to its trailing habit; *leukos*, white [Greek], *anthemon*, flower [Greek].

Identification: Small **creeper** with tips of stems erect, c. 50 cm long; covered in a tangled mass of soft **hairs**. **Leaves** alternate, oblong, base deeply cordate to accommodate the flowers that sit in the leaf axil, tip apiculate, margins entire, c. 4 cm (inc. c. 1 cm leaf stalk) × 1.8 cm. **Flowers** white occasionally with a blush of pink in throat, sessile in leaf axil, c. 7 mm dia. No latex in cut stems.

Habitat: Common in mopane woodland and in sandy areas of the floodplain.

Flowering: Middle to end of the main rains.

FZ Vol.8 pt1.

Convolvulaceae (morning glory family)

Ipomoea magnusiana Schinz var. *magnusiana*

Common names: Setswana motsididi.

Derivation: *Ips*, worm [Greek], *homoios*, like [Greek], referring to its trailing habit; *magnus*, great [Latin], big, *-iana*, belonging to, pertaining to [Latin].

Identification: Creeping twining herb with **annual stems** from a woody **perennial taproot**. **Stems** c. 1 m long, short harsh hairs. **Leaves** palmately compound with 5 deep lobes, margins wavy, tips apiculate; upper side with short harsh hairs; underside covered in a mass of dense woolly silvery **hairs**; veins on the underside of the leaves have harsh yellowish hairs, which make them distinct. **Flowers** in groups of up to three on c. 14 cm long stalks, usually white with faint pink markings in the throat but may vary through pale blue-mauve to magenta; corolla funnel-shaped; bracts just below the flower head. **Latex** is present.

Habitat: Found occasionally locally growing in full sun in deep sand.

Flowering: During the main rains.

FZ Vol.8 pt1, B Van Wyk Flowers p.200, Germishuizen p.85, Turton p.45.

Convolvulaceae (morning glory family)

Ipomoea plebeia R.Br. subsp. *africana* A.Meeuse

Derivation: *Ips*, worm [Greek], *homoios*, like [Greek], referring to its trailing habit; *plebeia*, of the people [Latin], i.e. ordinary or plain; *africana*, of Africa.

Identification: Small **climber** 1–2 m tall depending on rainfall and support. Sparse long **hairs** along stems. **Leaves** alternate, cordate with hairs along veins on both surfaces, margins slightly wavy, c. 7.2 cm (inc. c. 1.5 cm leaf stalk) × 3.2 cm. **Flowers** c. 1 cm dia., in pairs at leaf axil, white, funnel-shaped with acute tips to the lobes, 5 long sepals, short flower stalks (c. 0.4 cm). White **latex** when cut.

Habitat: Widespread in years of heavy rainfall in full sun and partial shade on sandy soil on the floodplain.

Flowering: Middle to end of the main rains.

FZ Vol.8 pt1.

Convolvulaceae (morning glory family)

Ipomoea sinensis (Desr.) Choisy subsp. ***blepharosepala*** (A.Rich.) A.Meeuse

Common names: Kalanga and **?Setswana** modandanyane; **English** purple-throated ipomoea.

Derivations: *Ips*, worm [Greek], *homoios*, like [Greek], referring to its trailing habit; *sinensis*, Chinese [Latin]; *blepharon*, eyelid [Greek], *sepala*, sepalled [Greek], sepals fringed like eyelashes. *Modandanyane*, possibly 'the climbing one', a generic name for plants that climb.

Identification: A creeping, climbing **annual herb**. **Leaves** soft, hairless, alternate and cordate with margins entire; leaf tips bristled. **Flowers** borne on long stalks, white with a magenta centre and slight magenta veining; pairs of small **bracts** c. 1 cm below the flowers.

Habitat: Widespread and common in clayey sand in areas with pans and waterways.

Flowering: Early in the main rains.

Uses and beliefs:
• Kalanga and Batswana mix the young leaves of this plant with others for *morogo*, a vegetable relish.
• The stems are rolled to form a pad, a *kgare*, to protect the head when carrying firewood.

FZ Vol.8 pt1, Flowers Roodt p.64. KS.

Convolvulaceae (morning glory family)

Ipomoea sinensis (Desr.) Choisy subsp. *sinensis*

Derivation: *Ips*, worm [Greek], *homoios*, like [Greek], referring to its trailing habit; *sinensis*, Chinese [Latin].

Identification: Creeping **annual herb** with stems c. 35 cm long; covered in short bristly **hairs**. **Leaves** alternate, deltoid in shape with a hastate base, leaf surface almost pleated, margins slightly toothed. **Flowers** borne singly or in pairs at the leaf axils; corolla tube white with a magenta throat and a greenish mid-lobe stripe, funnel-shaped with 5 star-shaped lobes. When the flower drops and the **fruit** matures, the leafy calyx grows to look almost like a hairy green flower.

Habitat: Locally uncommon in sandy disturbed areas of the floodplain.

Flowering: Middle to end of the main rains.

FZ Vol.8 pt1.

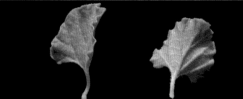

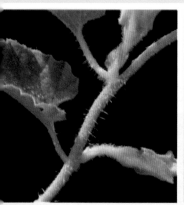

Convolvulaceae (morning glory family)

Merremia verecunda Rendle

Derivation: *Merremia* for Blasius Merrem (1761–1824), professor of natural sciences at Marburg in Germany; *verecunda*, modest [Latin].

Identification: Trailing, creeping **herb** with stems c. 1 m long. **Leaves** palmate, deeply lobed with up to 9 or 10 lobes, lobes almost linear and fold down the central vein. **Flowers** a white cone, usually with a maroon centre. They are larger than the flowers of many of the local *Convolvulaceae* (c. 3 cm dia.); **sepals** striped with maroon. **Latex** is sometimes present.

Habitat: Locally uncommon in full sun on open sandy ridges in the mopane woodland.

Flowering: Late in the main rains.

FZ Vol.8 pt1.

Cucurbitaceae (cucumber or gourd family)

Zehneria marlothii (Cogn.) R.Fern. & A.Fern.

Common names: Subiya ituhatuha, pekolola; **English** donkey flower.

Derivations: *Zehneria* for the Austrian botanical artist, Joseph Zehner; *marlothii* for the South African botanist Rudolph Marloth. *Ituhatuha* means perennial, a plant that grows again.

Identification: A vigorous fine-stemmed **semi-perennial climbing plant**. **Leaves** rather thin, palmate, trilobed, with a spike or bristle at the apex, margins serrated. **Tendrils** from the leaf axils. **Flowers** borne at the leaf axils, minute, white, with 5 pointed petals; **male flowers** c. 2 mm dia.; **female flowers** are smaller than the male flowers; the stalk of the female flower is twice the length of that of the male flower. **Fruit** small and spherical, c. 7 mm dia., ripening to bright red.

Habitat: Widespread in light shade in mixed woodland with sandy soil.

Flowering: Main rains.

Uses and beliefs:
* An attractive food plant for both animals and birds (especially donkeys, giraffe, elephants and grey lourie).

FZ Vol.4 pt0. CM, RM.

Fabaceae (pod-bearing family), Caesalpinioideae (cassia sub-family)

Bauhinia petersiana Bolle subsp. *macrantha* (Oliv.) Brummitt & J.H.Ross

Common names: Setswana dikgose, mochancha, mochope, mogose, mogotswe, mohuthi, mokoshi, mopondopondo, motaloga, mothlwa-o-jewa, motshentshe, motsope, motwakidja, mugutswe, nsekesa, cudwa, ndcwa, mohosi, muhotsi, tnotsantsa; **Lozi** mupondo; **Afrikaans** koffiebeesklou; **English** white bauhinia, wild coffee bean, camel's foot, coffee neat's foot.

Derivations: *Bauhinia* referring to the bilobed leaves as a symbol of the Swiss brothers, Jean (1541–1613) and Gaspard (1560–1624) Bauhin's interest in botany; *petersiana* in honour of the 19th century German botanist Professor W. Peters; *macrantha*, having large flowers [Latin]. The bush was known as 'wild coffee bean' because in the 19th century the beans were roasted and ground for use as a coffee substitute.

Identification: Semi-deciduous shrub, in this area rarely more than c. 1 m tall but may grow to c. 4 m. It can climb and scramble over surrounding vegetation, using hooks or spurs on its stems to attach itself to other plants. **Leaves** alternate, bilobed, with smooth margins, hairless. **Flowers** large, showy with crinkly white petals; open early in the day. **Fruit** hairless pods, opening explosively.

Habitat: Found occasionally locally growing in full sun in areas of deep sand.

Flowering: During the main rains.

Uses and beliefs:
- A substitute for coffee.
- Heavily browsed by game.

B&P van Wyk Trees p.372, Curtis & Mannheimer p.212, Hargreaves p.28, Palgrave p.284, Plowes & Drummond p.52, Setshogo & Venter p.70, WFNSA p.172.

Acacia arenaria Schinz

Common names: Setswana chimiwane, mogokatau, mophuratshukudu, mosokelateng, muparapara, pharaspikiri; **Afrikaans** sanddoring; **English** sand acacia, sand thorn.

Derivation: *Acacia* from *akantha*, thorn [Greek]; *arenaria*, relating to sand [Latin], hence growing in sandy places.

Identification: An untidy multi-stemmed **bush** or **small tree**, c. 3.5 m tall. **Bark** on the trunk is dark and rough. **Branches** have pairs of **thorns** of up to c. 6 cm in length that turn white with age. **Leaves** alternate, grey-green, feathery, bipinnate, c. 13 cm long with many small leaflets. **Flowers** in creamy white balls, c. 1 cm dia., borne in groups at leaf axils on the current year's wood; strongly **perfumed**. **Fruit** conspicuous long, slender, curved pods with slight restrictions between the seeds; becoming deep red-brown and splitting open to release the seeds when ripe.

Habitat: Widespread in areas of dry sandy bush.

Flowering: Early in the main rains.

FZ Vol.3 pt1, B&P van Wyk Trees p.490, Curtis & Mannheimer p.132, Palgrave p.230, Setshogo & Venter p.59.

Fabaceae (pod-bearing family), Mimosoideae (mimosa sub-family)

Acacia erubescens Welw. ex Oliv.

Common names: Setswana moloto, murengambo, omungongomwi; **G//ana** G//are; **!Kung** N!ā; **Afrikaans** blouhaak; **English** blue thorn.

Derivation: *Acacia* from *akantha*, the Greek for thorn; *erubescens*, blushing, becoming red [Latin].

Identification: A flat-crowned **deciduous tree**, c. 6 m tall. **Bark** of the main trunks and mature branches is fissured and peels in pale cream, rectangular sections. Younger branches have yellow bark. Pairs of hooked **thorns**, on opposite sides of the twigs, spiral up the stem. **Leaves** grey-green, bipinnately compound with the leaflets usually bending down. **Flowers** in cylindrical spikes that emerge before the leaves, fawny-white, with a delicate sweet **perfume**. **Fruit** flattened leathery pods, up to c. 13 cm long.

Acacia erubescens can be confused with *A. fleckii* as they both have short recurved thorns and cream-coloured flaky bark. The leaves of *A. erubescens* have longer stalks than those of *A. fleckii*, and *A. fleckii* has a large gland on its leaf stalk. *A. erubescens* flowers earlier than *A. fleckii*.

Habitat: Widespread in heavily grazed open woodland.

Flowering: During the cool dry period.

Uses and beliefs:
- The wood is used for firewood and fence posts.
- The bark is used to make thatching rope.

FZ Vol.3 pt1, Ellery p.84, Palgrave updated p.278, Setshogo & Venter p.10.

Fabaceae (pod-bearing family), Mimosoideae (mimosa sub-family)

Acacia fleckii Schinz

Common names: Setswana mhahu, mohahu, mfafu, mokoka, omutaurammmbuku, mukona, mokoko, mokokwane; **Hmbukushu** modiyangwe; **!xõ** n≠ahli; **!kung** n≠eng; **G//ana** /kane, gare; **Afrikaans** bladdoring; **English** blade thorn.

Derivation: *Acacia* from *akantha*, thorn [Greek]; *fleckii* for the German geologist E. Fleck who collected plants in Namibia in 1888.

Identification: A small bushy **tree** branching low down, c. 4 m tall. **Bark** grey to cream, peeling in small papery flakes. **Thorns** dark reddish-brown, hooked, in pairs. **Leaves** alternate, bipinnately compound with 8–20 pairs of crowded pinnae and 12–30 pairs of very small leaflets, greyish green and velvety/hairy. **Flowers** borne in axillary spikes, 4–5 cm long with stalk c. 1.5 cm, white, sweetly perfumed. **Fruit** small, thin, straight pods, c. 10 × 1.5 cm.

Can be confused with *A. erubescens* but *A. fleckii* has more pairs of leaflets and they are hairy.

Habitat: Found occasionally in dry sandy scrub, especially south of South Gate to the Moremi National Park.

Flowering: During the main rains.

Uses and beliefs:
- Hmbukushu ward off *harothi* (Hmbukushu) or *tokolosi* (Setswana) 'evil spirits' by burning branches of *Acacia fleckii*.
- Wayeyi children pick the flowers for use as ear-rings.

FZ Vol.3 pt1, Ellery p.85, Hargreaves p.25, Palgrave updated p.280, Tree Roodt p.23, Turton p.161. TK, SS, MT.

Fabaceae (pod-bearing family), Mimosoideae (mimosa sub-family)

Acacia hebeclada DC.
subsp. *chobiensis* (O.B.Mill.) A.Schreib.

Common names: Setswana setshi, sekhi, mogoka, mungcinda; **!xõ** n//ah; **G//ana** n//a; **Sheyeyi** setshe, setsee; **Afrikaans** trassiedoring; **English** candle pod acacia, candle thorn, Chobe candle acacia.

Derivations: *Acacia* from *akantha*, thorn [Greek], *hebe*, youth, pubescent [Greek], *klados*, branch [Greek], having twigs covered in soft hairs; *chobiensis* from Chobe. *Trassiedoring* is derived from the Hottentot *taras* meaning hermaphrodite or bisexual, alluding to the fact that the thorns are sometimes hooked and sometimes straight.

Identification: A small, spreading, tangled **shrubby tree**, c. 3 m tall, with some branches up to 4.5 m long. **Bark** dark-grey, longitudinally fissured and flaking. **Thorns** in pairs spiralling up the branches, sometimes hooked and sometimes straight. **Leaves** bipinnately compound, 6–9 pairs of primary leaflets with 11–13 pairs of secondary leaflets. Leaves and stems of the new growth have long **hairs** overall. New wood has white glandular spots. **Flowers** in pale fawn-grey balls, c. 1.5 cm dia., borne in clusters along the length of the twigs at leaf axils as the new growth of leaves appears. **Seed pods** covered in fine grey **hairs**, in erect clusters, c. 10 × 2.5 cm, flattened, broader and more rounded at the tip than those of the more widespread *Acacia hebeclada* subsp. *hebeclada*.

Habitat: Although on the endangered species list, this species is locally common and widespread in the areas from the Chobe along the Linyanti and Kwando Rivers and at the northern end of the Selinda Spillway.

Flowering: During the cool dry period.

Uses and beliefs:
• Subiya collect and store the pods as a treat for goats. They feed them to the flock if they want to keep them nearby in order to separate out animals for sale or slaughter.
• An important food for stock and game, especially giraffe, who enjoy picking the pods with their long tongues.

FZ Vol.3 pt1, Ellery p.87, Palgrave updated p.283, Setshogo & Venter p.61, Tree Roodt p.163, WFNSA p.162. BN, PN, BT.

Acacia luederitzii Engl. var. *luederitzii*

Common names: Setswana mokgwelekgwele, kangarangana, moka, mokala, moku, mooka, mooku, omungondo, mosaoka, mokha; **!Xo** !ola, **!Kung** g!u; **G//ana** and **Nharo** go; **Afrikaans** baster-haak-en-steek; **English** bastard umbrella thorn, Kalahari (sand) acacia, Kalahari sand thorn.

Derivation: *Acacia* from *akantha*, thorn [Greek]; *luederitzii* for the German explorer A. Luederitz, for whom the port in Namibia was also named.

Identification: Small multi-trunked **semi-deciduous tree** or **shrub** with a flat crown, c. 3 m tall. **Bark** on the main trunk dark brown to black, longitudinally deeply fissured. Younger branches ribbed. **Thorns** either small and hooked or long and straight. **Leaves** alternate, bipinnately compound, rarely more than 3 cm long. **Flowers** in rather loose pale creamy-white balls, c. 1 cm dia., growing in clusters in the leaf axils; with a honey-sweet strong **perfume**, very attractive to bees and other insects. **Fruit** straight flat pods, turning reddish-brown on ripening.

Habitat: Found occasionally locally in sandy areas of the floodplain turning to acacia scrub.

Flowering: During the main rains.

Uses and beliefs:
- Straight sections of the roots are hollowed out for use as quivers for arrows.
- The root cores are used to make pestles.
- The inner bark is pounded and coiled to make very good cord.

FZ Vol.3 pt1, B&P van Wyk Trees p.484, Curtis & Mannheimer p.158, Ellery p.89, P Van Wyk Kr Trees p.68, Palgrave p.243, Setshogo & Venter p.62.

Fabaceae (pod-bearing family), Mimosoideae (mimosa sub-family)

Acacia mellifera (Vahl) Benth. subsp. *detinens* (Burch.) Brenan

Common names: Setswana mongana, monka, more-o-mabele, unganda, mukona, nkogwana, nkoshwana, omusaona; **G//ana:** //wa, //kowa; **!Kung** !gou; **Nharo** and **!xo** //ha; **Afrikaans** swarthaak; **English** black thorn, wait-a-bit thorn, hook thorn.

Derivation: *Acacia* from *akantha*, thorn [Greek]; *mellis*, honey, *fera*, bearing [Latin]; *detinens*, detaining or holding [Latin], alluding to the hooked thorns.

Identification: Shrubby tree, c. 4 m tall. **Bark** smooth, grey. Black hooked **thorns** in close-set pairs spiral up the stems. **Leaves** blue green, bipinnately compound with 2 or 3 pairs of pinnae each bearing only 1 or 2 pairs of leaflets, individual leaflets comparatively large. **Flowers** appear before the leaves, in creamy-white elongated balls of stamens, c. 1.5 cm dia. The sweet scent of the flowers is diagnostic. **Fruit** flat, papery, broad pods, with few seeds, ripening to a rich beige colour. From a distance, the immature seed pods look like shiny new leaves.

Acacia mellifera may be confused with *Acacia nigrescens* but it does not have knobs on the stems and its leaflets do not have net veining.

Habitat: Common and widespread in areas where the soil is rich in calcium.

Flowering: Towards the end of the cool dry period.

Uses and beliefs:
- The bark is used to treat diarrhoea and the roots to treat stomach complaints.
- The dark heart-wood is used for carving and becomes black when it is oiled.
- The resin is used to mix and glue poison for arrows.
- The sweet flowers are highly attractive to bees.
- The flowers, leaves, twigs and seed pods are readily browsed by game.

FZ Vol.3 pt1, Blundell p.123, Ellery p.90, Hargreaves p.26, Palgrave updated p.289, Tree Roodt p.169, Setshogo & Venter p.11, Turton p.23.

Fabaceae (pod-bearing family), Mimosoideae (mimosa sub-family)

Acacia sieberiana DC.
var. *woodii* (Burtt Davy) Keay & Brenan

Common names: Setswana more-o-mosetlha, omuhengehenge, mughombe, morumosetlha; **Afrikaans** papierbasdoring; **English** paperbark thorn, flat-topped thorn, paperbark acacia.

Derivation: *Acacia* from *akantha*, thorn [Greek]; *sieberiana*, for Franz Sieber (1789–1844), a Bohemian botanist, traveller and plant collector; *woodii* for the South African botanist J.M. Wood.

Identification: Large, flat-topped **deciduous tree**, up to c. 18 m tall. **Bark** grey on the main trunk and golden yellow on the branches, sometimes corky and peels in papery strips, especially on the branches. **Thorns** straight, up to c. 7.5 cm long in pairs below the leaf axils, on both new and old wood. **Young twigs** and **leaf stalks** covered with a velvety down of hairs. **Leaves** bright green, bipinnately compound, c. 16 pairs of leaflets which have c. 36 pairs of pinnae. **Flowers** whitish balls, c. 1.5 cm dia., on c. 3 cm stalks from the leaf axils. **Fruit** long flat pods, c. 23 × 3 cm, margin clearly defined and seeds clearly delineated within the pod.

Habitat: Locally uncommon on termite mound islands in the floodplain.

Flowering: During the hot dry period prior to the early rains.

Uses and beliefs:
- The pods and leaves are grazed by stock and game.
- The caterpillars of *Anthene amarah amarah* (the black-striped hairtail) feed on the young shoots.

FZ Vol.3 pt1, Ellery p.93, Palgrave updated p.299, Setshogo & Venter p.65, Tree Roodt p.175.

Fabaceae (pod-bearing family), Mimosoideae (mimosa sub-family)

Albizia anthelmintica Brongn.

Common names: Setswana monoga; **Afrikaans** wurmbasvalsdoring; **English** worm-cure albizia, worm-cure false thorn, worm-bark false-thorn.

Derivation: *Albizia* for F. del Albizzi, a Florentine nobleman who introduced the first Albizia into cultivation in 1749; *anthelmintica*, acting against and expelling intestinal worms.

Identification: Open-crowned **deciduous tree** with slightly weeping branches, usually c. 5 m tall. **Bark** on young branches smooth, grey, with lenticels appearing as tiny raised whitish spots. **Leaves** bipinnately compound, 2–4 pairs of rounded leaflets with oblique bases. **Flowers** appear before the leaves, borne in clusters up the branches, showy with masses of shaggy half-balls of white stamens, c. 2 cm dia.

Habitat: Widespread in heavily grazed open bush.

Flowering: During the cool dry period.

Uses and beliefs:
• In Namibia, the bark is used as an effective treatment for tapeworm.

FZ Vol.3 pt1, Blundell p.124, Palgrave updated p.258, Setshogo & Venter p.66.

Fabaceae (pod-bearing family), Mimosoideae (mimosa sub-family)

Albizia harveyi E.Fourn.

Common names: Setswana mmola, molalakgaka; **Shiyeyi** orarakanga; **Afrikaans** bleekblaarboom; **English** common false-thorn, sickle-leaved albizia, sickle-leaved false-thorn.

Derivations: *Albizia* for F. del Albizzi, a Florentine nobleman who introduced the first Albizia into cultivation in 1749; *harveyi* for an Irish botanist, William Harvey (1811–66), who collected in S. Africa and elsewhere. *Molalakgaka* means guinea fowl roost.

Identification: Small tree with a flattened crown, c. 8 m tall. **Bark** dark grey, deeply fissured longitudinally; **main trunk** is almost braided. **New growth** covered in a golden down of short hairs. **Leaves** alternate, bipinnately compound, leaflets sickle-shaped with a sharp point; new leaves bright lime green, darker above and lighter below. **Flowers** borne on new growth from leaf axils and at the tips of branches. Inflorescences a ball of creamy-white florets each with 5 sepals and a mass of stamens. **Fruit** pinky-orange pods, occasional restrictions between the seeds.

Habitat: Widespread but not common on islands and in mixed woodland, preferring termite mounds.

Flowering: May produce some flowers throughout the rains but mainly during the early rains.

Uses and beliefs:
• This species is mixed with *Clerodendrum uncinatum* and *Euclea divinorum* to treat sexually transmitted diseases.

FZ Vol.3 pt1, Curtis & Mannheimer p.126, Ellery p.96, Hargreaves p.27, Palgrave p.220, P Van Wyk Kr Trees p.56, Setshogo & Venter p.66, Tree Roodt p.185. PN.

Fabaceae (pod-bearing family), Papilionoideae (pea sub-family)

Baphia massaiensis Taub. subsp. *obovata* (Schinz) Brummitt

Common names: Setswana isunde, monthe, motlhwakeja, nqoli, nqodi, munge, munde, motatija; **Afrikaans** sandkamhout; **English** jasmine pea, sand camwood.

Derivation: *Baphe*, a dye [Greek]; *massaiensis*, of the Massai.

Identification: A small **shrub**, usually up to c. 1 m tall locally but may grow up to c. 6 m in other areas. **Leaves** alternate, obovate with an apiculate tip to obcordate, leathery, with smooth margins. **Flowers** c. 1.5 cm across, growing in sprays at the apex of the branches, white to creamy pink, a yellow spot in the throat on the keel. **Fruit** long, narrow, woody pods, dark brown.

Habitat: Locally common in deep sand.

Flowering: In the Linyanti, these shrubs are grazed to the ground each year. They sprout again from a woody root stock and only come into flower on the new wood late in the rainy season. In other areas, they flower from before the early rains.

Uses and beliefs:
- Used for ropes and toothbrushes.

Hargreaves p.29, Palgrave p.302, Setshogo & Venter p.94.

Fabaceae (pod-bearing family), Papilionoideae (pea sub-family)

Crotalaria flavicarinata Baker f.

Derivation: *Krotalon*, rattle [Greek], as the seeds of many species rattle in the pod; *flavus*, yellow [Latin], *carinatus*, having a keel [Latin], yellow-keeled.

Identification: Branching **perennial herb** from a woody rooting stock, c. 60 cm tall. **Stems** ribbed covered in fine hairs. **Leaves** blue-grey, covered in fine hairs, alternate, compound, trifoliate, leaflets oval with a rounded tip, smooth margined. **Flowers** borne in rather short loose racemes at the apex of the branches; standard petal greenish-white striped on the reverse with red-brown streaks, wing petals yellow, keel yellowish green; stigma sharply bent. **Fruit** cylindrical with rounded tips, stigma retained on the fruit.

Habitat: Locally rare in deep sand in full sun.

Flowering: Late in the main rains. (This may be because it has been found growing in heavily grazed areas and must grow virtually from the root each year.)

Uses and beliefs:
• Much relished by black rhinoceros as a food plant.

FZ Vol.3 pt7. MI.

Fabaceae (pod-bearing family), Papilionoideae (pea sub-family)

Crotalaria heidmannii Schinz

Derivation: *Krotalon*, rattle [Greek], as the seeds of many species rattle in the pod.

Identification: Branching erect **perennial herb**, from a woody rooting-stock, c. 1 m tall. **Stem** slightly buttressed, often almost square. **Leaves** grey-green, compound, trifoliate, smooth margined. **Flowers** c. 2 cm; standard petals white and creamy pink with brown veining on the reverse, wing petals yellow, keel straight and green; stigma sharply bent. **Fruit** a cylindrical pod with bluntly rounded ends, held upright, stigma usually retained.

Habitat: Locally uncommon, growing in deep sand in full sun on the floodplain or in mixed scrub.

Flowering: During the main rains.

FZ Vol.3 pt7.

Fabaceae (pod-bearing family), Papilionoideae (pea sub-family)

Macrotyloma daltonii (Webb) Verdc.

Derivation: *Macro-*, long or big, *tyloma*, callus, lump or swelling [Greek], with large knobs or projections.

Identification: Annual or **perennial herbaceous climber,** up to c. 2 m tall depending on the height of its support. **Stems** tinged pink. **Leaf buds** covered in long soft hairs. **Leaves** compound, trifoliate, covered in long soft hairs, even the tips of the leaflets are hairy. **Flowers** produced in pairs in the leaf axils, stalkless, not flowering simultaneously, greenish-white with maroon marking on the standard.

Habitat: Locally rare on sandy ridges in *Combretum* scrub.

Flowering: During the main rains.

FZ Vol.3 pt5.

Gentianaceae (gentian family)

Enicostema axillare (Lam.) A.Raynal subsp. *axillare*

Common names: Setswana pelo botlhoko; **San** chotho.

Derivations: *Axillare* referring to the position of the flowers in the leaf axils. *Pelo botlhoko*, to heal your heart; the San name has a similar meaning.

Identification: Small, erect, short-lived **perennial**, up to c. 30 cm tall, usually a single stem but occasionally branching low down. **Stem** square with narrow wings. **Leaves** opposite and decussate, stalkless, linear, c. 6 × 0.7 cm, margins hairy, lower surface covered by mat of fine hairs. **Flowers** in dense clusters borne in the leaf axils, white with green centres, c. 0.7 cm dia., 5 lobes.

Habitat: Common and widespread in full sun on the floodplain.

Flowering: Throughout the rains.

Uses and beliefs:
- Hmbukushu and Wayeyi often use this plant mixed with *Aerva leucura*. The *A. leucura* acts as a medium through which it is possible to communicate with the ancestors whilst *Enicostema axillare* serves to 'heal the heart'. If there are problems or evil spirits, or if the ancestors are angry, these plants can help. For example, if a child is unlucky (e.g. he buys cattle and they die), his mother or father cuts the leaves and flowers and soaks them overnight. At sunrise, the child must face the sun while the parent takes a mouthful of the liquid and sprays it over the child towards the sun to bring good luck.
- If babies sleep restlessly, it is believed that there are *tokolosi*, bad spirits, in the yard. The leaves, flowers and roots are burned to drive away the evil spirits.
- *Aerva leucura* and *Enicostema axillare* may simply be rubbed together between the hands when seeking the answer to a problem. The ancestors will help.
- The whole *Enicostema axillare* plant, including the roots, is used to treat toothache. It is boiled and the mouth is washed out with the resulting decoction, which is very bitter and numbs the mouth.

FZ Vol.7 pt4. GM, IM.

Lapeirousia odoratissima Baker

Common names: English spider lily.

Derivation: *Lapeirousia* for Baron Philippe Picot de la Peirouse (1744–1818), a French botanist; *odoratissima* refers to the sweet scent of the flowers, which is particularly noticeable in the evening.

Identification: A small **perennial herb**, c. 20 cm tall arising from a corm. **Leaves** bright green, strap-like, folding at the central vein. **Flowers** delicate, white, like etiolated stars, 6 long lobes, c. 10 cm dia.; corolla tube up to c. 15 cm long, sweetly scented.

Habitat: Locally rare in sandy areas of the mopane woodland.

Flowering: Late in the main rains.

Uses and beliefs:
- The corms are part of the diet of the Khu tribe of Namibia and Botswana.

FZ Vol.12 pt4, Plowes & Drummond no 23.

Iridaceae (Iris family)

Lapeirousia schimperi (Asch. & Klatt) Milne-Redh.

Derivation: *Lapeirousia* for Baron Philippe Picot de la Peirouse (1744–1818), a French botanist; *schimperi* celebrates the German botanist W.G. Schimper, who collected in Ethiopia in the mid-19th Century.

Identification: Tall (c. 1 m) slender slightly lax **annual herb** arising from a bulb. **Bulb** at a depth of c. 17 cm, pear-shaped, c. 2.5 cm tall, dark brown fibrous robe. Multi-branching **stems** 4-angled or almost square with slight wings. **Leaves** grey-green, linear, sharply folded around the stem at the base, with a deep keel, fusing about one third of the way up and becoming flat; lower leaves up to c. 50 cm long, becoming shorter higher up the plant. **Flowers** borne on branching stems at the apex of the plant, delicate, white sometimes slightly tinged with violet, formed of a c. 15 cm corolla tube terminating in 6 lobes, c. 5 cm dia.

Habitat: Found occasionally locally in years of high rainfall, in damp clayey areas of mopane woodland.

Flowering: Late in the main rains.

Uses and beliefs:
- The corms are edible and are eaten raw or roasted in northern Namibia.

FZ Vol.12 pt4.

Lamiaceae (mint family)

Acrotome inflata Benth.

Common names: Setswana leatla, makhudugwane, seromo; **Ovambenderu** endumba; **English** acrotome, tumbleweed.

Derivation: *Acrotome*, cut off sharply at the tip [Greek]; *inflata*, inflated or swollen [Latin], referring to the spherical flower heads.

Identification: Much-branched **annual herb** almost forming a ball, c. 70 cm tall; covered in slightly coarse velvety **hairs**. **Stem** heavily buttressed or square. **Leaves** opposite and decussate, oval narrowing to form leaf stem, margins serrated, c. 18 cm (inc. c. 1.5 cm leaf stalk) × 3 cm. **Inflorescence** a ball of tubular white flowers borne at leaf axils; upper and lower lobes enlarged to form lips, emerging from a spiny green calyx. Makes use of its spherical shape as seed is often distributed by the whole plant being up-rooted and bowled along by the wind.

Habitat: Common and widespread in full sun throughout the floodplain and into island and treeline margins on clayey sand.

Flowering: Main rains through to the end of the cool dry period.

Uses and beliefs:
• Hmbukushu boil the leaves, flowers and fruit of this plant to make a hot compress with which they massage sore, tired or swollen legs.
• The boys pull the individual flowers and suck the nectar.

B Van Wyk Flowers p.64, Flowers Roodt p.95. BB, MM.

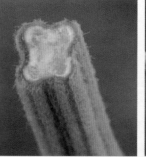

Lamiaceae (mint family) (formerly placed under *Verbenaceae*)

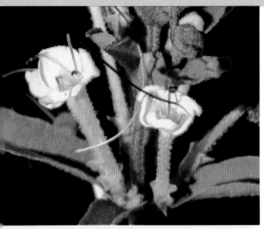

Clerodendrum ternatum Schinz

Common names: Setswana legonnyane?, sedupapula, lo-lonamagadi; **English** cat's claw.

Derivations: *Kleros*, chance [Greek], *dendron*, tree [Greek], thought to be an allusion to the variable medicinal qualities of the plants of this genus; *ternatus*, in groups of three [Latin]. *Sedupapula*, rain finder.

Identification: An erect **perennial herb**, c. 50 cm tall. **Leaves** lanceolate, in whorls of 3 to 5 up the stem, c. 8 × 1.5 cm, upper third of the leaf margin is serrated, surfaces densely covered with **hairs**; crushed leaves have a distinctive aromatic **smell**. **Flowers** borne in groups in leaf axils, delicately perfumed, white and furry with a long magenta style projecting c. 4 cm from the corolla tube, which terminates in 5 asymmetrical lobes. Stamens short, white, curved. **Calyx** very short, star-shaped. **Fruit** 4-lobed.

Habitat: Widespread in partial shade in sandy areas of islands and mopane woodland.

Flowering: Throughout the rains.

Uses and beliefs:
- For Batswana, the roots have many uses: they are boiled and the decoction is taken as a purgative for the stomach.
- The women use it as a douche; if they are suffering menstrual pain, they use it as long as the pain lasts.
- It is also used as a medicine for fevers.

Germishuizen p.340, Hargreaves p.58. KS.

Lamiaceae (mint family)

Hemizygia bracteosa (Benth.) Briq.

Common names: English white-tipped hemizygia.

Derivation: *Hemi*, half [Greek], *zygos*, yoke [Greek], possibly referring to the shape of the flowers; *bracteosa*, referring to the showy white bracts at the apex of the plant.

Identification: Erect, aromatic, annual **herb**, c. 55 cm tall. **Leaves** opposite and decussate, linear and stalkless, margins serrated. **Inflorescence** a c. 30 cm long spike of white and purple flowers, c. 7 × 6 mm, arranged in whorls around stem at the apex of plant and on side branches. Each flowering spike surmounted by a showy group of almost square white bracts, which are easily mistaken for flowers at first glance. **Flowers** tubular, with virtually no petals, white with purple markings, c. 1.2 cm long with slightly enlarged lower lip.

Habitat: Locally common in areas of deep sand, especially sandy ridges through the mopane woodland and on sandy banks in the floodplain.

Flowering: From the middle of the main rains into the cool dry period.

Lamiaceae (mint family)

Hoslundia opposita Vahl

Derivation: *Hoslundia*, for the Danish botanist Ole Haaslund-Schmidt (d. 1802), a naturalist, traveller and plant collector; *opposita*, opposite [Latin], possibly referring to the arrangement of the leaves.

Identification: A scrambling **herb**, more than 2 m tall, found growing on dense supports. **Leaves** opposite but all grow to face the same direction, grey-green, covered in hairs, margins toothed. **Flowers** borne in panicles at the apex of the stems, small, greenish white. **Fruit** like miniature orange pumpkins but soft and fleshy, c. 8 mm dia. The **plant** has an aromatic minty smell when crushed.

Habitat: Locally rare growing on dense supporting foliage such as juvenile *Hyphaene petersiana*, preferring termite mounds in sandy areas.

Flowering: During the main rains.

Uses and beliefs:
- The fleshy fruit are edible.

WFNSA p.356.

Amaryllidaceae (*Amaryllis* or vlei-lily family)

Pancratium tenuifolium Hochst. ex A.Rich.

Common names: English vlei lily.

Derivation: *Pancratium*, bulbous plant [Greek]; *tenuis*, slender, thin [Latin], *folius*, leaf [Latin], slender-leaved.

Identification: Attractive small **perennial herb** from a **bulbous rooting** stock, up to c. 25 cm tall. **Bulb** c. 2.5 cm dia., with a papery brown tunic. **Leaves** c. 16 × 0.8 cm, linear, narrow forming a basal rosette, very distinctive as they grow in spirals like rather lax cork-screws. **Flowers** c. 11 cm long, showy, white, with 6 linear white tepals (sepals that look like a second row of petals) and a large cup-shaped funnel with a dentate margin. Flowers open only at night and disappear soon after sunrise. There are 2 linear **bracts**, c. 5 cm long, at the base of the flower.

Habitat: Widespread in sandy areas of the floodplain and on sand ridges in the mopane woodland.

Flowering: During the early rains.

Germishuizen p.31.

Malvaceae (hibiscus or mallow family)

Hibiscus lobatus (Murray) Kuntze

Derivation: *Hibiscus* was the Greek name for the marsh mallow, possibly from the ibis which feeds on certain varieties of this genus; *lobatus*, lobed [Latin].

Identification: Erect, **annual herb**, up to c. 60 cm tall; sparsely covered in silvery **hairs**. **Leaves** slightly palmate, trilobed, margins toothed, spiralling up the stem, c. 12 cm (inc. c. 3 cm stalk) × 7 cm, slightly sticky, pairs of **stipules** at the leaf axils. **Flowers**, c. 2 cm dia., brilliant white with 5 petals.

Habitat: Found occasionally locally in the deep shade of the riverine forest. [Introduced.]

Flowering: During the main rains.

FZ Vol.1 pt2.

Kostelezkya buettneri Gürke

Derivation: *Kosteletzkya*, for Vincenz Franz Kosteletzky (1813–1866) of Prague, a medical botanist; *buettneri*, for Büttner, who first collected this species in Zaïre.

Identification: Erect, harshly hairy **annual** or **perennial herb,** branching at a low level, up to c. 1 m tall, growing from a rhizomous **rooting** system. **Leaves** alternate, stalkless, oblong to linear, some slightly hastate, margins serrated, with net veining and often 3 veins from the base of the leaf; densely covered in star-shaped **hairs** on both surfaces. **Flowers** borne singly at leaf axils and at the apex of the plant, white with yellow markings at the base of the 5 petals, a yellow stamen tube bending to one side. Flower dries to yellow.

Habitat: Locally rare growing in the margin of fast-flowing channels sheltered by grasses and sedges.

Flowering: During the main rains.

FZ Vol.1 pt2, Ellery p.167.

Malvaceae (hibiscus family)

Pavonia clathrata Mast.

Derivation: *Pavonia* for José Antonio Pavon (1754–1840), a Spanish botanist; *clathrata*, latticed or pierced with openings like a grating or trellis [Latin].

Identification: Erect, branching **annual herb,** c. 80 cm tall; coarsely **hairy**. **Leaves** alternate, palmate with 5 deep almost linear lobes with deeply toothed margins, sticky on the underside. **Flowers** white tinged with pink on the underside and turning pink as the flower matures; petals are long and narrow. Below the flowers, the **epicalyx** is a ring of long narrow linear bracts.

Habitat: Found occasionally locally in full sun on open sandy ridges in the mopane woodland.

Flowering: Late in the main rains.

FZ Vol.1 pt2.

Malvaceae (hibiscus family)

Sida alba L.

Common names: Afrikaans stekeltaaiman; **English** spiny sida.

Derivation: *Sida*, water plant [Greek]; *alba*, white [Latin].

Identification: Slender erect **annual herb,** up to c. 40 cm tall. **Leaves** soft green, leaf stalks almost as long as the leaf blades, alternate and broadly ovate with toothed margins. Both the leaves and the calyx are outlined in dark red. **Flowers** white to pale cream, 5 petals with obtuse tips; stamens bright yellow. **Fruit** 5 sections, seeds light brown and wedge-shaped.

Habitat: Found only occasionally in full sun near seasonal pans and on the spillway.

Flowering: During the main rains.

FZ Vol.1 pt2, B Van Wyk Flowers p.160, Botweeds p.84.

Menyanthaceae (bogbean family)

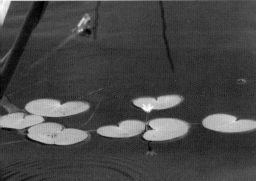

Nymphoides indica (L.) Kuntze subsp. *occidentalis* A.Raynal

Common names: Subiya isotho, **English** water gentian, floating heart, water snowflake.

Derivation: *Nymphoides* from *nymphaia*, a water nymph [Greek], resembling a water lily; *indica*, of India; *occidentalis*, western [Latin].

Identification: Erect **perennial aquatic herb** with rhizomous **rooting** system. **Stems** hollow and therefore float, so that the leaves and flowers are not affected by changes in the depth of the water. **Leaves** kidney-shaped with a slight indentation at the outer edge, c. 10 cm dia., upper surface green, the lower surface reddish-brown. **Flowers** on long stalks, in clusters, at the leaf axil just below the leaf blade, feathery, white with a yellow centre, 5 lobes, c. 2 cm dia. These **plants** can withstand considerable changes in water level and even the drying out of pools. They have a delicate almond **perfume**.

Habitat: Widespread in shallow lagoons with little current.

Flowering: Early in the main rains.

Uses and beliefs:
• Wayeyi and Subiya consider this a delicious food plant. The roots are boiled, mixed with meat, fish or bubble fish (cat fish).
FZ Vol.7 pt4, Ellery p.146, WFNSA p.300. GM.

Molluginaceae (mollugo family)

Glinus bainesii (Oliv.) Pax

Common names: Setswana baswabile.

Derivations: *Glinus*, origin obscure as *glinos* is Greek for the maple tree; *bainesii*, named for John Thomas Baines (1820–75), painter, naturalist and explorer of southern Africa. *Baswabile* means 'Let them be ashamed'.

Identification: A rather lax spreading **herbaceous annual**, c. 20 cm tall. **Stems** covered in short tangled white hairs. Stems and leaf buds brownish red. **Leaves** more prominent than the flowers, slightly thick and appearing almost succulent, occasionally with a bristled tip, borne in whorls up the stems. **Flowers** borne individually on long stalks in clusters at the leaf axils, very small petals, white with dark red anthers.

Habitat: Found occasionally in damp areas of seasonally drying lagoon.

Flowering: Early in the main rains.

Uses and beliefs:
- Batswana traditional doctors mix *Glinus bainesii* with other plants, including *Abrus precatorius* and *Neptunia oleracea*, to make good luck charms.
- San either wash themselves with an infusion of the plants or apply a mixture of the crushed leaves with fat or petroleum jelly to the whole body for a good result in a court case.
- Subiya and Wayeyi have similar traditions with regard to court cases. Also, after using *Glinus bainesii*, if they ask a girl to marry them, she will look shy and agree. For the charm to be particularly effective, they also mix *G. bainesii* with *Neptunia oleracea* and *Abrus precatorius*.

BT.

Molluginaceae (mollugo family)

Glinus oppositifolius (L.) Aug. DC. var. *oppositifolius*

Derivation: *Glinus*, origin obscure as *glinos* is Greek for the maple tree; *oppositus*, opposite [Latin], *folius*, leaf [Latin], referring to the arrangement of the leaves.

Identification: Mat-forming **perennial herb,** c. 1.3 m in diameter. Multi-branched **stems** forming a dense system that attains a height of c. 12 cm. **Leaves** borne in whorls or opposite pairs, lanceolate to obovate, smooth margined. **Flowers** borne on long stalks in groups at leaf axils, white, sometimes browny-pink on the reverse of the petals, 5 petals, prominent yellow stamens.

This plant frequently **hybridises** with *Glinus lotoides* (p. 104) and a variety of forms may be found if both plants exist in an area.

Habitat: Found occasionally near seasonal pans.

Flowering: During both early and main rains.

FZ Vol.4 pt0, Ellery p.162.

G. oppositifolius × *G. lotoides*

Mulluginaceae (mollugo family)

Limeum fenestratum (Fenzl) Heimerl

Common names: Subiya umbolo; **English** window seed.

Derivation: *Limeum* meaning pest, to the point of ruin [Greek], referring to the toxicity of this genus; *fenestra*, window [Latin], referring to the transparent wings of the fruit.

Identification: Sparse free-standing, stiffly branching **annual herb**, up to c. 1 m tall. **Leaves** alternate, linear, stalkless, c. 5 cm long, margins entire, semi-succulent. **Flowers** stalkless (sessile), minute (c. 3 mm dia.), white, with 5 petals. **Fruit** flat, disc-like with circular translucent wings, c. 1 cm dia. **Seeds** with a sharp spine on each side, which sticks into the skin.

Habitat: Widespread on sandy areas of the floodplain.

Flowering: During the main rains.

Uses and beliefs:
• Young men take the root, boil it and drink the resulting decoction 3 times per day for about a week to cure venereal diseases.

FZ Vol.4 pt0, Hargreaves p.9, Turton p.85, WFNSA p.132. GM.

Molluginaceae (mollugo family)

Limeum sulcatum (Klotzsch) Hutch.

Common names: Subiya insekwasekwa; **Afrikaans** klosaarbossie.

Derivation: *Limeum*, a pest to the point of ruin [Greek], referring to the toxicity of these plants; *sulcatum*, furrowed [Latin].

Identification: Low-growing, much branched, slightly trailing, **annual herb**, c. 25 cm tall. **Leaves** alternate, linear and hairless, c. 32 × 1 mm, with a strong midrib that is prominent on the lower surface. **Flowers** carried in dense balls at leaf axils, white with a broad green stripe. **Fruit** fawny-yellow, spiky overall, especially at the edges.

Habitat: Found occasionally locally on islands and seasonal pan margins.

Flowering: Throughout the rainy period.

Uses and beliefs:
• Subiya eat the roots and use the stems to make small mats for decoration.

FZ Vol.4 pt0, WFNSA p.132. GM.

Molluginaceae (mollugo family)

Limeum viscosum (J.Gay) Fenzl var. *kraussii* Friedrich

Common names: Afrikaans klosaarbossie.

Derivation: *Limeum*, a pest to the point of ruin [Greek], referring to the toxicity of these plants; *viscosus*, sticky [Latin], referring to the fact that the plant is sticky over all; *kraussii* probably for the German botanist and traveller Christian Ferdinand Friedrich von Krauss (1812–1890).

Identification: Small **annual herb** with single **taproot**, c. 25 cm tall; covered in sticky glandular **hairs**. **Stems** pink. **Leaves** alternate, linear, up to c. 6.5 × 1.1 cm, smooth margined, folding at the mid-vein. **Flowers** borne in groups at the leaf axils, white, with 5 petals, c. 3 mm dia., many yellow stamens. When the flowers fall, they leave a pendant green-yellow calyx of 5 sepals, c. 7 mm dia., which is retained on the plant and looks like a small green flower.

These plants are highly variable and may easily be confused with other species of *Limeum*.

Habitat: Locally rare in areas of open grass in mixed bush in seasons of high rainfall.

Flowering: During the main rains.

FZ Vol.4 pt0, Botweeds p.6.

Molluginaceae (mollugo family)

Limeum viscosum (J.Gay) Fenzl var. *viscosum*

Common names: Afrikaans klosaarbossie.

Derivation: *Limeum*, a pest to the point of ruin [Greek], referring to the toxicity of these plants; *viscosus*, sticky [Latin], referring to the fact that the plant is sticky over all.

Identification: An erect, branching, low-growing, fleshy **annual herb**, c. 50 cm tall, may also be prostrate; covered in sticky glandular **hairs**. **Leaves** alternate, oval with a rounded tip, margins smooth. **Flowers** borne in clusters on long stalks opposite the leaf axil, with 5 petals, white turning yellow with age.

These plants are highly variable and may easily be confused with other species of *Limeum*.

Habitat: Found occasionally in lightly shaded areas of the riverine woodland.

Flowering: During the main rains.

FZ Vol.4 pt0, B Van Wyk New p.76, Botweeds p.6, WFNSA p.132.

Mulluginaceae (mollugo family)

Mullugo cerviana (L.) Ser. ex DC.

Derivation: *Mollis*, soft [Latin], alluding to the plant's soft herbaceous habit; *cerviana*, for J. Cervi (b. 1663), personal physician to Philip V of Spain.

Identification: A small **annual herb** up to c. 15 cm tall with a prolifically branching structure of light stems; hairless. **Stems** pink or brown. **Leaves** pale grey-green, linear, stalkless, borne in whorls. **Flowers** white with 5 petals (c. 1 mm dia.); borne singly or in twos and threes at the leaf axils, or singly on fine stiff stalks that are as long as the leaves or longer.

Habitat: Found occasionally locally in full sun in open sandy areas of degraded mopane woodland.

Flowering: During the main rains.

FZ Vol.4 pt0, Botweeds p.8, WFNSA p.132.

Nyctaginaceae (bougainvillea family)

Commicarpus plumbagineus (Cav.) Standl.

Common names: Setswana bogoma; **Hmbukushu** kaXhee; **Shiyeyi** okaXhee; **English** tattoo plant, false plumbago.

Derivations: *Kommi*, gum [Greek], *-carpus*, -fruited [Greek], referring to the sticky fruit; *plumbagineus*, plumbago-like, from the genus of that name. *Bogoma*, a plant whose parts readily attach themselves to passing animals.

Identification: Climbing, scrambling **herb** from a woody root stock, up to 5 m tall. Stems and leaves are slightly **hairy** overall. **Leaves** opposite and decussate, ovate, with a cordate base, margins wavy. **Flowers** white, 7 mm dia., growing in groups at the apex of the plant and on up to c. 10 cm stalks from the axils. Stamens protruding from the flowers, anthers pink. **Fruit** with rings of stalked glands that exude a glutinous substance when ripe. This causes the pods to stick to animals, aiding dispersal.

Habitat: Locally common in shaded areas of islands and riverine forest, preferring moister parts.

Flowering: Main rains through to the end of the cool dry period.

Uses and beliefs:
- Both Wayeyi and Hmbukushu skin the root, cut it into a design and tape it onto the skin, a black scar is left on the skin.
- Spring hares like to eat the roots.

FZ Vol.9 pt1, Blundell p.55, Hargreaves p.8. GM, PN, MT.

Oleaceae (olive family)

Jasminum fluminense Vell. subsp. *fluminense*

Common names: Setswana motsweketsane; **Hmbukushu** mudhangwe; **English** wild jasminum.

Derivations: *Jasminum*, the latinised form of the Persian name, yasmin, for these sweetly perfumed shrubs and climbers; *fluminense* in the naming of plants means, of Rio de Janeiro, Brazil, which is on the Flumen (river in Latin) Januarii. *Motsweketsane* means that it tangles around other plants.

Identification: A **perennial herbaceous climber,** up to c. 3 m tall. **Stems** hairy. **Leaves** grey-green, both surfaces velvety, opposite and trifoliate with wavy edges; leaflets ovate, the terminal leaflet is the largest. **Flowers** white, star-shaped, with c. 8 lobes fused to form a tube. **Fragrance** sweet. **Fruit** globose, c. 5 mm dia., ripening to a shiny black.

Habitat: Widespread, found occasionally locally in light shade on islands and in mixed woodland.

Flowering: Almost throughout the year.

Uses and beliefs:
- Fed on by the oleander hawk moth, *Deliphia nerif*.

FZ Vol.7 pt1, Ellery p.165, Flowers Roodt p.125, Hargreaves p.58, WFNSA p.296.

Oleaceae (olive family)

Jasminum stenolobum Rolfe

Common names: English shrub jasmine.

Derivation: *Jasminum*, the latinised form of the Persian name, yasmin, for these sweetly perfumed shrubs and climbers; *steno*, narrow [Latin], *lobum*, a lobe [Latin], possibly referring to the narrow lobes of the corolla.

Identification: Erect, occasionally climbing, **shrub**, up to c. 3 m tall. **Leaves** elliptical, slightly hairy overall, margins wavy, borne opposite and decussate or in whorls on new growth. **Flowers** appearing on new growth almost before the leaves, creamy-white, c. 2.2 cm dia., with long narrow lobes recurved across their breadth. **Fragrance** a strong sweet jasmine smell. **Calyx** a tube of long needle-like sepals that is retained around the fruits. **Fruit** a curved drop shape, fleshy, often maturing in pairs curving towards each other, ripening to a shiny black.

Habitat: Widespread in lightly shaded woodland and on islands.

Flowering: During the early rains and into the beginning of the main rains.

FZ Vol.7 pt1, WFNSA p.296.

Passifloraceae (passion flower family)

Basananthe pedata (Baker f.) W.J.de Wilde

Derivation: *Pedata*, like a bird's foot [Latin], i.e. with a few divisions radiating from the same centre, referring to the form of the leaves.

Identification: Erect, usually unbranched **annual or biennial herb**, c. 40 cm tall. **Leaves** grey-green, alternate, palmate with 3–7 deeply incised main lobes, c. 5 × 3 cm, margins slightly dentate. **Flowers** growing in pairs on long stalks from the leaf axils, white, c. 5 mm dia., 5-keeled, sepals white on the upper surface and pale silvery-green below. **Fruit** ellipsoid, pendant, single-seeded, with persistent sepals. **Aromatic** when crushed.

Habitat: Locally rare in lightly shaded mopane woodland.

Flowering: Early in the main rains.

FZ Vol.4 pt0.

Plumbaginaceae (sea lavender or plumbago family)

Plumbago zeylanica L.

Common names: Setswana bogoma, masigomabe; **English** white plumbago, wild plumbago.

Derivations: *Plumbago*, leadwort a name derived from *plumbum*, lead or lead-like [Latin]; *zeylanicus*, of Ceylon. *Bogoma*, a plant whose parts readily attach themselves to passing animals. Various plants of different families have this name.

Identification: Perennial **herbaceous climber**, up to c. 1.5 m tall. **Leaves** hairless, alternate, oval, c. 8 × 3.5 cm, smooth margined, with net veining. **Flowers** white, c. 1.5 cm dia.; borne in short spikes from the leaf axils and stem apex; 5 lobes with a small sharp point are fused to form a long (c. 2 cm) corolla tube. **Calyx** tubular, covered with club-shaped sticky glandular hairs.

Habitat: Widespread, found occasionally locally in the treeline bordering the floodplain and on islands.

Flowering: Late in the main rains.

FZ Vol.7 pt1, B Van Wyk Flowers p.78. GM.

Polygonaceae (buckwheat family)

Oxygonum alatum Burch.

Common names: Setswana letswai-la-khudu, motswe; **Hmbukushu** mongwa, kashe.

Derivations: *Oxy*, sharp, pointed or sour [Greek], *gonum*, about the fruit [Greek], this might refer to either the shape or the taste of the fruit; *alata*, winged [Latin], referring to the shape of the fruit. *Mongwa*, salt, referring to the use of the plant as a salt substitute.

Identification: A lax, occasionally branching, **annual herb**. **Stems** up to c. 70 cm long, mainly from a whorl at the base. **Leaves** alternate, linear (with long teeth) to oval (with deeply incised margins), short **hairs** on both surfaces. Whorls of bristled **stipules** at the leaf axils. **Flowers** in long widely spaced spikes, star-shaped, white.

Habitat: Common and widespread in full sun in deeply sandy areas of the floodplain and in clearings in mopane woodland.

Flowering: During the main rains.

Uses and beliefs:
* Hmbukushu use the seeds as a salt-like seasoning when cooking meat. The seeds are sometimes dried and ground but are preferred fresh.
* A strong decoction of the leaves and roots is given to babies who are sick after drinking breast milk. They are given a tablespoon twice a day for 2 or 3 days to stop the vomiting.

MM, MT.

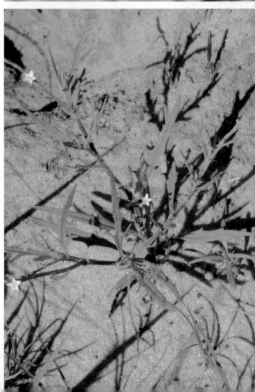

Polygonaceae (buckwheat family)

Oxygonum sinuatum (Meisn.) Dammer

Common names: Afrikaans dubbeltjie.

Derivation: *Oxy*, sharp, pointed or sour [Greek], *gonum*, fruit [Greek], this might refer either to the shape or the taste of the fruit; *sinuatum*, sinuous, strongly wavy [Latin], referring to the leaf margins.

Identification: Lax, occasionally branching, **annual herb**. **Stems** c. 20 cm long, mainly from a whorl at the base. **Leaves** oval, hairless, wavy margined, narrowing at the base to form a stalk. **Flowers** white, star-shaped with 5 petals in long widely spaced spikes, re-flowering beside the fruit. **Fruit** trilobed with an unpleasant spike on each lobe.

Habitat: Found only once in this locality in the shelter of a log near a seasonal pan.

Flowering: During the main rains.

B Van Wyk Flowers p.80, Blundell p.38, Germishuizen p.130.

Rubiaceae (coffee family)

Gardenia volkensii K.Schum. subsp. *spatulifolia* (Stapf & Hutch.) Verdc.

Common names: Setswana morala, monyapula, sulu, kabunga, nnala, moravi; **Subiya** kavangu; **Afrikaans** bosveldkatjiepiering; **English** common gardenia, Transvaal gardenia, bushveld gardenia, savanna gardenia, woodland gardenia.

Derivation: *Gardenia*, for Dr Alexander Garden (1730–91), a Scottish physician who worked in South Carolina and corresponded with Linnaeus; *volkensii*, for Georg Volkens (1855–1917), a German botanist.

Identification: Small sturdy **tree**, up to c. 6 m high with a dense crown. **Bark** pale grey, smooth. **Leaves** and **flowers** arranged on short, stubby, side branches. **Leaves** broadly spathulate, hairy margined, with net veining, occasional warts at the junction between the main and side veins; upper surface shiny but rough to touch, lower surface matt and covered with short hairs. **Leaves**, **branches** and **stipules** all in whorls of 3, branches occasionally in fours. **Flowers** (c. 4 cm dia.) creamy white with a green stripe at one edge of the lower side of each lobe, darkening to yellow as they mature, 6–7 twisted and recurved lobes fused to form a tube; stigma yellow-green and protruding; delicately perfumed. **Fruit** have a rich winey smell when ripe.

Habitat: Widespread but locally uncommon, growing in sandy areas on the edge of the floodplain.

Flowering: Throughout the cool dry period.

Uses and beliefs:
- The fruit is poisonous and causes severe vomiting. An infusion of roots and fruit is taken as an emetic.
- Subiya soak the roots in the daily bath water of new born babies. They believe that the tree is very strong and cannot be broken and that this will make the baby strong.
- The trees are believed to be a protection from lightning.
- The wood is white and fine-grained and is suitable for carving small items.
- May be used to source a black dye for baskets.
- Animals eat the fruit only when they have fallen from the tree and become soft.

Blundell p.152, Ellery p.117, Flowers Roodt p.117, Hargreaves p.65, Setshogo & Venter p.114, WFNSA p.400. GM, KS.

Rubiaceae (coffee family)

Kohautia virgata (Willd.) Bremek.

Common names: Setswana motlhala-wa-pitse.

Derivations: *Kohautia* for Francis Kohaut who collected in Senegal in 1822; *virgata*, twiggy [Latin], referring to the habit of the plant. *Motlhala-wa-pitse*, tracks of the zebra.

Identification: Delicate, branching **annual herb**, c. 30 cm tall. **Leaves** narrowly linear, c. 1.5 cm long, in whorls. **Stipules** in groups at the branching of the flower stalks. **Flowers** small (c. 3 mm dia.), 4 pointed lobes fused to form a corolla tube (c. 3 mm long); lobes dark pink on the underside and white above. **Fruit** cup-shaped with 4 pointed sepals retained on the rim.

Habitat: Found occasionally in open grassland in damp areas of the spillway and on the margins of seasonal pans.

Flowering: During the main rains.

FZ Vol.5 pt1, Germishuizen p.400. GM.

Oldenlandia capensis L.f. var. *capensis*

Derivation: *Oldenlandia*, after Heinrich Bernhard Oldenland (1663–97), a German botanist who travelled in southern Africa; *capensis*, of the Cape of Good Hope.

Identification: Rather lax small **annual herb** with plants growing together to form a loose mat. Mature **roots** yellow. Flowering **stems** erect, c. 13 cm long; rather hairy overall. **Leaves** opposite, linear, c. 20 × 2 mm, folded on the central vein, smooth margined. Bristly **stipules** at the leaf axils. **Flowers** borne singly or in small groups at the leaf axils, white, with 4 lobes, c. 2.5 mm dia., with a pungent **scent**.

Habitat: Widespread in sandy areas of the floodplain.

Flowering: Throughout the rains.

FZ Vol.5 pt1.

Rubiaceae (coffee family)

Oldenlandia corymbosa L. var. *caespitosa* (Benth.) Verdc.

Common names: English false spurry.

Derivation: *Oldenlandia,* for Heinrich Bernhard Oldenland (1663–97), a German botanist who travelled in southern Africa; *corymbosa,* the flowers are arranged in a corymb [Latin]; *caespitosa,* growing in dense clumps [Latin].

Identification: Very slim, erect, multi-branched **annual herb**, c. 20 cm tall with fine stems. **Leaves** stalkless, opposite and decussate, finely linear, smooth margined. A papery **stipular** sheath at the leaf axils. **Flowers** borne singly on long hair-like stalks from the leaf axils, minute (c. 3 mm dia.), white with slight pink markings in the throat, 4 pointed lobes fused to form a tube.

Habitat: Found only occasionally in heavily grazed grassland near seasonal pans.

Flowering: Early in the main rains.

FZ Vol.5 pt1, B Van Wyk New p.82, Blundell p.153.

Rubiaceae (coffee family)

Pavetta cataractarum S.Moore

Common names: Afrikaans Zambezi-bruidsbos; **English** Zambezi brides-bush.

Derivation: *Pavetta*, the name given to this genus in Malabar; *cataractarum*, of waterfalls or cataracts [Latin], referring to the flowers.

Identification: Shrub or **small tree**, usually 1–2 m tall but occasionally up to 4.5 m. **Leaves** leathery, oval, smooth margined, the base narrowing to form the leaf stalk, c. 12 × 4 cm. Leaf undersides have bacterial nodules that form small black dots, which are diagnostic for this genus. **Flowers** (c. 2 cm dia.) in dense terminal heads, showy, white, 4 petals fused to form a corolla tube. **Fruit** round, c. 7 mm dia., crowned with the remains of the calyx lobes.

Habitat: Locally uncommon occurring in the deep shade of the riverine forest.

Flowering: Early in the main rains, fruiting towards the end of the rains.

Palgrave updated p.1115.

Rubiaceae (coffee family)

Spermacoce senensis (Klotzsch) Hiern

Common names: Setswana phesana-yangwana, matlebilo, phesana tsa bathwana; **English** buttonweed.

Derivation: *Spermus* and *coce* both from the Greek for seed; *senex*, old man [Latin], possibly referring to the white hairs on this plant.

Identification: Erect **annual herbaceous** plant that branches mainly from the base, up to c. 50 cm tall. **Stem** square. **Leaves** stalkless, lanceolate to narrowly ovate, c. 4 × 0.7 cm, smooth margined, covered in bristly hairs overall, borne in clusters up the stem. A **stipular** sheath with bristled margin, c. 3 mm long, at the leaf axils. **Flowers** small (c. 3 mm), white with purple veining in the throat, 4 lobes fused to form a tube; borne in clusters in the leaf axils.

Habitat: Locally uncommon except in years of heavy rainfall when these plants are widespread on the floodplains.

Flowering: During the main rains.

Uses and beliefs:
- An infusion of the boiled roots is used to treat irregular menstruation.

FZ Vol.5 pt1, Botweeds p.96.

Sapindaceae (sandolive or litchi family)

Cardiospermum halicacabum L.

Common names: Afrikaans blaasklimop, opblaaboontjie; **English** balloon vine, black winter cherry, heart seed, heart pea.

Derivation: *Kardia*, heart [Greek], *sperma*, seed [Greek], referring to the heart-shaped spot on the seeds; *halikakaban* is the Greek name for the genus *Physalis*, the fruit of this species are similar to those of *Physalis*.

Identification: Herbaceous climber, c. 2 m tall depending on support. **Leaves** compound, c. 9.5 × 12 cm, trifoliolate, oval and deeply incised, margins decurrent. Leaf stalk tends to bend at a right-angle to the leaf blade. Single **male flowers** in leaf axils; minute white **female flowers** in groups on a c. 10 cm stalk, 2 tendrils at each side of the flower cluster (c. 8 mm dia.). **Fruit** 3-winged, c. 2.2 cm long, swelling like a balloon as they ripen.

Habitat: Usually found only occasionally locally but widespread in years of heavy rainfall. *Cardiospermum halicacabum* appears to grow wherever it can find suitable support: in both light and heavy shade in the treeline or on isolated termite mounds in the floodplain or even in tall grasses (e.g. *Phragmites australis*) in wetland areas.

Flowering: During the main rains.

FZ Vol.2 pt2, WFNSA p.238.

Solanaceae (potato family)

Solanum tarderemotum Bitter

Derivation: *Solanum*, the ancient Roman name for one plant in this family, probably this one as it is widespread in Europe, may derive from the Latin *solamen* referring to the soothing or narcotic properties of some species; *tarde*, slowly [Latin], *motum*, movement [Latin].

Identification: Large **bushy herb**, c. 1 m tall. **Stems** covered in short white **hairs**. **Leaves** alternate, oval, narrowing at the base to form the leaf stalk, margins sparsely toothed. **Flowers** white, borne in clusters on a short stalk mid-way between the leaf axils; small (c. 5 mm dia.), star-shaped with 5 pointed lobes. Stamens protuberant, bright yellow. **Fruit** small, spherical, becoming shiny and black when ripe.

Habitat: Uncommon locally, growing in light shade in moist areas on islands in the floodplain. [Introduced.]

Flowering: During the main rains.

Verbenaceae (verbena family)

Phyla nodiflora (L.) Greene

Common names: English fog grass, daisy lawn.

Derivation: *Phyla*, race, tribe or class [Greek]; *nodiflora*, flowering at the nodes [Latin].

Identification: Small mat-forming **marginal herb**, c. 30 cm tall; hairless overall. **Leaves** opposite and decussate, spathulate with margin dentate in the upper half, c. 25 mm (inc. c. 3 mm leaf stalk) × 13 mm. **Flowers** on long stalks (c. 5 cm) arising from the leaf axils, a dark brown composite head with a fringe of white florets with magenta or yellow throats, c. 5 mm dia.

Habitat: A locally common moisture-loving species growing in marshy areas, along water courses and around pans.

Flowering: Main rains into the cool dry period where moisture persists.

Grasses

Introduction to the grasses

The key features used to identify grasses are overall structure, life history (annual or perennial), inflorescence structure, root structure, form of the ligule (a structure at the junction of the leaf sheath) and form of the leaf blade. In this field guide, we have used the term 'inflorescence' to include the entire structure supporting the spikelets at the top of the stem and any subtending leaf or structure.

The grasses are arranged in this book according to the features contained in the inflorescence, then by genus and species. We are very indebted to Tom Cope of the Royal Botanic Gardens, Kew for providing a system of classification of grasses developed from the work of Bor in the Flora of Iraq.

When attempting to identify a grass, be sure to select a specimen with a fully developed inflorescence structure, preferably with stamen or anthers showing. Taking an immature specimen can lead to misidentification of, say, an open panicle as a contracted one. Further down the plant, the stem needs to be in prime condition and not too young so that structures such as the ligule are properly developed. Long feathery awns can obscure the structure of a grass, so care is needed to tease out the details of the structure under examination.

Brief illustrated descriptions of the inflorescence structures used in this book follow.

Simple spike

Spikelets stalkless (sessile) and all alike. They are arranged singly along an axis (rhachis) either in one row (uniseriate) or two alternating rows (biseriate), in a single, unbranched, spike.

Spike of spikes

Spikelets arranged in spikes, as in a simple spike, but several to many such spikes are arranged along an elongated axis. This axis is longer than the spikes themselves. Sometimes the spikelets occur in pairs, one of them shortly stalked, but the spikelets in a pair are always otherwise identical.

Digitate spike

A spike of spikes that are not arranged along an elongated axis, instead they radiate from the tip of the stem like the outspread fingers of the hand. Sometimes one or more spikelets are set below the tip (semi-digitate), but the distance between the lowest of these and the tip of the stem is always much less than the length of the individual spikes. True digitate and semi-digitate forms tend to intergrade and are thus treated as a single group.

False raceme

Spikelets appear to be arranged in simple spikes but, on closer inspection, are found to be in pairs at each joint (node) of the rhachis. The spikelets comprising a pair are different in size and structure: one of them is sessile, the other has a stalk (pedicel). These false spikes may be solitary, paired or in groups of three or more, and are usually subtended by a modified leaf (spathe). Several of these structures may be gathered together into a leafy false panicle.

Plumose panicle

A richly branched panicle resembling a small feather duster.

Open panicle

The primary and secondary branches are widely spreading. The spikelets are often on conspicuous stalks (pedicels) that are also spreading, but the pedicels may be rather short and stiff.

Subcontracted panicle

Not always absolutely distinct from an open panicle and some species may occur in both groups depending on growth conditions. The primary branches are spreading, as in an open panicle, but the secondary branches (if any) and spikelets are condensed about them. Not to be confused with a spike of spikes because either the spikelets clearly have stalks or secondary branches are present.

Contracted panicle

Not only are the spikelets condensed about the primary branches, but the primary branches are themselves condensed about the main axis.

Spike-like panicle

The nature of the spike-like panicle is obscure, but it is always tight and cylindrical. Either the spikelets are in deciduous groups (*Tragus*) or they are surrounded by an involucre of bristles. These bristles may persist after the spikelets have fallen (*Setaria*) or fall with them (*Cenchrus* or *Pennisetum*).

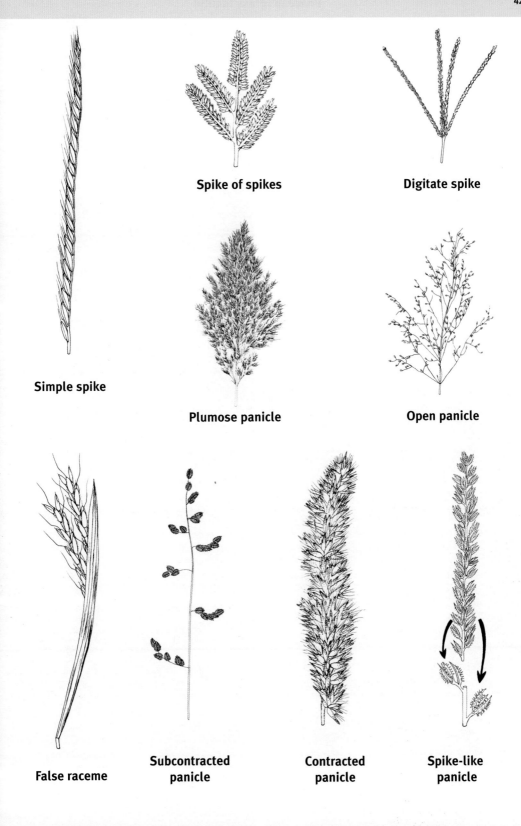

Poaceae (grass family)

Simple spike

Enteropogon macrostachyus (Hochst. ex A.Rich.) Munro ex Benth.

Common name: English mopane grass.

Derivation: *Entos*, within or inside, *pogon* [Greek], beard [Greek], possibly alluding to the hairy ligule; *macro*, long or big [Greek], *stachys*, an ear of wheat [Greek], referring to the long spike of the inflorescence.

Identification: Tufted branching **short-lived perennial grass**, c. 1.1 m tall with a rhizomatous **rooting** system. **Roots** fleshy and white. **Nodes** a narrow ring of dark red. **Ligules** an untidy ring of long hairs. **Leaves** c. 40 × 7 cm, may be flat or tightly rolled with rasping **hairs** on both surfaces. **Inflorescence** a single arching spike, c. 15 cm long, **spikelets** arranged alternately on the upper side of the spike.

Habitat: Locally rare in partial shade mainly in mopane woodland, occasionally on islands along the floodplain.

Flowering: During the main rains.

FZ Vol.10 pt2, Ellery p.188, Van Oudtshoorn 2 p.74.

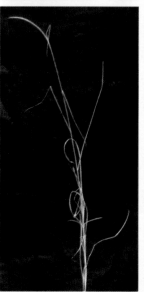
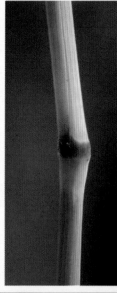

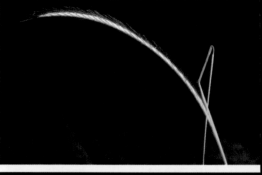

Poaceae (grass family)

Simple spike

Microchloa caffra Nees

Common names: Afrikaans elsgras; **English** pincushion grass.

Derivation: *Micro*, small [Greek], *chloe*, grass [Greek], a small grass; *caffra*, from Kaffraria, the old name for the Eastern Cape of South Africa.

Identification: Small inconspicuous densely tufted **perennial grass**, c. 55 cm tall. **Nodes** dark red. **Ligule** a ring of short hairs. **Leaves** fine and wiry, up to c. 60 cm × 1 mm, with occasional sparse long hairs. **Inflorescence** a single curved raceme, c. 16 cm tall, with the **spikelets** arranged on one side.

Habitat: Locally uncommon on the margin of seasonal pans.

Flowering: Early in the main rains.

FZ Vol.10 pt2, B Van Wyk New p.317, Van Oudtshoorn 2 p.70.

Poaceae (grass family)

Simple spike/spike of spikes

Vossia cuspidata (Roxb.) Griff.

Common names: Setswana mojakubu, monxidi; **English** hippo grass.

Derivations: *Vossia* named for either the German poet, Johann Heinrich Voss (1751–1826) or the Dutch humanist theologian Gerhard Johann Voss (1577–1649); *cuspidata* refers to long flattened awn-like projections on the spikelets. *Mojakubu*, grass for hippopotamus.

Identification: Gigantic **perennial grass** with spongy floating branching stems up to c. 8 m long, stems above water c. 1 m tall. **Leaves** with razor sharp edges, c. 80 × 1.5 cm, flat with a conspicuous broad white mid-rib. **Ligule** a narrow membrane topped with a ring of hairs. **Inflorescence** sessile spikelets, rarely with 2–6 racemes on a short axis, c. 24 cm long.

Habitat: Widespread in fast-flowing channels.

Flowering: Throughout the main rains and into the cool dry period.

Uses and beliefs:
- A species of wild rice.

FZ Vol.10 pt4, Ellery p.206.

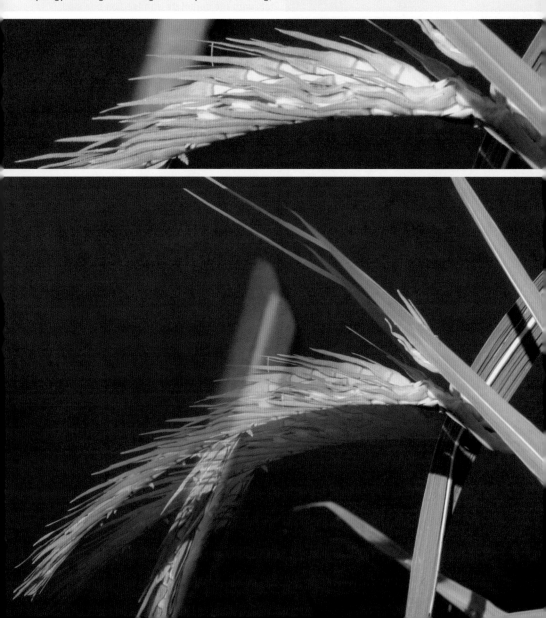

Poaceae (grass family)

Spike of spikes

Brachiaria dura Stapf

Derivation: *Brachiatus*, having arm-like branches [Latin]; *dura*, hard or harsh [Latin].

Identification: Tufted **perennial grass**, c. 1.35 m tall. **Leaves** narrow, rolled. **Nodes** dark red. **Inflorescence** c. 50 cm tall, usually with 2 branches, sometimes 3. **Spikelets** arranged singly up the branches. Anthers brilliant orange.

Habitat: Locally rare, growing in full sun in deeply sandy areas of the floodplain.

Flowering: During the main rains.

FZ Vol.10 pt3.

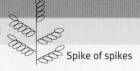 Spike of spikes

Brachiaria grossa Stapf

Derivation: *Brachiatus*, having arm-like branches [Latin]; *grossa*, very large [Latin].

Identification: Much-branched tufted **annual grass**, c. 75 cm tall. **Stems** often bend at the nodes. Stems and leaves covered in unpleasantly coarse **hairs**. **Nodes** green, **roots** produced from the lower nodes. **Leaves** blue-green, c. 30 × 1.3 cm, flat. **Inflorescence** appearing initially as a single spike but when mature irregular branches open from a single axis. **Spikelets** arranged irregularly around branches. Anthers orange, stigma maroon.

Habitat: Rare, found in open areas in mopane woodland by track-sides.

Flowering: During the main rains.

FZ Vol.10 pt3.

Poaceae (grass family)

Spike of spikes

Brachiaria humidicola (Rendle) Schweick.

Common names: English creeping signal grass, creeping false paspalum.

Derivation: *Brachiatus*, having arm-like branches [Latin]; *humidus*, damp or moist [Latin], *colo*, to dwell or inhabit [Latin], alluding to its preference for a damp habitat.

Identification: Perennial grass, c. 1.2 m tall, with a stoloniferous **rooting** system. **Nodes** a double ring of dark red-brown colouration, **stems** often bend and root at the nodes. **Leaves** c. 28 × 0.5 cm, folding at a strong mid-rib and having widely spaced tooth-like bristles on the margin. **Inflorescence** c. 15 cm tall, with 2–3 branches to one side of a central axis. **Spikelets** covered in coarse hairs and arranged on one side of the branch. **Seeds** have a special protective coating which aids dispersal by allowing them to pass through the gut of animals without being digested.

Habitat: Common and widespread throughout the floodplain.

Flowering: During the main rains.

Uses and beliefs:
- A fairly palatable grass.

FZ Vol.10 pt3, Ellery p.183.

Poaceae (grass family)

Spike of spikes

Brachiaria nigropedata (Ficalho & Hiern) Stapf

Common names: Setswana motetene; **English** black-footed grass, spotted signal grass, black-footed brachiaria.

Derivations: *Brachiatus*, having arm-like branches [Latin]; *nigro*, black [Latin], *-pedatus*, -footed [Latin]. *Black-footed grass*, alludes to the short stalks of the mature spikelets which are dark purple to black in colour.

Identification: Densely tufted **hairy perennial grass**, c. 80 cm tall. **Nodes** a hairy dark red ring. **Leaves** c. 35 × 0.8 cm, flat, covered in velvety hairs, tending to curl when they dry. **Inflorescence** c. 15 cm tall. Branches arranged at more or less regular intervals up a central axis, 2 rows of spikelets on the lower side. Spikelets covered in soft hairs.

Habitat: Locally uncommon in deep sand in mopane woodland.

Flowering: During the main rains.

Uses and beliefs:
• A palatable grass that quickly disappears if overgrazed.

FZ Vol.10 pt3.

Poaceae (grass family)

Spike of spikes

Echinochloa colona (L.) Link

Common names: English jungle rice, awnless barnyard grass, bird's grass, marsh grass.

Derivation: *Echinos*, hedgehog [Greek], *chloe*, grass [Greek]; *colona*, of farms [Latin].

Identification: Annual grass growing either in or on the edge of standing water, up to 1 m tall; **hairless** overall. **Stem** has 2 distinctly curved sides. **Ligule** absent. **Leaves** flat, c. 16 × 0.5 cm. Young leaves often have purple stripes. **Inflorescence** spike of spikes with the racemes more or less evenly spaced along a central axis and sloping slightly down, c. 13 cm tall. **Spikelets** neatly arranged in fours along the underside of the branches.

Habitat: Common and widespread in shallow water on the margin of seasonal pans.

Flowering: Throughout the main rains.

Uses and beliefs:
• A palatable grass that may be used as a cereal in times of famine.

FZ Vol.10 pt3, Van Oudtshoorn 2 p.207.

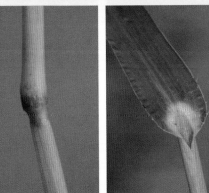

Poaceae (grass family)

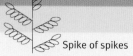
Spike of spikes

Echinochloa jubata Stapf

Derivation: *Echinos*, hedgehog [Greek], *chloe*, grass [Greek]; *jubata*, crested [Latin], alluding to the long awns on the spikelets.

Identification: Rather lax rambling **annual grass**, c. 90 cm tall. **Nodes** green. **Ligule** a ring of short hairs. **Leaves** c. 12 × 0.5 cm, folding at the mid-rib, margins occasionally wavy towards the base. **Inflorescence** spike of spikes with branches arranged alternately and more or less evenly up stem, c. 12 cm tall. **Spikelets** with long awns, tightly packed on one side of the branches in triple rows.

Habitat: Found occasionally locally in dried-out areas on the edge of lagoons and in damp parts of the floodplain.

Flowering: During the main rains.

FZ Vol.10 pt3.

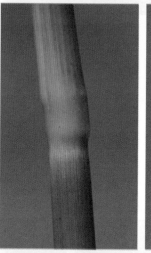
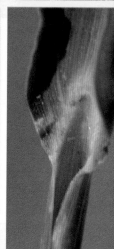

Poaceae (grass family)

Spike of spikes

Echinochloa stagnina (Retz.) P.Beauv.

Common names: English long-awned water grass, water grass, hippo grass.

Derivation: *Echinos*, hedgehog [Greek], *chloe*, grass [Greek]; *stagnina*, growing in standing water [Latin].

Identification: Creeping **annual** or **short-lived perennial grass**, c. 95 cm tall. **Stems** pithy, which helps them to float, bending at and rooting from the nodes. **Nodes** green. **Ligule** absent or a ring of short hairs. **Leaves** c. 30 × 1 cm, folding slightly at the mid-rib, with occasional indistinct purple blotched stripes. The plant is hairless except for the upper leaf surface which has rasping hairs. **Inflorescence** with alternate branches held closely to a central axis, c. 20 cm tall. **Spikelets** with awns c. 2 cm long.

Habitat: Locally rare on the margin of seasonal pans.

Flowering: During the main rains.

Uses and beliefs:
- A palatable grass.

FZ Vol.10 pt3, Ellery p.188.

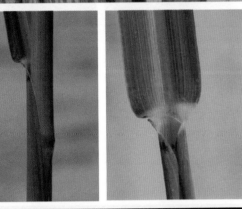

Poaceae (grass family)

 Spike of spikes

Leptochloa fusca (L.) Kunth

Derivation: *Lepto-*, thin or slender [Greek], *chloe*, grass [Greek]; *fusca* [Latin], brown, dusky [Latin].

Identification: Aquatic or semi-aquatic **perennial grass**, c. 1.1 m tall. Often branching and rooting from the lower nodes, with a **rhizomatous rooting** system. **Nodes** pale green. **Ligule** a pointed membrane. **Leaves** c. 40 × 0.3 cm, folding at the midrib which is white. **Inflorescence** a spike of spikes, c. 40 cm tall, with long slender branches radiating from a central axis. **Spikelets** alternately spaced. Stamens dusky pink.

Habitat: Locally rare on the margin of seasonal pans.

Flowering: During the main rains.

FZ Vol.10 pt2.

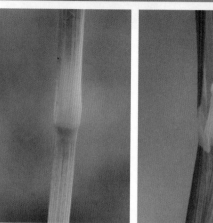

Poaceae (grass family)

Spike of spikes

Oplismenus burmannii (Retz.) P.Beauv.

Derivation: *Hoplismos*, equipment for war [Greek], alluding to the awns on the spikelets; *burmannii* for the Dutch botanist and physician Johannes Burman (1707–79), professor of botany at Amsterdam, a close friend of Linnaeus.

Identification: Small trailing **annual grass**, c. 20 cm long. **Nodes** and mature **leaf sheaths** slightly **hairy**. Narrowly ovate **leaves**, c. 5.5 × 1.5 cm, with an attractive ripple in their surface. **Inflorescence** spike of spikes, c. 5 cm long. **Spikelets** arranged in pairs, awns c. 2 cm long. Stigmas dark pink.

Habitat: Locally uncommon in deeply shaded riverine forest.

Flowering: Late in the main rains.

FZ Vol.10 pt3.

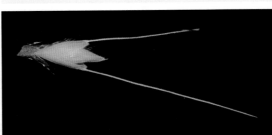

Poaceae (grass family)

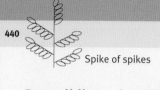
Spike of spikes

Paspalidium obtusifolium (Delile)
N.D.Simpson

Derivation: *Paspalidium*, a diminutive form of the generic name *Paspalum* from *paspalos*, millet [Greek]; *obtusus*, blunt [Latin], *folius*, leaf [Latin], blunt-leaved.

Identification: Rather lax **perennial grass** deeply rooted in the water margin, c. 90 cm tall. **Stems** hollow, with a distinct restriction at the **nodes**. **Leaves** c. 30 × 1.5 cm, rolled, may be obtuse or acute at the tip. **Inflorescence** spike of spikes, c. 24 cm tall, short side branches arranged alternately and held tightly against the main stem. **Spikelets** arranged alternately on one side of the branch. Anthers orange, stigmas maroon.

Habitat: Found occasionally locally in the margin of slow-flowing channels.

Flowering: Late in the main rains.

FZ Vol.10 pt3.

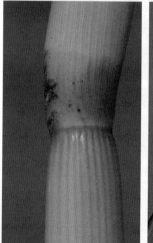

Poaceae (grass family)

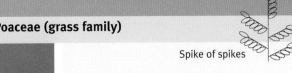

Spike of spikes

Paspalum scrobiculatum L.

Common names: Setswana puka; **English** veld paspalum, creeping paspalum, ditch grass.

Derivations: *Paspalos*, millet [Greek]; *scrobiculus*, little ditch [Latin], referring to the fact that ditch grass thrives in wet places. *Puka*, guinea fowl, because the seeds drop and are eaten by guinea fowl.

Identification: Rank, untidy, mat-forming short-lived **perennial grass**. **Stems** up to c. 1.1 m long. **Nodes** red. **Ligule** a small pinkish membrane. **Leaves** c. 20 × 0.7 cm, flat. **Inflorescence** spike of spikes, usually with 2 branches, c. 7 cm long, with 2 rows of overlapping rounded spikelets on the lower side. **Spikelets** quickly ripen to black and soon fall. Anthers creamy yellow turning orange with age, stigmas white.

Habitat: Locally rare in areas of the floodplain where water collects in years of heavy rain.

Flowering: Late in the main rains.

FZ Vol.10 pt3, Van Oudtshoorn 2 p.245.

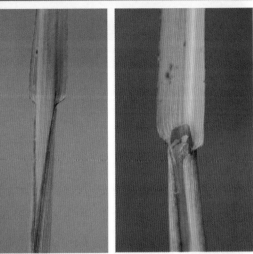

Poaceae (grass family)

Spike of spikes

Pogonarthria fleckii (Hack.) Hack.

Common names: Afrikaans eenjarige denneboomgras; **English** herringbone grass.

Derivation: *Pogon*, beard [Greek], *arthron*, joint [Greek], alluding to the hairs on the central stalk of the spikelet; *fleckii* for the German geologist E. Fleck who collected plants in Namibia in 1888.

Identification: Tufted **annual grass** with occasionally branching stems, c. 60 cm tall. **Stems** with a tendency to bend at the lower nodes. **Nodes** green. **Ligule** is 2 tufts of long white hairs and a ring of short hairs. **Leaves** slightly rolled, c. 20 × 0.6 cm, with **sparse long hairs** on the outer surface and towards the base of the inner surface; leaf sheath covered in erect white hairs. **Inflorescence** spike of spikes, c. 30 cm tall, with unbranching side branches regularly spaced along a central axis.

Habitat: Common and widespread in sandy areas in full sun.

Flowering: Late in the main rains.

Uses and beliefs:
- A pioneer grass of little nutritional value.
- May be an indicator of drought or over-grazing.

FZ Vol.10 pt2, Müller p.212.

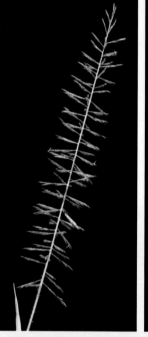

Poaceae (grass family)

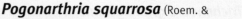
Spike of spikes

Pogonarthria squarrosa (Roem. & Schult.) Pilg.

Common names: Setswana lepheto, seloka; **Afrikaans** sekelgras, meerjarige denneboomgras; **English** herringbone grass, cross grass, sickle grass.

Derivation: *Pogon*, beard [Greek], *arthron*, joint [Greek], alluding to the hairs on the central stalk of the spikelet; *squarrosa*, with parts spreading or recurved at the ends [Latin].

Identification: Vigorous tufted **short-lived perennial grass**, c. 90 cm tall. **Nodes** yellowish green. **Ligule** a ring of short hairs. **Leaves** c. 20 × 0.4 cm, rolled, inner surface with **rasping hairs**. **Inflorescence** a brown spike of spikes, c. 30 cm tall, with branches arranged in whorls around the stem at more or less regular intervals.

Habitat: Widespread in sandy areas in full sun.

Flowering: Late in the main rains.

Uses and beliefs:
- A pioneer grass of little nutritional value.
- An indicator of poor, sandy soils.

FZ Vol.10 pt2, B Van Wyk Flowers p.326, Ellery p.199, Müller p.214, van Oudtshorn p.201, WFNSA p.24.

Poaceae (grass family)

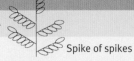
Spike of spikes

Trichoneura grandiglumis (Nees) Ekman

Common names: Afrikaans kleinrolgras; **English** rolling grass, tumbleweed.

Derivation: *Thrichos*, hair [Greek], *neura*, vein or nerve [Greek], with hair-like veins; *grandis*, large [Latin], *glumis*, glume or spikelet sheath [Latin], alluding to the fact that the sheaths are longer than the spikelet.

Identification: Tufted **perennial grass**, c. 90 cm tall, with occasionally branching stems. **Nodes** green above and pink below a dark brown ring. **Ligule** a wide membrane. **Leaves** c. 11 × 0.6 cm, flat. **Inflorescence** a large spike of spikes, long horizontal side branches arranged at more or less regular intervals up a central axis, c. 40 cm tall. When mature, the inflorescence breaks off and rolls away in the wind, aiding dispersal. **Spikelets** spaced by almost their own length up the branches.

Habitat: Found occasionally locally on the floodplain.

Flowering: Late in the main rains.

Uses and beliefs:
- An unpalatable grass with low leaf production.

FZ Vol.10 pt2, B Van Wyk New p.338, Van Oudtshoorn 2 p.211.

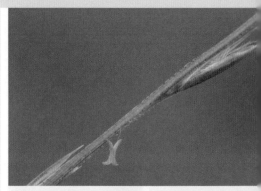
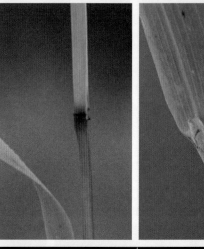

Poaceae (grass family)

Spike of spikes

Urochloa brachyura (Hack.) Stapf

Common names: Setswana and **Shiyeyi** phuka; **Setswana** phoka; **Ovambenderu** oruejo; **Afrikaans** merjarige beesgras; **English** gonya grass.

Derivations: *Oura*, tail [Greek], *chloe*, young green corn or grass [Greek], referring to the sharply pointed inner spikelet sheaths; *brachys*, short [Greek], *oura*, tail [Greek], possibly because this species has shorter spikelet sheaths. *Phoka*, guinea fowl, because the seeds drop and are eaten by guinea fowl.

Identification: Hairy **annual grass**, c. 75 cm tall, with branching stems. **Nodes** pink above, green below, with a ring of hairs. **Ligule** a ring of sparse long hairs. **Leaves** softly hairy, c. 10 × 1 cm, flat or folding at the mid-rib. **Inflorescence** spike of spikes, with slightly crooked branches arranged at roughly regular intervals up the central axis, up to c. 8 cm tall.

Urochloa brachyura may easily be confused with *U. trichopus* but it has narrower spikelets and is generally more hairy. It is identical to *U. oligotricha* except that it is a perennial.

Habitat: Found occasionally locally in hard sandy clay on the open floodplain.

Flowering: During the main rains.

Uses and beliefs:
• Wayeyi women collect the high-protein seed as a food. It can be stored for a long time. It is pounded like mealie. Mixed with water, it can be eaten cold or cooked and tastes like millet.
• San also grind the seeds to make flour.
• A highly palatable grass much appreciated by game.

FZ Vol.10 pt3, Müller p.272. TK, ET.

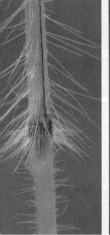

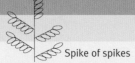
Spike of spikes

Poaceae (grass family)

Urochloa trichopus (Hochst.) Stapf

Common names: Setswana and **Shiyeyi** phuka; **Setswana** phoka; **Ovambenderu** oruejo; **English** signal grass.

Derivations: *Oura*, tail [Greek], *chloe*, young green corn or grass [Greek], referring to the sharply pointed inner spikelet sheaths; *thrix*, hair [Greek], *-pus*, footed [Greek]. *Phoka* or *phuka*, guinea fowl, because the seeds drop and are eaten by guinea fowl.

Identification: Coarse **annual grass**, c. 40 cm tall. **Stem** red-brown. **Node** a hairy ring. **Ligule** a ring of hairs. **Leaves** softly hairy, sometimes folding at the mid-rib, c. 40 × 1.5 cm. **Inflorescence** spike of spikes, with up to 10 side branches regularly spaced along a central axis, c. 10 cm tall, may be slightly hairy.

Urochloa trichopus may easily be confused with *U. brachyura* but has rounder spikelets. It is almost identical to *U. mosambicensis* except that it is a perennial and usually has a slightly taller slimmer inflorescence.

Habitat: Found occasionally locally in almost any habitat from deep shade of the riverine forest to full sun on the floodplain.

Flowering: Throughout the main rains.

Uses and beliefs:
- Wayeyi women collect the high-protein seed as a food. It can be stored for a long time. It is pounded like mealie. Mixed with water, it can be eaten cold or cooked and tastes like millet.
- San also grind the seeds and use them as flour.
- A highly palatable grass much appreciated by game.

FZ Vol.10 pt3, Van Oudtshoorn 2 p.249. TK, ET.

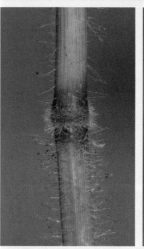
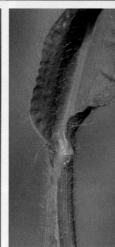

Poaceae (grass family)

Digitate spike

Chloris virgata Sw.

Common names: Setswana tshitladingwetsi; **Afrikaans** witpluimgras; **English** feather top chloris, feathertop grass, blue grass, feather finger grass, hay grass, old-land's grass.

Derivation: *Chloris*, dedicated to Chloris, the Greek goddess of flowers and the personification of spring; *virgata*, twiggy [Latin].

Identification: Tufted **annual** or **short-lived perennial grass**, c. 40 cm tall. **Leaves** c. 20 × 0.5 cm, folding at the mid-rib. **Inflorescence** digitate spikes from a single point, c. 9 cm tall. **Spikelets** feathery, arranged along one side of spikes, often turning black late in the season.

Habitat: Widespread and common on the floodplain and along track-sides.

Flowering: During the main rains.

Uses and beliefs:
• Valuable grazing in arid areas where few palatable grasses grow.

FZ Vol.10 pt2, B Van Wyk New p.323, Van Oudtshoorn 2 p.231.

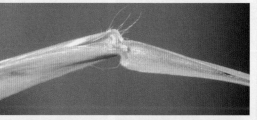

Poaceae (grass family)

Digitate spike

Cynodon dactylon (L.) Pers.

Common names: Setswana motlhwa, motlho, nganje, ngxaio; **Afrikaans** kweek; **English** couch grass, quick grass, Bermuda grass.

Derivation: *Kyon*, dog [Greek], *odontos*, tooth [Greek], alluding to the scale-like leaves at the end of the creeping stems which resemble dog's teeth; *dactylos* a finger [Greek], referring to the finger-like inflorescence.

Identification: Small **perennial grass**, up to c. 44 cm tall, with rhizomatous and stoloniferous rooting systems. **Nodes** usually green but occasionally have red stripes above and below. **Ligule** of some long hairs and an inconspicuous membrane or ring of short hairs. **Leaves** c. 9 × 0.3 cm, flat. **Inflorescence** digitate spikes, c. 5 cm tall, with branches from a single point. Anthers creamy pink, stigma maroon.

Habitat: Common and widespread on the floodplain margin and on the banks of rivers and lagoons. Prefers termite mounds and areas of high salinity.

Flowering: Throughout the rainy period.

Uses and beliefs:
- The leaves are rich in vitamin C and are edible.
- Also used medicinally.
- Underground stems, which grow up to 1 m deep, may cause hydrocyanic poisoning to livestock if eaten when wilting.

FZ Vol.10 pt2, B Van Wyk New p.324, Ellery p.185, Van Oudtshoorn 2 p.229.

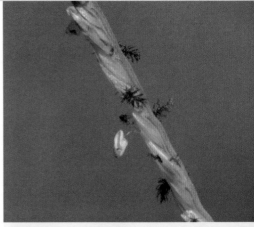

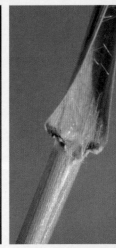

Poaceae (grass family)

Digitate spike

Dactyloctenium aegyptium (L.) Willd.

Common names: Hmbukushu dingofo; **English** crowfoot, common crowfoot, duck grass, coast button grass, Egyptian finger grass.

Derivation: *Daktylos*, finger [Greek], *ktenos*, comb [Greek], alluding to the comb-like fingers of the inflorescence; *aegyptium*, Egyptian.

Identification: Small tufted **annual grass** c. 30 cm tall rooting from the lower nodes. **Leaves** hairy, up to c. 11 × 0.7 cm, tightly rolled, with hairy margins and prominent mid-ribs. **Inflorescence** a digitate spike from a single point with relatively short thick branches.

Habitat: Widespread but found only occasionally locally in damp areas on the banks of channels and lagoons.

Flowering: During the main rains.

Uses and beliefs:
• A reasonably palatable grass that can quickly colonise disturbed areas but is not particularly valuable for grazing.

FZ Vol.10 pt2, Van Oudtshoorn 2 p.236.

Poaceae (grass family)

Digitate spike

Dactyloctenium giganteum B.S.Fisher & Schweick.

Common names: Setswana ngharara, tadwa; **Hmbukushu** kangungwe; **English** giant crowfoot.

Derivations: *Daktylos*, finger [Greek], *ktenos*, comb [Greek], alluding to the comb-like fingers of the inflorescence; *giganteum*, giant [Latin], referring to the large size of this grass. *Giant crowfoot*, the branches of the inflorescence curl upwards on ripening to form the shape of a crow's foot.

Identification: Strong tufted **annual grass**, up to c. 1.3 m tall, with **stoloniferous rooting** from the lower nodes. **Stems** branching. **Nodes** green. **Ligule** a small membranous ring with a hairy margin. **Leaves** c. 40 × 1.1 cm, flat, both surfaces smooth with bristled margins. **Inflorescence** a digitate spike, branching from a single point on a central axis, with 3–6 branches of c. 8 cm in length. **Spikelets** arranged in a double row on the lower side of the branch.

Habitat: Common and widespread on the open floodplain and in lightly shaded areas of woodland.

Flowering: Throughout the main rains.

Uses and beliefs:
• A palatable grass especially when young. Makes good hay when cut early.
• Hmbukushu harvest the grain, pound it and cook it like mealie.

FZ Vol.10 pt2, Ellery p.186, Van Oudtshoorn 2 p.235. JC, MT.

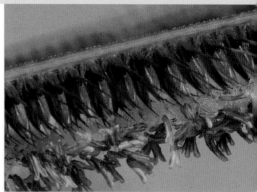

Poaceae (grass family)

Digitate spike

Digitaria debilis (Desf.) Willd.

Common names: English finger grass.

Derivation: *Digitus*, finger [Latin]; *debilis*, weak, feeble [Latin].

Identification: Erect tufted **perennial grass**, c. 1 m tall, with occasionally branching stems and a stoloniferous **rooting** system, often deeply rooted in the water margin. **Stems** red from the first node below the inflorescence and green where they are covered by the leaf sheath. **Nodes** dark red with a ring of hairs. **Ligule** a membranous ring. **Leaves** c. 13 × 0.7 cm, folding at the mid-rib; leaves and leaf sheaths hairy. **Inflorescence** a digitate spike, c. 15 cm tall, branching close together from a single axis.

Habitat: Widespread found occasionally locally in damp areas of the floodplains and often rooted in water.

Flowering: Late in the main rains and into the beginning of the cool dry period.

FZ Vol.10 pt3, Ellery p.186.

Poaceae (grass family)

Digitate spike

Digitaria milanjiana (Rendle) Stapf

Common names: Setswana namele, moseka?; **Hmbukushu** nondothinde, tswa; **Afrikaans** makarikari vingergras, milanje vingergras, panvingergras; **English** milanje finger grass, makarikari finger grass, milanje grass.

Derivation: *Digitus*, finger [Latin]; *milanjiana*, from Milanje in South Africa.

Identification: Erect **perennial grass**, c. 1.5 m tall, with a rhizomatous **rooting** system with occasional stolons. **Stems** and leaf sheaths **hairy**. **Nodes** green. **Ligule** an insignificant membrane with a hairy margin. **Leaves** c. 30 × 0.8 cm, flat. **Inflorescence** c. 25 cm tall, digitate spike, with long slim branches arising close together from a single axis. **Spikelets** arranged in pairs along the branches.

Habitat: Widespread in areas of deep sand on the floodplain and throughout lightly shaded areas of mixed woodland.

Flowering: Late in the main rains.

Uses and beliefs:
• Used by Hmbukushu to weave bangles.

FZ Vol.10 pt3. MT.

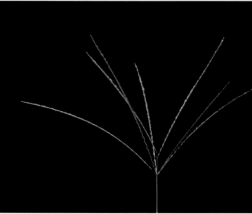

Poaceae (grass family)

Digitate spike

Digitaria sanguinalis (L.) Scop.

Common names: English crab finger grass, crab grass, crop grass, land grass.

Derivation: *Digitus*, finger [Latin]; *sanguineus*, blood-red [Latin].

Identification: Sparsely tufted **annual grass**, up to c. 1.1 m tall. **Stems** hairless and occasionally branching, tending to bend at and may **root** from the lower nodes. **Nodes** tinged brown, paler green above than below. **Ligule** a membranous sheath with 2 tufts of long hairs. **Leaves** c. 25 × 1.2 cm, flat, hairy on both surfaces, sheath covered in short erect hairs. **Inflorescence** c. 18 cm tall, digitate spike with branches arising from almost a single point.

Spikelets arranged alternately on the lower side of the branches, one with a stalk the next without.

Habitat: A widespread naturalised weed found occasionally locally in lightly shaded mixed woodland.

Flowering: Throughout the main rains.

Uses and beliefs:
- Of little value for grazing because it has little foliage.

FZ Vol.10 pt3, Van Oudtshoorn 2 p.225.

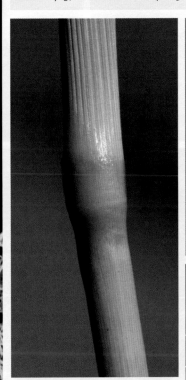

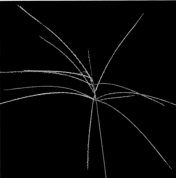

Poaceae (grass family)

Digitate spike

Digitaria velutina (Forssk.) P.Beauv.

Common names: English flaccid finger grass, long plumed finger grass, finger grass.

Derivation: *Digitus*, finger [Latin]; *velutina*, velvety [Latin].

Identification: Almost creeping **annual grass**, c. 60 cm tall, bending and rooting at the nodes. **Stems** branching. **Nodes** green. **Ligules** membranous with 2 tufts of long hairs, c. 2.5 mm long. Leaves and stems covered in dense **hairs**. **Leaves** up to c. 13 × 1.3 cm, flat. **Inflorescence** a digitate spike, radiating singly or more or less in pairs from a central axis, c. 13.5 cm tall.

Habitat: Common and widespread in damp patches of deep shade in clearings within mixed woodland.

Flowering: Throughout the main rains.

Uses and beliefs:
• A palatable pioneer grass but of little use for grazing because of its low foliage production.

FZ Vol.10 pt3, Van Oudtshoorn 2 p.215.

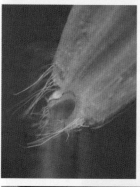
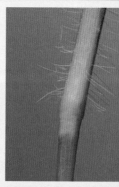

Poaceae (grass family)

Digitate spike

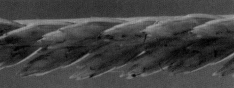
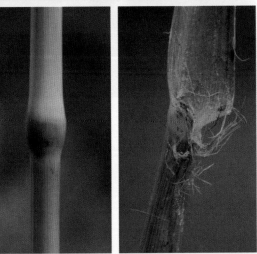

Eleusine indica (L.) Gaertn.

Common names: Setswana selekangwetsi; **English** goose grass, Bermuda grass, crab grass, crowfoot grass, dog's tail, Indian millet, land grass.

Derivation: *Eleusine* for the city of Eleusis in Greece where the temple of Ceres stood; *indica*, of India, this grass is a native of southern Africa but many plants that reached Europe on board an 'indiaman' were given the epithet *indica* wherever they originated.

Identification: Strong tufted **perennial grass**, c. 95 cm tall. **Stems** branching, often oval. **Nodes** a darker green ring with paler green above and below. **Ligule** a ring of sparse short hairs with a mass of long tousled hairs on the outer surface. **Leaves** c. 30 × 0.8 cm, folding at the mid-rib, with sparse long hairs on the inner surface. **Inflorescence** a digitate spike, c. 20 cm tall, with branches c. 17 cm long almost all arising from a single point. **Spikelets** flattened, arranged in 2–3 rows on the underside of the branches.

Habitat: Widespread but only found occasionally locally in damp and disturbed areas on the floodplain.

Flowering: During the main rains.

Uses and beliefs:
• A highly nutritious grain. Improved forms were the staple diet of southern Africa several thousand years ago.

FZ Vol.10 pt2.

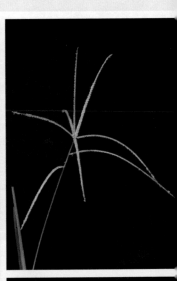

Poaceae (grass family)

Digitate spike

Eulalia aurea (Bory) Kunth

Derivation: *Eulalia* for the painter Eulalie Delile who illustrated the work of the French naturalist Victor Jacquemont; *aurea*, golden [Latin].

Identification: Tufted **perennial grass** scrambling through other grasses and vegetation, c. 1.8 m tall. Lower **nodes** often bending and tending to root where they touch the ground. **Ligule** a rather uneven membrane, fringed with very short hairs. **Leaves** c. 14 × 0.7 cm, flat, hairless. **Inflorescence** digitate spikes with all the golden brown branches springing almost from a single point.

Habitat: Found occasionally on seasonally flooded floodplain.

Flowering: During the main rains.

FZ Vol.10 pt4.

Poaceae (grass family)

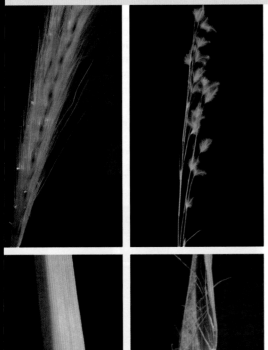

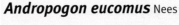
False raceme

Andropogon eucomus Nees

Common names: Afrikaans veergras; **English** snowflake grass, silver thread grass, old man's beard.

Derivation: *Andro*, male [Greek], *pogon*, beard [Greek], alluding to the hairy inflorescence; *eu*, well [Greek], *kome*, hair [Greek], implying a beautiful head, again referring to the inflorescence.

Identification: Tufted **perennial grass**, up to c. 1 m tall. **Stems** branching, hairless, pink; leaf sheaths green, giving the impression of alternate pink and green sections of stem. **Ligule** membranous. **Leaves** short and fine, tightly folded at the mid-vein, c. 10 × 0.4 cm. **Inflorescence** false racemes, c. 25 cm tall, with conspicuous fluffy white spikelets arranged digitately.

Habitat: Widespread but only found occasionally and then in large stands of plants in damp areas of the floodplain.

Flowering: Throughout the rainy period.

Uses and beliefs:
• An unpalatable grass that is rarely grazed.

FZ Vol.10 pt4, B Van Wyk New p.327, Ellery p.182, Van Oudtshoorn 2 p.48.

Poaceae (grass family)

False raceme

Andropogon gayanus Kunth

Common names: English blue grass, Rhodesian blue grass.

Derivation: *Andro*, male, *pogon*, beard [Greek], alluding to the hairy inflorescence; *gayana* for J.E. Gay, an 18th century French civil servant who worked in Senegal.

Identification: Large tufted **perennial grass**, c. 2.2 m tall, with branching stems. **Leaves** c. 60 × 2 cm, narrow and folding to form what appears to be a leaf stalk. Leaf blade c. 35 cm long, with a pronounced white mid-rib. A **waxy coating** on the leaves gives them their **blue colour**. **Inflorescence** a false raceme, c. 70 cm long. Anthers yellow.

Habitat: Found only occasionally, in partial shade in mixed woodland and scrub.

Flowering: Throughout the rainy period.

Uses and beliefs:
- Good grazing but of low nutritional value.

FZ Vol.10 pt4, Hargreaves p.68, Van Oudtshoorn 2 p.58.

Poaceae (grass family)

False raceme

Cymbopogon caesius (Hook. & Arn.) Stapf
(syn. *Cymbopogon excavatus* (Hochst.) Burtt Davy)

Common names: Setswana mokamakama; **Afrikaans** breëblaarterpentyngras; **English** broad-leaved turpentine grass, common turpentine grass, ginger grass, lemon grass.

Derivation: *Kymbe*, boat [Greek], *pogon*, beard [Greek], alluding to the shape of the spikelets; *caesius*, light blue [Latin], referring to the colour of the leaves.

Identification: Erect tufted **perennial grass**, up to c. 2 m tall. **Stems** appear horizontally striped because the stem is cream and the leaf sheaths are green. **Ligule** a rounded membrane. **Leaves** appearing blue-grey because of their waxy surface, flat, usually twice as wide at the base as the stem, c. 30 × 1.5 cm. **Inflorescence** a false raceme, up to c. 70 cm tall, with short side branches. Stamens pale yellow. A pungent **smell** of turpentine when the plant is crushed.

Habitat: Common and widespread across the floodplain in lower areas. Interestingly, when driving across the floodplain at night, the temperature usually drops upon entering stands of this grass.

Flowering: During the main rains.

Uses and beliefs:
- An unpalatable grass because of its strong turpentine taste. It is only grazed when nothing else is available.
- Widely used as a thatching grass.

FZ Vol.10 pt4, B Van Wyk New p.328, Ellery p.185, Van Oudtshoorn 2 p.52.

Poaceae (grass family)

False raceme

Heteropogon contortus (L.) Roem. & Schult.

Common names: Setswana seloka; **Afrikaans** assegaaigras; **English** (common) spear grass, tanglehead, stick grass.

Derivation: *Hetero*, various, diverse [Greek], *pogon*, beard [Greek], referring to there being awned and awnless spikelets in the same raceme; *contortus*, twisted, contorted [Latin].

Identification: Easily recognised later in the season by its tangled, knotted heads. A vigorous **tufted perennial**, c. 1 m tall, often with **branching** stems. **Leaves** c. 25 × 0.7 cm, folding at the mid-rib and often with a blunt tip. **Ligule** an inconspicuous membrane. **Inflorescence** a false raceme, c. 12 cm tall, with awns twisting and bending as they mature.

Habitat: Locally common and widespread in damp disturbed areas of the floodplain along the tracksides.

Flowering: During the main rains.

Uses and beliefs:
- Good grazing when young.
- When older, the seed corkscrews into flesh (as well as into the ground) when dampened; it is thus painful to people and animals, and damages animal hides.

FZ Vol.10 pt4, B Van Wyk New p.316, Ellery p.191, Van Oudtshoorn 2 p.66.

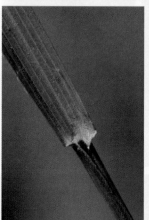

Poaceae (grass family)

False raceme

Hyperthelia dissoluta (Steud.) Clayton

Common names: English yellow thatching grass.

Derivation: *Hyper-*, above, *thele*, nipple [Greek] or *thelys, thelia*, female [Greek], alluding to the male spikelets overtopping the female; *dissoluta*, dissolved [Latin], possibly referring to the rather sparse inflorescence.

Identification: Noticeably large tufted **annual grass**, c. 2 m tall and usually standing above the surrounding grasses. **Stems** yellow, leaf sheaths green, giving the appearance of alternate green and yellow segments on the stems. **Ligule** a membranous ring with distinct ears on either side. **Leaves** unusual as they are sometimes revolute, up to c. 22 × 0.7 cm. **Inflorescence** a sparse false raceme, c. 80 cm long.

Habitat: Widespread in dry areas of the spillway floodplain and in mixed *Combretum* scrub.

Flowering: Late in the main rains.

Uses and beliefs:
- Widely used as a thatching grass.
- Grazed only when young.

FZ Vol.10 pt4, Van Oudtshoorn 2 p.59.

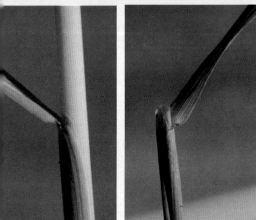

Poaceae (grass family)

False raceme

Schizachyrium jeffreysii (Hack.) Stapf

Common names: English silky autumn grass.

Derivation: *Schizo*, to divide, to split [Greek].

Identification: Very fine lax trailing and branching **perennial grass**, up to c. 2 m tall given sufficient support. **Ligule** membranous. **Leaves** flat, c. 13 × 0.4 cm, lower part with sparse **long white hairs** (c. 5 mm long) along the margin. **Inflorescence** a false raceme with long branches of spikelets alternating up the stem, covered in long white hairs. Stamens pale yellow, stigma maroon.

Habitat: Locally rare on sandy islands on the spillway margin.

Flowering: Late in the main rains and into the cool dry period.

Uses and beliefs:
• A harsh unpalatable grass.

FZ Vol.10 pt4, Van Oudtshoorn 2 p.60.

Poaceae (grass family)

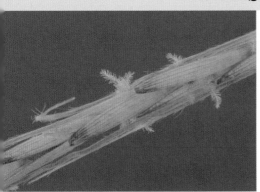

False raceme

Trachypogon spicatus (L.f.) Kuntze
(syn. *Trachypogon capensis* (Thunb.) Trin.)

Common names: Afrikaans bokbaardgras; **English** giant spear grass, grey tussock grass.

Derivation: *Trachys*, rough [Greek], *pogon*, beard [Greek], alluding to the rough hairy awns; *spicatus*, spiked [Latin], referring to either the inflorescence or the spikelets.

Identification: Perennial grass growing in large clumps, c. 1.6 m tall. **Node** a hairy ring. **Ligule** a membrane with 2 pronounced lobes. **Leaves** c. 75 × 0.3 cm, folding at a heavy central mid-rib. **Inflorescence** c. 25 cm tall, a semi-digitate false raceme with up to 4 long spikes. Anthers yellow, stigma white.

Habitat: Widespread but found occasionally locally in damp areas of the floodplain.

Flowering: During the main rains and into the cool dry period.

Uses and beliefs:
- Moderately palatable when young.

FZ Vol.10 pt4, B Van Wyk New p.320, Ellery p.204, Van Oudtshoorn 2 p.67.

Poaceae (grass family)

Plumose panicle

Phragmites australis (Cav.) Trin. ex Steud.

Common names: Setswana letlhaka; **Shiyeyi** setou; **Afrikaans** fluitjiesriet; **English** common reed.

Derivation: *Phragma*, hedge or fence [Greek], referring to the fact that it grows in stands that look like hedges; *australis*, southern [Latin], probably referring to its distribution.

Identification: Perennial grass growing in dense stands, c. 3 m tall. **Roots** long and rhizomatous. **Ligule** a ring of long tangled white hairs. **Leaves** c. 40 × 2.5 cm, flat near the stem and rolled at the tip. **Inflorescence** a plumose panicle, c. 50 cm tall. **Spikelets** dark brown, covered in silky hairs.

Habitat: Common and widespread in dense stands along waterways and in damp areas.

Flowering: During the main rains.

Uses and beliefs:
- Used as firewood on islands where there are no trees.
- Used to make the grids for smoking fish.
- Also used in construction.

FZ Vol.10 pt1, B Van Wyk Flowers p.336, Ellery p.198, Van Oudtshoorn 2 p.178.

Poaceae (grass family)

Open panicle

Aristida meridionalis (Stapf) Henrard

Common names: Setswana riri-satau, seriri, seriri-sa-tau, bojang-jwa-tau; **English** giant three awn, coppery three awn.

Derivation: *Arista*, needle [Latin], alludes to the awns of the spikelets; *meridionalis*, of noonday, blooming at noon-time.

Identification: Attractive, tall, tufted **perennial grass**, up to c. 1.6 m tall. **Stems** suffused with red towards the base. **Nodes** dark red-brown. **Ligules** with a tuft of hairs on either side. **Leaves** up to c. 55 × 0.5 cm, tightly rolled. **Inflorescence** an open panicle, a delicate shining bronze as they shimmer in the wind.

Aristida meridionalis may easily be confused with the annual *A. stipoides* but the latter is perennial and the shaft of the awns is shorter.

Habitat: Common and widespread in damp sandy areas of the floodplain and on track-sides.

Flowering: Late in the main rains.

Uses and beliefs:
- A hard grass grazed only when young.

FZ Vol.10 pt1, Hargreaves p.68, Van Oudtshoorn 2 p.162.

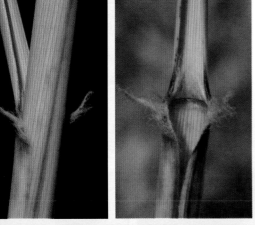

Poaceae (grass family)

Open panicle

Aristida scabrivalvis Hack.

Common names: Setswana seloka; **Afrikaans** besemgras; **English** purple three-awn.

Derivation: *Arista*, needle [Latin], alludes to the awns of the spikelets; *scabra*, rough [Latin], *-valvis*, valved [Latin], referring to the rough valves of the fruit.

Identification: Tufted **annual grass**, c. 50 cm tall. **Stems** occasionally branching. **Nodes** tinged red. **Leaves** c. 25 cm × 3 mm, flat. **Inflorescence** a large purple open panicle, approximately half the overall height of the grass.

Habitat: Found occasionally locally in sandy clay on the margin of seasonal pans.

Flowering: During the main rains.

Uses and beliefs:
- Little value for grazing.

FZ Vol.10 pt1, B Van Wyk New p.327, Van Oudtshoorn 2 p.164.

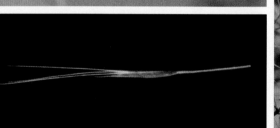

Poaceae (grass family)

Open panicle

Aristida stipoides Lam.

Common names: Setswana seloka, mogatla-wa-tau; **English** large fountain bristle grass.

Derivation: *Arista*, needle [Latin], alludes to the awns of the spikelets; *-oides*, looks like or similar to [Greek], looks like members of the genus *Stipa*.

Identification: Attractive, tall, tufted **annual grass** up to c. 1.6 m tall. **Stems** suffused with red towards the roots. **Ligules** tinged with red, with 2 distinct hairy tufts. **Nodes** dark red-brown. **Leaves** up to c. 55 × 0.5 cm and tightly rolled. **Inflorescences** shimmering bronze, open panicles, c. 40 cm tall.

Aristida stipoides may easily be confused with the perennial *A. meridionalis* but it is annual and the shaft of the awns is longer.

Habitat: Locally widespread in damp sandy areas of the floodplain and occasionally in mixed scrub.

Flowering: Late in the main rains.

Uses and beliefs:
• A hard grass grazed only when young.

FZ Vol.10 pt1.

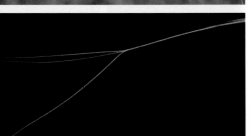

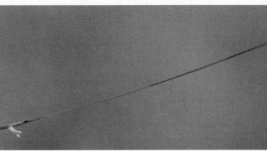

Open panicle

Eragrostis aspera (Jacq.) Nees

Common names: English rough love grass, large plume eragrostis.

Derivation: *Eros*, love [Greek], *agrostis*, grass [Greek], the reason for this name is not known; *aspera*, rough [Latin].

Identification: Tufted **annual grass**, c. 80 cm tall. **Nodes** green. **Ligule** a ring of long hairs. **Leaves** c. 25 × 1.1 cm, flat, with **rasping hairs** on both surfaces. **Inflorescence** an open panicle with evenly placed single white to reddish purple spikelets on the tips of fine stalks.

Habitat: Widespread but found only occasionally locally in partial shade on islands in the floodplain.

Flowering: Late in the main rains.

Uses and beliefs:

- Poor grazing as it has little leaf growth.

FZ Vol.10 pt2, Hargreaves p.70, Van Oudtshoorn 2 p.166.

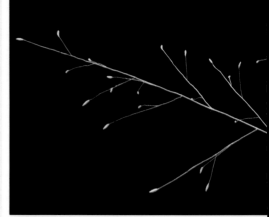
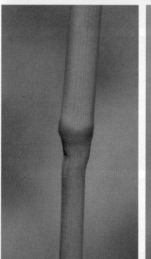

Poaceae (grass family)

Open panicle

Eragrostis cilianensis (All.) Vignolo ex Janch.

Common names: English stink grass, stink love grass, stink eragrostis, grey love grass.

Derivation: *Eros*, love [Greek], *agrostis*, grass [Greek], the reason for this name is not known; *cilianensis* for a village in Italy called Ciliana.

Identification: Tufted **annual grass** with branching **stems**, up to c. 80 cm tall. **Stems** dark red-brown above the nodes, growing at an angle and bending at the lower nodes. **Ligules** with 2 clumps of straggly long hairs and a ring of shorter ones. **Leaves** c. 25 × 0.6 cm flat at the base of the plant when young but rolled later. **Inflorescence** a grey open panicle with rather short branches, up to c. 20 cm tall. This plant gives off a strong rotting sweet green **smell** when crushed.

Habitat: Common and widespread locally in disturbed damp sandy places.

Flowering: Mainly late in the main rains.

Uses and beliefs:
- A pioneer species in disturbed areas.
- Generally considered poor grazing but locally heavily grazed.

FZ Vol.10 pt2, Ellery p.189, Van Oudtshoorn 2 p.153, WFNSA p.24.

Open panicle

Eragrostis cylindriflora Hochst.

Common names: Setswana segobe.

Derivation: *Eros*, love [Greek], *agrostis*, grass [Greek], the reason for this name is not known; *cylindricus*, long and round, cylindrical [Latin], *-flora*, -flowered [Latin], with cylindrical flowers.

Identification: Tufted branching **perennial grass** with fleshy **roots**, up to c. 1.1 m tall. Very variable in size depending on the weather and where it is growing. **Stems** and lower leaf surfaces **hairless**, inner leaf surfaces covered in **rasping hairs**. **Nodes** light brown to red, tending to bend and droop to the ground where they root. **Ligule** has 2 tufts of long straight hairs growing at right angles to the leaf and a ring of shorter hairs. **Leaves** c. 25 × 0.7 cm, tightly rolled. **Inflorescence** an open panicle, c. 25 cm tall. Anthers lemon yellow, stigmas white.

Habitat: Common and widespread throughout the floodplains in full sun.

Flowering: During the main rains.

FZ Vol.10 pt2, Van Oudtshoorn 2 p.265.

Poaceae (grass family)

Open panicle

Eragrostis lappula Nees

Common names: Setswana makoholi; **English** love grass.

Derivation: *Eros*, love [Greek], *agrostis*, grass [Greek], the reason for this name is not known; *lappula*, burr-like [Latin], possibly alluding to the spikelets.

Identification: Tufted **perennial grass**, c. 1.2 m tall. **Nodes** pinkish brown occasionally with a yellow ring. **Ligule** a mass of straggly hairs, a distinct tuft on either side. **Leaves** up to c. 25 × 0.7 cm, slightly rolled, with straggling sparse long hairs on the inner surface. **Inflorescence** a soft grey open panicle, up to c. 23 cm tall. **Spikelets** covered in long grey hairs.

Habitat: Widespread but not common near to or in seasonally flooded areas of the floodplain and island margins. May be associated with termite mounds.

Flowering: Throughout the rains.

Uses and beliefs:
• A palatable grass when young.

FZ Vol.10 pt2, Ellery p.189, Van Oudtshoorn 2 p.154.

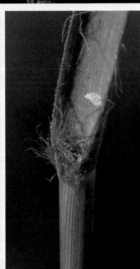

Poaceae (grass family)

Open panicle

Eragrostis pallens Hack.

Common names: Setswana motsikiri, motshikiri, moseka; **English** broom love grass, thatching grass.

Derivation: *Eros*, love [Greek], *agrostis*, grass [Greek], the reason for this name is not known; *pallens*, pale [Latin].

Identification: Vigorous densely tufted **perennial grass**, c. 1.5 m tall. **Nodes** greenish, paler above than below. **Ligule** a slight membrane with a hairy margin. **Leaves** up to c. 45 × 0.7 cm, tightly rolled with short harsh hairs on the inner surface. **Inflorescence** an open panicle up to c. 50 cm tall. Anthers pale lemon yellow, stigmas white.

Habitat: Common and widespread in seasonally flooded areas of the floodplain.

Flowering: Late in the main rains.

Uses and beliefs:
- Widely used as a thatching grass.

FZ Vol.10 pt2, Van Oudtshoorn 2 p.131.

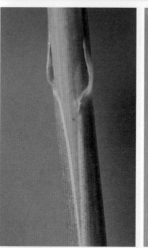

Poaceae (grass family)

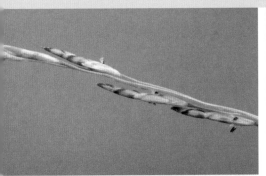

Open panicle

Eragrostis pilosa (L.) P.Beauv.

Common names: English Indian love grass, slender meadow grass.

Derivation: *Eros*, love [Greek], *agrostis*, grass [Greek], the reason for this name is not known; *pilosa*, covered with long soft hairs.

Identification: Very delicate **annual grass** with branching stems, c. 40 cm tall. **Nodes** variable and may be tinged red above a ring of green and a ring of red. **Ligules** a ring of hairs with a tuft of long straight hairs on either side. **Leaves** up to c. 10 × 0.3 cm, tightly rolled or flat. **Inflorescence** an open panicle of whorls of fine branches, c. 25 cm tall.

Habitat: Found occasionally locally in sandy clay on the margin of seasonal pans.

Flowering: Late in the main rains.

Uses and beliefs:
• Tef (*Eragrostis tef*), which is widely grown in Ethiopia and surrounding countries for its grain, is probably an improved variety of *Eragrostis pilosa*.

FZ Vol.10 pt2, Van Oudtshoorn 2 p.131.

Poaceae (grass family)

Open panicle

Eragrostis porosa Nees

Derivation: *Eros*, love [Greek], *agrostis*, grass [Greek], the reason for this name is not known; *poros*, an opening or pore [Greek], *porosus*, pierced with small holes.

Identification: Untidy lax tufted **perennial grass**, up to c. 1.1 m tall with **branching** stems. **Nodes** red above a green ring, **rooting** from the nodes where they touch the ground. **Ligule** a ring of hairs. **Leaves** c. 15 × 0.4 cm, rolled or flat. **Inflorescence** an open panicle, up to c. 23 cm tall.

Habitat: Common and widespread in full sun and light shade on the floodplain and in mixed woodland.

Flowering: Throughout the main rains.

FZ Vol.10 pt2.

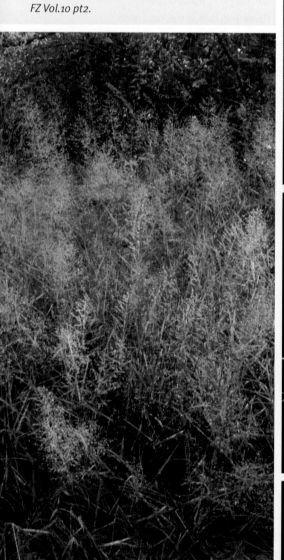

Poaceae (grass family)

Open panicle

Eragrostis pusilla Hack.

Derivation: *Eros*, love [Greek], *agrostis*, grass [Greek], the reason for this name is not known; *pusilla*, very small [Latin].

Identification: Small pink-tinged tufted **annual grass**, c. 40 cm tall, with branching stems. **Nodes** green. **Ligule** a small membrane. **Leaves** bright green, c. 15 × 0.7 cm, flat with some harsh hairs. **Inflorescence** an open pink panicle, c. 18 cm tall.

Habitat: Common and widespread on the seasonally flooded floodplain.

Flowering: Late in the main rains.

FZ Vol.10 pt2.

Poaceae (grass family)

Open panicle

Eragrostis viscosa (Retz.) Trin.

Common names: Setswana tlatlana; **English** sticky love grass.

Derivation: *Eros*, love, *agrostis*, grass [Greek], the reason for this name is not known; *viscosus*, sticky, clammy, viscous [Latin], alluding to the overall stickiness of this plant.

Identification: Small **annual grass**, c. 50 cm tall, with **branching** stems. Long straggly **hairs** (c. 4 mm) on the stems and on both surfaces of the leaves. **Nodes** dark red. **Ligule** a mass of long white hairs, c. 7 mm long. **Leaves** c. 12 × 0.5 cm, flat. **Inflorescence** an open panicle of pink spikelets, c. 20 cm tall. Plants very **sticky** overall and have a delicate **spicy aroma** when crushed.

Habitat: Widespread but not common in sandy areas of lightly shaded islands and mopane woodland.

Flowering: During the main rains.

Uses and beliefs:
- An unpalatable grass.
- Indicative of poor growing conditions.

FZ Vol.10 pt2, Ellery p.190, Van Oudtshoorn 2 p.142.

Poaceae (grass family)

Open panicle

Melinis kallimorpha (Clayton) Zizka
(syn. *Rhynchelytrum kallimorphon* Clayton)

Derivation: *Meli*, honey [Greek], with the colour of new honey; *kallos*, beauty [Greek], *morphe*, form or shape [Greek], beautifully shaped.

Identification: Small tufted **annual** or **short-lived perennial grass**, c. 40 cm tall. **Stems** branching and rooting from the lower nodes, lower stems suffused with dark red. **Nodes** dark red. **Ligule** a few short hairs, virtually non-existent. **Leaves** flat, c. 9 × 0.6 cm. **Inflorescence** an open panicle, c. 9 cm long. Anthers golden. Stigmas white.

Habitat: Found occasionally in mixed woodlands on sandy soil.

Flowering: During the main rains.

FZ Vol.10 pt3.

Poaceae (grass family)

Open panicle

Melinis repens (Willd.) Zizka subsp. ***grandiflora*** (Hochst.) Zizka
(syn. *Rhynchelytrum grandiflorum* Hochst.)

Common names: Setswana senyane, sanyane, lenapa; **Afrikaans** natalse rooipluim; **English** Natal red top, red top grass, fairy grass.

Derivation: *Meli*, honey [Greek], with the colour of new honey; *repens*, creeping [Latin]; *grandis*, big, showy [Latin], *flora*, flower [Latin], large flowered.

Identification: Attractive tufted short-lived perennial grass, c. 80 cm tall, with **branching stems**. Stems and lower parts of the leaves are covered in long straggly **hairs**. **Node** a protuberant green ring. **Ligule** a ring of short hairs. **Leaves** c. 1.5 × 0.8 cm may be flat or rolled with a prominent mid-rib. **Inflorescences** shining, may be dark pink to white, open panicles, c. 10 cm tall. **Spikelets** covered with long silky pink or white hairs.

Habitat: Common and widespread in open sandy areas of the floodplain.

Flowering: During the main rains.

Uses and beliefs:
- A palatable grass but produces few leaves.
- Used for basket and hat weaving.

FZ Vol.10 pt3, B Van Wyk New p.337, Hargreaves p.72, Van Oudtshoorn 2 p.149, WFNSA p.22.

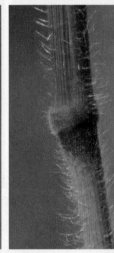

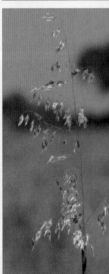

Poaceae (grass family)

Open panicle

Panicum fluviicola Steud.

Derivation: *Panicum*, millet [Latin], *fluvium*, river [Latin], *-cola*, dweller [Latin], preferring watery places.

Identification: Fine tall, tufted, **perennial grass** with a **stoloniferous rooting** system and a branching structure, c. 1.2 m tall. **Roots** dense, fleshy and retaining moisture. **Nodes** red above and green below. **Ligule** membranous with hairs along the margin. **Leaves** c. 35 × 0.6 cm, some partly rolled along their lengths with other parts flat. **Inflorescence** a dark pink open panicle, c. 24 cm tall. Stigmas light maroon.

Habitat: Locally rare in damp areas of the floodplain.

Flowering: During the main rains.

FZ Vol.10 pt3.

Poaceae (grass family)

Open panicle

Panicum hirtum Lam.

Derivation: *Panicum*, millet [Latin], *hirtum*, hairy [Latin].

Identification: Erect branching tufted **annual grass**, up to c. 35 cm tall, with a tendency to bend at the lower nodes. **Ligule** a vestigial membrane. **Leaves** bright green, broad and flat, with a rippling surface, up to c. 8 × 0.8 cm, margins have long white **hairs** getting longer towards the base. **Inflorescence** an open panicle, c. 10 cm tall, with small drop-like **spikelets**. Anthers pale maroon.

Habitat: Locally rare in Chobe riverine bush.

Flowering: Late in the main rains.

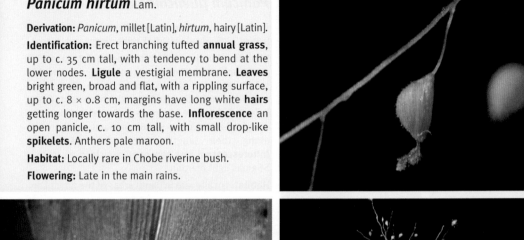

Poaceae (grass family)

Open panicle

Panicum kalaharense Mez

Common names: English Kalahari panicum.

Derivation: *Panicum*, millet [Latin]; *kalaharense*, of the Kalahari.

Identification: Tufted **perennial grass** with a **rhizomatous rooting** system c. 1.55 m tall. **Nodes** green. **Ligule** a fringe of erect c. 6 mm hairs. **Leaves** c. 55 × 0.8 cm, usually folded at the mid-rib but becoming flattened as they age. **Inflorescence** an open panicle, c. 26 cm tall but wider than its height. Anthers orange, stigmas pink.

Habitat: Widespread in mixed woodlands on sandy soil.

Flowering: During the main rains.

FZ Vol.10 pt3.

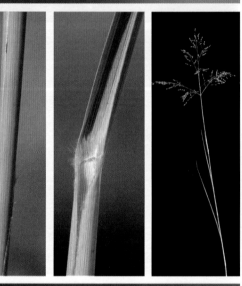

Poaceae (grass family)

Open panicle

Panicum maximum Jacq.

Common names: Setswana mhaha?, puka, mfhafha, mphaga, mfafa, mohaha; **Afrikaans** blousaad, blousaadsoetgras, buffelsgras; **English** guinea grass, brown top buffel-grass, bush buffalo grass, common buffalo grass.

Derivations: *Panicum*, millet [Latin], *maximum*, largest [Latin]. *Puka*, guinea fowl, because the seeds drop and are eaten by guinea fowl.

Identification: Large tufted **perennial grass**, c. 2.5 m tall with branching **stems** and occasionally stoloniferous **rooting. Stems** vary in colour, sometimes having dark red joints. Leaves and stems often coarsely **hairy. Nodes** sometimes hairy, green or dark red. **Ligule** a membrane with a hairy margin. **Leaves** up to c. 60 × 2 cm, flat. **Inflorescence** open panicle, white or very pale pink, up to c. 40 cm tall, lower branches arranged in whorls.

Habitat: Locally common and widespread in light shade in woodland margins and on islands in the floodplain.

Flowering: During the main rains.

Uses and beliefs:
• In times of famine, the Wayeyi eat the seed.
• A highly palatable grass, one of the most valuable grasses for game and livestock.

FZ Vol.10 pt3, B Van Wyk New p.335, Ellery p.195, Hargreaves p.72, Van Oudtshoorn 2 p.179, WFNSA p.24. MMo.

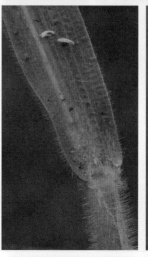

Poaceae (grass family)

Open panicle

Panicum subalbidum Kunth

Common names: Afrikaans elmboogbuffelsgras; **English** elbow buffalo grass.

Derivation: *Panicum*, millet [Latin]; *sub-*, rather, slightly [Latin], *albidus*, whitish [Latin].

Identification: Robust **annual** or **short-lived perennial grass** with branching stems, c. 1.2 m tall. **Stems** tinged red on the lower $^2/_3$ of the plant. Mature stems often bend sharply away from the new inflorescence at the first node. **Nodes** pink above a green ring, then red-brown below. **Ligule** a ring of long erect hairs. **Leaves** c. 13 × 0.6 cm, tightly wrapped around the stem and with a distinct white line at the mid-rib. **Inflorescence** an open panicle, c. 25 cm tall. Anthers bright orange, stigmas dark maroon.

Habitat: Locally uncommon on the margin of flowing channels, rooted in the edge of the water.

Flowering: During the main rains.

FZ Vol.10 pt3.

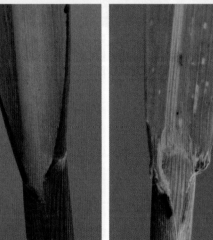

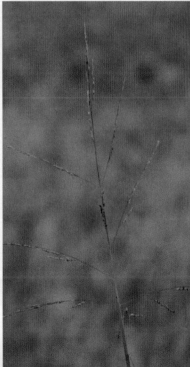

Poaceae (grass family)

Open panicle

Sorghastrum nudipes Nash

Derivation: *Sorghastrum*, derived from *suricum granum*, meaning grain from Syria [Latin]; *nudus*, bare, naked [Latin], *pes*, foot [Latin], bare-footed.

Identification: Tufted **perennial grass** with a **rhizomatous rooting** system, c. 1.2 m tall. **Nodes** slightly hairy, tinged with pink, often **rooting** from the lower nodes. **Ligule** 2 tufts of white hairs and a membrane with a hairy margin. **Leaves** up to c. 20 × 0.4 cm, folding at the mid-rib. **Inflorescence** a red-gold normally open (but in this case compact) panicle, up to c. 28 cm tall. Anthers yellow, stigmas dark pink.

Habitat: Found occasionally locally in damp areas of the spillway in full sun.

Flowering: During the main rains.

FZ Vol.10 pt4.

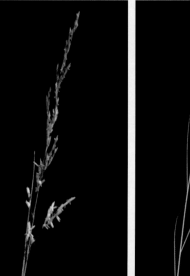

Poaceae (grass family)

Open panicle

Sporobolus cordofanus (Steud.) Coss.

Derivation: *Spora*, seed [Greek], *ballein*, to throw [Greek], this genus of grasses is commonly known as dropseed because the pericarp swells when damp causing the seed to be forcibly ejected.

Identification: Small tufted **annual grass**, c. 30 cm tall with branching stems. **Nodes** usually green but often flushed with red. **Ligule** a ring of short hairs. **Leaves** c. 3.5 × 0.5 cm, flat. **Inflorescence** an open panicle, c. 8 cm tall.

Habitat: Locally uncommon growing on pans that are seasonally flooded but may be baked hard and dry.

Flowering: Late in the main rains.

FZ Vol.10 pt2.

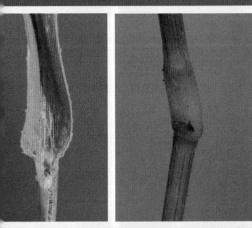

Poaceae (grass family)

Open Panicle

Sporobolus ioclados (Trin.) Nees

Common names: Afrikaans panfynsaadgras; **English** pan dropseed.

Derivation: *Spora*, seed [Greek], *ballein*, to throw [Greek], this genus of grasses is commonly known as dropseed because the pericarp swells when damp causing the seed to be forcibly ejected; *ion*, purple [Greek], *klados*, branch [Greek], alluding to the branches of the inflorescence.

Identification: Densely tufted short-lived **perennial grass**, up to c. 90 cm tall with **stoloniferous rooting** system and occasional rhizomes. **Stems** branching, occasionally drooping to the ground. **Nodes** green. **Ligule** a ring of hairs. **Leaves**, arranged near the base of the plant, up to c. 15 × 0.6 cm, folding at the midrib, long **hairs** on both surfaces and regularly spaced on the margins. **Inflorescence** an open branching panicle, up to c. 15 cm, tall with branches arranged in whorls. Anthers cream, stigma white.

Habitat: Widespread in areas of damp shaded woodland and beside seasonal pans, often where the soil is very saline.

Flowering: During the main rains.

Uses and beliefs:
* A palatable grass providing important grazing if sufficiently plentiful.

FZ Vol.10 pt2, Ellery p.203, Müller p.236, Van Oudtshoorn 2 p.171.

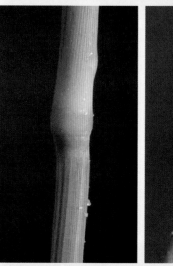

Poaceae (grass family)

Open panicle

Sporobolus macranthelus Chiov.

Derivation: *Spora*, seed [Greek], *ballein*, to throw [Greek], this genus of grasses is commonly known as dropseed because the pericarp swells when damp causing the seed to be forcibly ejected; *makros*, long or big [Greek], *anthela*, the flower of a sedge [Greek], possibly referring to the height of the inflorescence.

Identification: Branching tufted **perennial grass**, c. 1.5 m tall, with a **rhizomatous rooting** system. **Nodes** fawn above a dark red ring, then green below. **Ligule** a vestigial membrane. **Leaves** c. 30 × 0.4 cm, folding at the mid-rib, ribbed longitudinally. **Inflorescence** an open panicle, c. 40 cm long. Anthers lemon yellow, stigmas white.

Habitat: Locally rare occurring on an island in the Selinda Spillway.

Flowering: During the main rains.

FZ Vol.10 pt2.

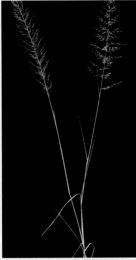

Open panicle

Sporobolus panicoides A.Rich.

Common names: Afrikaans grootsaad-sporobolus; **English** famine grass, Christmas tree grass.

Derivation: *Spora*, seed [Greek], *ballein*, to throw [Greek], this genus of grasses is commonly known as dropseed because the pericarp swells when damp causing the seed to be forcibly ejected; *panicoides*, resembling the genus *Panicum*.

Identification: Branching, delicate, rather lax, tufted **annual grass**, c. 75 cm tall. **Nodes** green. **Ligule** a ring of short hairs. **Leaves** c. 30 × 0.5 cm, flat, hairless, with shiny bristled margins; leaf sheaths covered in sparse, erect, long hairs. **Inflorescence** an open panicle, c. 15 cm tall, bearing distinctive single spherical **spikelets**.

Habitat: Found occasionally locally in lightly shaded woodland.

Flowering: During the main rains.

Uses and beliefs:
• Of little value for grazing but the spherical seeds are popular with birds.

FZ Vol.10 pt2, Müller p.240, Van Oudtshoorn 2 p.170, WFNSA p.22.

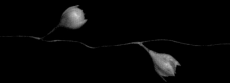

Poaceae (grass family)

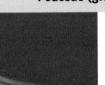

Open panicle

Stipagrostis uniplumis (Roem. & Schult.) de Winter

Common names: Setswana mogatla-watau, tshikitshane; **Afrikaans** blink(h)aarboesmangras; **English** bushman grass, silky bushman grass, tall bushman grass, small bushman grass.

Derivation: *Stipes*, stalk or stem [Latin], *agrostis*, a kind of grass [Greek], referring to the stalk at the base of the 3-awned branches; *unus*, one [Latin], *pluma*, feather [Latin], one feather referring to the single-plumed awn.

Identification: Conspicuously **shining** (when in seed), densely tufted, short-lived **perennial grass**, c. 1 m tall. **Stems** sometimes bending at the nodes. **Nodes** yellowish with pale green rings above and below, sometimes with a ring of hairs. **Ligule** a circle of long white hairs, c. 5 mm long. **Leaves** c. 25 × 0.3 cm, flat but curling towards the tip, light grey-green above and darker green below. **Inflorescence** an open panicle, c. 35 cm tall. Anthers yellow, stigma white.

Habitat: Common and widespread on the sandy floodplain.

Flowering: During the main rains and into the cool dry period.

Uses and beliefs:
• A highly palatable grass that offers good grazing in dry areas.

FZ Vol.10 pt1, B Van Wyk New p.330, Ellery p.204, Müller p.260, Van Oudtshoorn p.132.

Poaceae (grass family)

Open panicle

Tricholaena monachne (Trin.) Stapf & C.E.Hubb.

Common names: Setswana mofala; **Afrikaans** blousaadgras; **English** blue-seed tricholaena.

Derivation: *Thrichos*, hair [Greek], *chlaena*, cloak or covering [Greek], referring to the hairy spikelets of the first specimen collected; *monos*, single [Greek], *achyron*, chaff [Greek], alluding to the form of the spikelets.

Identification: Rather lax, branching, **annual** to **short-lived perennial grass**, c. 1.2 m tall. **Stems** tending to bend and root at the nodes. **Nodes** green and rather irregularly formed. **Ligule** a ring of hairs. **Leaves** blue-green, up to c. 20 × 0.5 cm, usually folded at the mid-rib. **Inflorescence** an open panicle, c. 15 cm tall, with distinctive single spikelets on the branches. **Spikelets** have long thin twisting stalks. Anthers buttery yellow, stigma white.

Habitat: Widespread in sandy clay on the floodplain and in mopane woodland.

Flowering: During the main rains.

Uses and beliefs:
• Of little nutritional value, although it can be important grazing in dry areas.

FZ Vol.10 pt3, Müller p.264, Van Oudtshoorn 2 p.148.

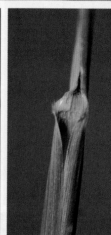
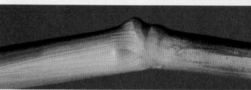

Poaceae (grass family)

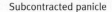
Subcontracted panicle

Brachiaria deflexa (Schumach.) Robyns

Common names: English false signal grass.

Derivation: *Brachiatus*, having arm-like branches [Latin]; *deflexa*, bent abruptly downwards [Latin].

Identification: Tufted annual grass with occasionally branching stems, c. 70 cm tall. **Ligule** a ring of short hairs. **Leaf blade** covered in sparse **velvety hairs** above the ligule. **Leaves** c. 17 × 1 cm, flat. **Inflorescences** subcontracted panicles, up to c. 20 cm tall. **Spikelets** usually arranged in pairs.

This grass may easily be confused with grasses of the genus *Panicum*. The main difference is that *Brachiaria* usually has spikelets arranged in pairs.

Habitat: Found occasionally in shaded areas of mixed and mopane woodland.

Flowering: During the main rains.

Uses and beliefs:
- A palatable grass that is a useful pioneer species.

FZ Vol.10 pt3, Van Oudtshoorn 2 p.183.

Poaceae (grass family)

Subcontracted panicle

Chrysopogon nigritanus (Benth.)
Veldkamp

Common names: English vetiver grass.

Derivation: *Chrysos*, gold [Greek], *pogon*, beard [Greek], alluding to the gold-coloured inflorescence; *nigritanus*, from Nigeria, where it was first collected.

Identification: Perennial grass growing in large tufts, c. 1.9 m tall. **Stems** branching occasionally. **Nodes** towards base of the plant, a yellow-green ring. **Ligule** a ring of short bristles. **Leaves** c. 75 × 0.9 cm, folding at the mid-rib and having a harshly bristled margin. **Inflorescence** a subcontracted panicle with whorls of branches around a central axis, c. 26 cm tall.

Habitat: Locally uncommon, growing at the margin of seasonal pans.

Flowering: During the main rains.

FZ Vol.10 pt4.

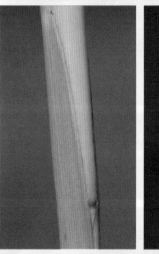
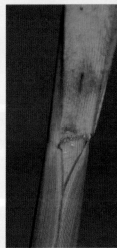

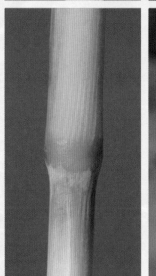

Poaceae (grass family)

Subcontracted panicle

Eragrostis echinochloidea Stapf

Common names: English tick grass.

Derivation: *Eros*, love [Greek], *agrostis*, grass [Greek], possibly referring to a graceful inflorescence. *Echinos*, hedgehog [Greek], *chloe*, grass [Greek], hedgehog grass, possibly resembling grasses of the genus *Echinochloa*.

Identification: Annual or short-lived tufted **perennial grass** with occasionally branching **stems**, c. 65 cm tall. **Nodes** tinged brownish purple. **Ligules** a nondescript ring of hairs. **Leaves** vary up to c. 30 × 0.7 cm (but usually c. 15 × 0.3 cm), tightly rolled. **Inflorescence** a subcontracted panicle, c. 10 cm tall. **Spikelets** with sharp tips, packed close together on the branches.

Habitat: Common and widespread in light shade or full sun on islands and on the floodplain.

Flowering: Late in the main rains.

Uses and beliefs:
- Tick grass is reasonably palatable when young but produces sparse foliage.

FZ Vol.10 pt2, Hargreaves p.70, Van Oudtshoorn 2 p.184.

Subcontracted or open panicle

Eragrostis inamoena K.Schum.

Common names: English tite grass.

Derivation: *Eros*, love [Greek], *agrostis*, grass [Greek], possibly referring to a graceful inflorescence; *inamoena*, unlovely, gloomy [Latin].

Identification: Tufted **perennial grass**, up to c. 70 cm tall, with short **rhizomes**. **Nodes** and leaf sheath often tinged red and slightly bending. **Ligule** a ring of hairs. **Leaves** c. 10 × 0.3 cm, slightly folding at the mid-rib. **Inflorescence** an open or subcontracted panicle, up to c. 13 cm tall. **Spikelets** smooth, blue-grey to purple.

Habitat: Common and widespread, growing on dry land close to water.

Flowering: During the main rains.

Uses and beliefs:
• Rarely grazed, even in areas where everything else has been heavily eaten.

FZ Vol.10 pt2, Van Oudtshoorn 2 p.154, WFNSA p.22.

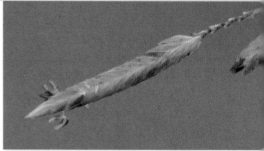
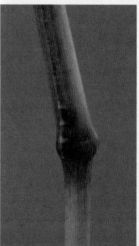

Poaceae (grass family)

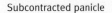
Subcontracted panicle

Eragrostis superba Peyr.

Derivation: *Eros*, love [Greek], *agrostis*, grass [Greek], possibly referring to a graceful inflorescence; *superba*, shining or proud [Latin], possibly referring to the spikelets.

Common names: Setswana mogamapodi, totsane; **Afrikaans** weeluisgras; **English** saw-tooth love grass, flat-seeded love grass, heart-seed love grass.

Identification: Perennial grass, c. 1 m tall, growing in tufts. **Leaf blades** flat, c. 40 cm long. **Inflorescence** a subcontracted panicle with a central axis and several side branches. **Spikelets** loosely arranged on the lower side of the branch, attractive, large, oval, flat, creamy-pink with toothed margins.

Habitat: Found occasionally in damp areas along track-sides, and in disturbed sandy areas in lightly shaded woodland or on the floodplain where scrub is encroaching.

Flowering: Intermittently throughout the rainy seasons.

Uses and beliefs:
- A palatable grass for animals.

FZ Vol.10 pt2, B Van Wyk Flowers p.334, Ellery p.190, Van Oudtshoorn 2 p.127.

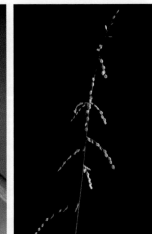

Poaceae (grass family)

Subcontracted panicle

Leersia hexandra Sw.

Common names: Setswana mokanja; **Afrikaans** wilderysgras; **English** rice grass, rasp grass.

Derivation: *Leersia* for Johann Georg Daniel Leers (1727–74), a German botanist and apothecary; *hexa*, 6 in compound words, *andros*, man [Greek], alluding to the 6 stamens in each spikelet.

Identification: Perennial grass, c. 90 cm tall, with a **rhizomatous rooting** system and often with branching stems. Some **roots** white and spongy, holding water. **Nodes** protuberant and covered in white velvety **hairs**. **Ligule** a horn-shaped membrane. **Leaves** c. 15 × 1.2 cm, lightly rolled, with sparse rasping hairs. **Inflorescence** a pinkish-brown subcontracted panicle, c. 14 cm tall. **Spikelets** flat with stiff hairs around the margin.

Habitat: Found occasionally locally deeply rooted in water or on the water margin.

Flowering: During the main rains.

Uses and beliefs:
- A palatable grass.

FZ Vol.10 pt1, B Van Wyk New p.334, Ellery p.193, Van Oudtshoorn 2 p.191.

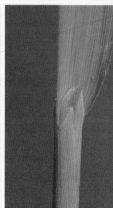

Poaceae (grass family)

Subcontracted or open panicle

Panicum coloratum L. var. *coloratum*

Common names: Setswana mhaha?, puka; **Shiyeyi** empunga; **English** white buffalo grass, small buffalo grass.

Derivations: *Panicum*, millet [Latin], *coloratum*, coloured [Latin], possibly for the colourful anthers and stigmas of the species. *Puka*, guinea fowl, because the seeds drop and are eaten by guinea fowl.

Identification: Tufted **perennial grass**, c. 1.1 m tall with a **rhizomatous rooting** system. **Stems** hollow and occasionally branching. **Nodes** red. Stems and leaves **hairy** overall. **Leaves** c. 20 × 0.7 cm slightly folded at the mid-rib. **Inflorescence** a sparsely branched subcontracted or open panicle, c. 20 cm tall, often with a single side branch at the base. Stigmas purple, anthers bright orange.

Habitat: Widespread in damp areas of the floodplain, sometimes growing in water.

Flowering: During the main rains.

Uses and beliefs:
• Good grazing but never plentiful in the wild. Now widely cultivated for grazing.

FZ Vol.10 pt3, Van Oudtshoorn 2 p.180. PN.

Poaceae (grass family)

Subcontracted panicle

Schmidtia pappophoroides Steud. ex J.A.Schmidt
(syn. *Schmidtia bulbosa* Stapf)

Common names: Setswana molalaphage, tshwang, tsube, bojang-jwa-dipitse, khezane; **Afrikaans** Kalahari sand kweek; **English** staggers grass, Kalahari sand quick grass.

Derivation: *Schmidtia*, for the German botanist M. Schmidt; *-oides*, looks like or similar to [Greek], looks like members of the genus *Pappophorum*.

Identification: Perennial grass with a short creeping **rhizome**, often with long surface **stolons** up to c. 80 cm tall. Both the **stems**, including the **nodes**, and the **leaves** covered in short erect **hairs**. **Nodes** yellowish green. **Ligule** a ring of short hairs. **Leaves** c. 15 × 0.5 cm, loosely rolled. **Inflorescence** beginning as a tight spike of spikelets but opening to a subcontracted panicle, c. 13 cm tall. Anthers pale lemon yellow, stigma white.

Habitat: Common and widespread on the open floodplain and in lightly shaded woodland.

Flowering: Late in the main rains.

Uses and beliefs:
• A palatable grass that is reasonably nutritious and drought resistant.

FZ Vol.10 pt1, Ellery p.200, Müller p.226, Van Oudtshoorn 1 p.129.

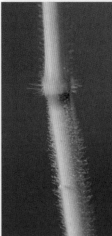

Poaceae (grass family)

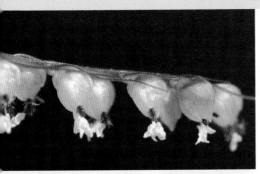

Subcontracted panicle

Setaria sagittifolia (A.Rich.) Walp.

Common names: English arrow grass.

Derivation: *Seta*, bristle [Latin], alluding to the bristles on the spikelets; *sagitta*, arrow, *folia*, leaf [Latin], with arrow-shaped leaves.

Identification: Soft **annual tufted grass**, c. 40 cm tall. **Ligule** membranous. **Leaves** c. 15 × 1 cm, flat, leaf base sagittate. **Inflorescence** subcontracted panicle, c. 7.5 cm tall, side branches have a single row of spikelets. **Spikelets** often a rich red-brown, arranged along the underside of the side branches.

Habitat: Common and widespread in heavily shaded areas of islands and woodland, especially in years of heavy rainfall.

Flowering: During the main rains.

Uses and beliefs:
- The spherical spikelets are popular with birds.
- Probably good grazing but not plentiful enough to have any effect in the area.

FZ Vol.10 pt3, Ellery p.200, Van Oudtshoorn 2 p.200.

Subcontracted or open panicle

Sporobolus fimbriatus (Trin.) Nees

Common names: Setswana modi, moshanje; **Afrikaans** fynsaadgras; **English** bushveld dropseed, common dropseed, dropseed grass.

Derivation: *Spora*, seed [Greek], *ballein*, to throw [Greek], this genus of grasses is commonly known as dropseed because the pericarp swells when damp causing the seed to be forcibly ejected; *fimbriatus*, fringed [Latin], alluding to the hairy margin of the leaf sheath.

Identification: Tufted **perennial grass**, up to c. 1.9 m tall, with branching stems. Noticeable because it is taller than the surrounding grasses. **Ligule** a ring of hairs, some long and quite tousled. **Leaves** borne mainly at the base of the plant, c. 50 × 0.6 cm, either flat or folding at the mid-vein, which is often white and quite prominent. **Inflorescence** either a subcontracted or an open panicle, c. 40 cm tall.

Habitat: Common and widespread in lightly shaded woodland and occasionally on the open floodplain.

Flowering: During the main rains.

Uses and beliefs:
• A palatable grass much grazed when young.

FZ Vol.10 pt2, Ellery p.202, Müller p.236, Van Oudtshoorn 2 p.144.

Poaceae (grass family)

Subcontracted panicle

Sporobolus pyramidalis P.Beauv.

Common names: English white herring-bone grass, cat's tail grass, cat's tail dropseed, narrow-plumed dropseed.

Derivation: *Spora*, seed [Greek], *ballein*, to throw [Greek], this genus of grasses is commonly known as dropseed because the pericarp swells when damp causing the seed to be forcibly ejected; *pyramidalis*, pyramid-shaped [Latin], alluding to the inflorescence.

Identification: Pale-coloured tufted **perennial grass**, c. 1.1 m tall, without rhizomes or stolons. **Stems** branching. **Nodes** green, paler green above. **Ligule** a ring of short hairs. **Leaves** c. 30 × 0.5 cm, flat. **Inflorescence** a subcontracted panicle, c. 27 cm tall, with short more or less regularly spaced side branches.

Habitat: Found occasionally locally in damp areas on the edge of the floodplain.

Flowering: During the main rains.

Uses and beliefs:
• An unpalatable grass.

FZ Vol.10 pt2, Van Oudtshoorn 2 p.195.

Poaceae (grass family)

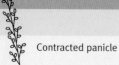
Contracted panicle

Aristida adscensionis L.

Common names: Setswana seloka; **Afrikaans** eenjarige steekgras; **English** annual bristle grass, annual three-awn.

Derivation: *Arista*, needle [Latin], alludes to the awns of the spikelets; *adscendens*, alludes to Ascension Island where Linnaeus collected the original specimen.

Identification: Tall, graceful, **annual grass**, c. 20–50 cm tall. **Nodes** red above, green below. Branching stems also occasionally flushed red at the **joints**. **Leaves** c. 17–35 × 0.2 cm, flat, upper surface covered with sparse coarse **hairs**, fewer hairs on the lower surface. **Inflorescence** a single fine, arching, contracted panicle of 3-awned spikelets, up to c. 30 cm long.

Habitat: A widespread grass of disturbed areas and track-sides, often found in light shade.

Flowering: During the main rains.

Uses and beliefs:
• This grass is not good grazing because it has few leaves.

FZ Vol.10 pt4, Müller p.62, Van Oudtshoorn 2 p.111.

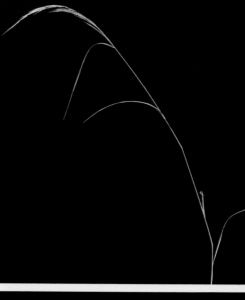

Poaceae (grass family)

Contracted panicle

Aristida hordeacea Kunth

Derivation: *Arista*, needle [Latin], alludes to the awns of the spikelets; *hordeum*, barley [Latin], possibly alluding to the inflorescence, which is like that of barley.

Identification: Annual or **short-lived perennial grass**, up to 80 cm tall. **Ligule** a ring of short hairs with a tuft of longer hairs on either side. **Leaves** c. 25 cm, tightly rolled. **Inflorescence** contracted panicle occasionally with less contracted branches near the base, c. 15 cm tall.

Habitat: Locally widespread in dry sandy areas of the floodplain.

Flowering: During the main rains.

FZ Vol.10 pt4.

Poaceae (grass family)

Contracted panicle

Aristida junciformis Trin. & Rupr.

Common names: English gongoni three-awn, wire grass, bristle grass.

Derivation: *Arista*, needle [Latin], alludes to the awns of the spikelets; *juncus*, a rush or reed, *formis*, having the shape or form of [Latin], reed-like.

Identification: Tufted **perennial grass**, up to c. 1.5 m tall. **Nodes** white above and green below. **Ligules** a short membrane. **Leaves** c. 28 × 0.3 cm, rolled. A slender branching hairy **inflorescence**, c. 30 cm tall, with branches held tightly to the stem.

Habitat: Widespread on shallowly sloping areas of the floodplain.

Flowering: During the main rains.

Uses and beliefs:
* An unpalatable grass only grazed late in the season.
* Widely used to make brooms.

FZ Vol.10 pt1, Van Oudtshoorn 2 p.110.

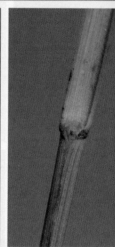

Poaceae (grass family)

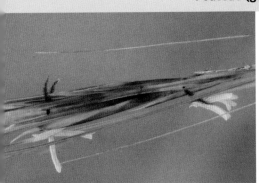

Contracted or subcontracted panicle

Aristida stipitata Hack.

Common names: Setswana seloka; **English** long-awned grass.

Derivation: *Arista*, needle [Latin], alludes to the awns of the spikelets; *stipitata*, stipitate [Latin], i.e. provided with a stipe or little stalk.

Identification: Tufted **perennial grass,** up to c. 1.3 m tall. **Branching** at almost every node in its upper parts. **Ligules** a circle of short hairs and 2 tufts of long hairs, c. 3 mm long. **Leaves** c. 50 × 0.5 cm, tightly rolled. **Inflorescence** a contracted panicle of spikes in a curving plume, c. 32 cm long. Anthers green-yellow, stigmas purple. **Spikelets** strongly twisted along their shafts.

Habitat: Widespread in deeply sandy disturbed areas and in mixed woodland.

Flowering: During the main rains.

Uses and beliefs:
* A grass with little value for grazing, often an indicator of overgrazing.

FZ Vol.10 pt1, Van Oudtshoorn 2 p.122.

Poaceae (grass family)

Contracted panicle

Enneapogon cenchroides (Roem & Schult.) C.E.Hubb.

Common names: Setswana molekangwetsi, bojang-jwa-mahupu; **Afrikaans** vaalsuurgras; **English** (common) nine-awned grass, grey sour grass.

Derivation: *Ennea-*, nine (in compound words) [Greek], *pogon*, beard [Greek], nine bearded, referring to the awns on the spikelets; *cenchroides*, like grasses of the genus *Cenchrus*.

Identification: Tufted **annual** or **short-lived perennial grass**, c. 60 cm tall. **Nodes** dark brown and hairy, lower nodes bent and knee-like. **Ligule** a ring of short hairs. **Leaves** c. 13 × 0.5 mm, rolled, covered in softly velvety hairs. **Inflorescence** a contracted panicle, occasionally with 1–2 short side branches at the base, c. 8 cm tall, purplish-grey.

Habitat: Common and widespread in compacted areas of poor soil on the floodplain and islands.

Flowering: Late in the main rains.

Uses and beliefs:
- A rather unpalatable pioneer grass.

FZ Vol.10 pt1, B Van Wyk New p.315, Van Oudtshoorn 2 p.113.

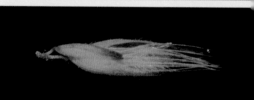

Poaceae (grass family)

Spike-like panicle

Cenchrus biflorus Roxb.

Common names: English burgrass.

Derivation: *Cenchrus*, from *kenchros*, millet [Greek]; *bi*, two, *-florus*, -flowered [Latin], probably alluding to the pairs of spikelets within each burr.

Identification: Unpleasantly spiky **annual grass**, c. 50 cm tall. **Rooting** from the nodes where they touch the soil. **Stems** branching. **Nodes** with a central green ring, dark brown above and below. **Leaves** c. 8 × 0.7 cm, rolled. **Inflorescence** a single spike-like panicle, c. 7 cm tall.

Habitat: A locally uncommon introduced noxious weed, growing in open sandy areas of the spillway.

Flowering: During the main rains.

FZ Vol.10 pt3.

Poaceae (grass family)

Spike-like panicle

Cenchrus ciliaris L.

Common names: Setswana molekangwetsi, modikangwetsi, mosela-watshwene, mosekangwetsi; **Afrikaans** bloubuffelsgras, **English** blue buffalo grass, foxtail buffalo grass, African foxtail, buffalo grass, pearl millet.

Derivation: *Cenchrus* from *kenchros*, millet [Greek]; *ciliaris*, fringed with hairs (*cilium*, eyelids and lashes [Latin]), probably alluding to the long hairs on the leaves.

Identification: Erect, bush-forming, branching and tufting **perennial grass**, c. 90 cm tall. Stoloniferous **rooting system**. **Nodes** sometimes tinged brown above. **Ligule** a ring of short bristles. **Leaves** flat or folding at the mid-rib, up to c. 30 × 0.7 cm, sheaths and lower leaf surface have sparse long **hairs**, upper leaf surface has short rasping hairs. **Inflorescence** a single dense spike-like panicle, c. 10 cm tall, ranging in colour from pale straw to purple. **Stamens** yellow.

Habitat: Common and widespread in lightly shaded areas of woodland and islands, with a particular liking for the well-drained soil of termite mounds.

Flowering: Throughout the rainy period.

Uses and beliefs:
- Good grazing and used for hay.

FZ Vol.10 pt3, B Van Wyk New p.315, Hargreaves p.68, Van Oudtshoorn 2 p.88.

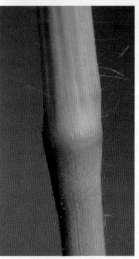

Poaceae (grass family)

Spike-like panicle

Pennisetum macrourum Trin.
(syn. *Pennisetum glaucocladum* Stapf & C.E.Hubb.)

Common names: Setswana lebelebele, mahango; **English** riverbed grass, riverbank pennisetum.

Derivation: *Penna*, feather or plume [Latin], *saeta*, bristle or hair [Latin], alluding to the bristly spikelets; *macro-*, long or big [Latin], *-urum*, tailed [Latin], alluding to the large tail-like inflorescence.

Identification: Erect densely tufted **perennial grass**, c. 2.3 m tall with a **rhizomatous rooting** system. **Nodes** inconspicuous and green. **Ligule** a ring of hairs. **Leaves** c. 70 × up to c. 1.2 cm, mainly flat and rolled towards the tip, harshly hairy especially on the edges. **Inflorescence** a single erect spike, c. 20 cm tall. **Spikelets** surrounded by bristles of varying length. Anthers yellow.

Habitat: Common and widespread in dense stands at the water's edge.

Flowering: During the main rains.

Uses and beliefs:
• Used for thatching.

FZ Vol.10 pt3, Ellery p.197, Van Oudtshoorn 2 p.86.

Poaceae (grass family)

Spike-like panicle

Perotis patens Gand.

Common names: Setswana mojakubu, bojang-jwa-phiri; **Afrikaans** katstertgras; **English** bottle brush grass, cat's tail grass, hyena grass, purple spike grass.

Derivation: *Per*, very or throughout [Latin]; *otos*, ear [Greek], referring to the ear-shaped base of the leaves; *patens*, spreading [Latin].

Identification: Semi-perennial tufting **grass**, up to c. 90 cm tall. Stoloniferous **rooting system**. **Stems** branching. **Nodes** pale green and pink. **Ligule** an inconspicuous membrane. **Leaves** blue-green, c. 60 cm × 1.5 cm, flat with a twist in the blade, margins are lined with stiff erect hairs, leaf base rounded. **Inflorescence** a single spike of long-awned spikelets, c. 20 cm in length.

Habitat: Common and widespread in sandy areas of the floodplain.

Flowering: During the main rains.

Uses and beliefs:
• A pioneer grass of over-grazed or drought areas.

FZ Vol.10 pt2, B Van Wyk Flowers p.318, Ellery p.198, Müller p.210, Van Oudtshoorn 2 p.82.

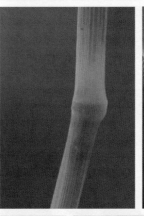

Poaceae (grass family)

Spike-like panicle

Setaria pumila (Poir.) Roem. & Schult.
(syn. *Setaria ustilata* de Wit)

Derivation: *Seta*, bristle [Latin], alluding to the bristles on the spikelets; *pumila*, dwarf [Latin].

Identification: Small tufted **annual grass**, c. 55 cm tall, with branching stems. **Nodes** green flushed pink above and below. **Ligule** membranous. **Hairless** except for the leaf margins, which are hairy. **Leaves** c. 10 × 1 cm, flat. **Inflorescence** a single spike, c. 5 cm tall. Anthers, stigma and bristles on the **spikelets** mahogany brown.

Habitat: Locally uncommon growing in partial shade on islands in the floodplain.

Flowering: Late in the main rains.

FZ Vol.10 pt3.

Poaceae (grass family)

Spike-like panicle

Setaria sphacelata (Schumach.) M.B.Moss

Common names: Setswana mabele, motawaphesa, mogatapeba, mogatla-wapeba; **Afrikaans** katstertmannagras; **English** golden bristle grass, common bristle grass, dog's tail, buffalo grass, golden timothy grass, golden tumbling grass, land grass, old-land's grass, golden setaria.

Derivation: *Seta*, a bristle [Latin], alluding to the bristles on the spikelets; *sphacelata*, withered as if dead [Latin].

Identification: Tufted perennial grass, up to c. 1.8 m tall, with short **rhizomes** and branching stems. **Nodes** pale green. **Ligule** a ring of hairs. **Leaves** mostly at the base of the plant, up to c. 24 × c. 0.8 cm, folding at the midrib. **Inflorescence** rounded and unbranched, a single erect spike resembling a small bottle brush, up to c. 24 cm tall. The bristles remain on the stem when the spikelets have fallen. This grass is very variable, witness the number of common names. Its size may vary considerably and some plants may be creeping rather than tufted.

Habitat: Found occasionally locally in damp sandy and peaty areas, usually near water.

Flowering: From the main rains into the cool dry period.

Uses and beliefs:
• A highly palatable grass.

FZ Vol.10 pt3, B Van Wyk Flowers p.319, Ellery p.201, Van Oudtshoorn 2 p.90, WFNSA p.24.

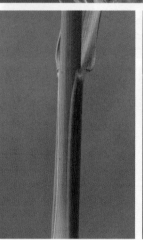

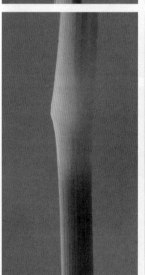

Poaceae (grass family)

Spike-like panicle

Setaria verticillata (L.) P. Beauv.

Common names: Setswana bogoma; **Afrikaans** klitsgras, klettgrass; **English** bur bristle grass, bur grass, sticky bristle grass, bristly foxtail, cat's tail.

Derivations: *Seta*, bristle [Latin], alluding to the bristles on the spikelets; *verticillatus*, whorled [Latin], referring to the lower part of the inflorescence. *Bogoma*, a plant parts of which readily attach themselves to passing animals. Various plants of different families have this name.

Identification: Rather lax **annual grass**, growing in loose tufts up to c. 1 m tall. **Stems** triangular to D-shaped in section and branching. **Nodes** green tinged brown. **Leaves** flat and soft, c. 25 × 2.5 cm. **Inflorescence** a single spike-like panicle, c. 5 cm long. Individual **spikelets** have unpleasant bristles covered in barbs which catch onto themselves and onto any other plant or animal, forming tangled masses.

Habitat: Common, growing thickly in light shade, mainly on islands and in mixed woodland. Flourishes on disturbed, over-grazed soils.

Flowering: Early in the main rains.

Uses and beliefs:
• The seed heads are used by Hmbukushu to trap birds (even francolin or guinea fowl). Grain is added to the seed heads to attract birds, which become entangled.
• A very palatable grass that retains its nutritional value even when it has dried out.

FZ Vol.10 pt3, Ellery p.201, B Van Wyk Flowers p.320, Müller p.232, Van Oudtshoorn 2 p.104. MM.

Poaceae (grass family)

Spike-like panicle

Tragus berteronianus Schult.

Common names: Setswana mogamapodi, segowe, nama, bogoma; **Afrikaans** wortelsaadgras; **English** (common) carrot-seed grass, small carrot-seed grass, burgrass.

Derivation: *Tragos*, goat [Greek], *berteronianus* for Carlo Giuseppe Bertero (1789–1831) an Italian physician who studied botany in the West Indies and South America.

Identification: Small erect **annual grass**, up to c. 40 cm tall. **Stoloniferous rooting** system. **Stem** flushed red below the node. **Nodes** dark red. **Ligule** a ring of short hairs. **Leaves** slightly rolled, c. 3.5 cm × 0.5 cm, margin fringed with widely spaced bristles. **Inflorescence** a spike-like panicle, c. 9 cm tall. **Spikelets** covered in hooks to catch passing animals. Anthers and stigma are indigo in colour.

This species may easily be confused with *Tragus racemosus*, which has larger spikelets and more prominent erect hairs on the leaf margin.

Habitat: Common and widespread on sand and compacted sandy clay on the floodplain.

Flowering: Throughout the main rains.

Uses and beliefs:
- Grows in poor soil where other plants cannot survive.
- An indicator of drought or overgrazing.

FZ Vol.10 pt2, B Van Wyk New p.321, Ellery p.205, Van Oudtshoorn 2 p.80.

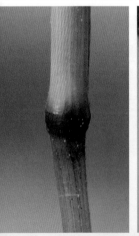

Poaceae (grass family)

Spike-like panicle

Tragus racemosus (L.) All.

Common names: Setswana bogoma; **English** large carrot-seed grass.

Derivation: *Tragos*, goat [Greek]; *racemosus*, with flowers in racemes [Latin].

Identification: Small, erect **annual grass**, up to c. 25 cm tall, with a stoloniferous **rooting** system. Tending to bend at the nodes. **Nodes** dark red, the stem is flushed red below the node. **Ligule** a ring of short hairs. **Leaves** c. 3 × 0.5 cm, flat with a slightly powdery surface, margin fringed with erect long hairs. **Inflorescence** a spike-like panicle, up to c. 9 cm tall. **Spikelets** covered in hooks to catch passing animals. Anthers and stigma dusky pink.

Easily confused with *Tragus berteronianus*, which is slightly taller and has smaller, more densely packed spikelets.

Habitat: Common and widespread on sand and compacted sandy clay on the floodplain and in mopane woodland.

Flowering: Throughout the main rains.

Uses and beliefs:
- Grows in poor soil where other plants cannot survive.
- An indicator of drought or overgrazing.

FZ Vol.10 pt2.

Sedges
and sedge-like plants

Introduction to sedges and sedge-like plants

The word sedge as used below is taken to include rushes, reeds, sedges and sedge-like plants.

Sedges and sedge-like plants can be found in varying habitats. They are most frequently found in damp to waterlogged locations but can grow in the sandy areas and in mopane woodland far from pans and water courses.

The sedges and sedge-like plants are arranged in this book according to the structure of the inflorescence, then by genus and species.

Sedges and sedge-like plants frequently have triangular stems with an angular cross-section, and normally have linear leaves arising in a basal rosette. Bracts, which can be long, are often found on the stem immediately below the inflorescence. The flowers are wind pollinated.

When attempting to identify a sedge or sedge-like plant, be sure to select a specimen whose inflorescence is fully developed, ideally with flowers showing. Taking an immature specimen can lead to misidentification of say an umbel as a cluster. Further down the plant, the stem needs to be in prime condition and not too young for the features to be properly developed.

We are indebted to Dr David Simpson of the Royal Botanic Gardens Kew for his advice on this section, including a classification of sedges by structure. A list of the inflorescence structures together with brief descriptions follows:

Branched pseudolateral

An inflorescence in the form of a branched cluster that appears to emerge from the side of the stem below its apex (pseudolateral). However the apparent continuation of the stem is in fact a bract.

Panicle

An inflorescence emerging from more than one point up the stem. Usually each part of the inflorescence has its own supporting bract at its point of emergence.

Umbel-like with single spikelets

An inflorescence with branches arranged in an umbel-like structure having single spikelets each with its own clearly defined stalk at the tips of the branches.

Capitate

An inflorescence where the spikelets are arranged in a group at the apex of the stem above the bract or bracts.

Unbranched pseudolateral

An inflorescence in the form of an unbranched cluster that appears to emerge from the side of the stem below its apex (pseudolateral). However, the apparent continuation of the stem is in fact a bract.

Umbel-like clusters

An inflorescence with branches arranged in an umbel-like structure having the spikelets arranged in clusters at the tips of the branches.

Terminal elongate

An inflorescence arranged as a single elongated terminal spikelet at the apex of the stem.

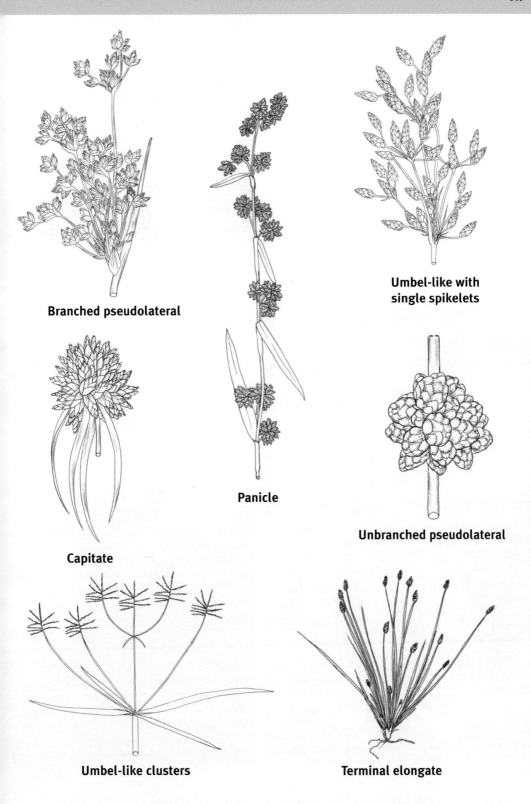

Cyperaceae (sedge family)

Capitate

Cyperus dubius Rottb.
(syn. *Mariscus dubius* (Rottb.) Kük. ex C.E.C.Fisch.)

Derivation: *Cuperos* [Latin], *kypeiros* [Greek], sedge or rush; *dubius*, doubtful in the sense of not conforming to pattern [Latin].

Identification: Annual tufted sedge, c. 60 cm tall. **Stem** triangular with rounded corners. **Leaves** grass-like, basal, about half the height of the flowering plant. The capitate **inflorescence** is a compact ball of flattened greenish white **spikelets**, c. 1.5 cm dia. Supporting bracts below the inflorescence are up to c. 25 cm long.

Habitat: Found occasionally locally in sandy areas of woodland.

Flowering: During the main rains.

Cyperaceae (sedge family)

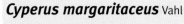
Capitate

Cyperus margaritaceus Vahl

Common name: Setswana monakaladi; **Afrikaans** witbiesie.

Derivation: *Cuperos* [Latin], *kypeiros* [Greek], sedge or rush; *margarita*, pearl [Latin], hence pearly, alluding to the inflorescence.

Identification: Tufted **annual sedge** of sandy areas, c. 50 cm tall. **Stem** rounded. **Inflorescence** a ball of wide flattened **spikelets** that are pearly white with a green margin; basal **bracts** supporting the inflorescence stiff and narrow, sloping downwards.

Habitat: Found occasionally locally in sandy mixed woodland.

Flowering: During the main rains.

Hargreaves p.67.

Cyperaceae (sedge family)

Capitate

Cyperus pectinatus Vahl
(syn. *Cyperus nudicaulis* Poir.)

Derivation: *Cuperos* [Latin], *kypeiros* [Greek], sedge or rush; *pectinatus*, comb-like [Latin], referring to the close arrangement of the segments of the spikelets.

Identification: Slender tufted **perennial sedge** with long fine rather lax stems, up to c. 80 cm tall. **Stems** ribbed and oval, often curving over and producing **roots** from near the inflorescence. **Leaves** absent but there are small leaf **sheaths** at the base. **Inflorescence** capitate. Basal **bracts** supporting the inflorescence are slightly longer than the inflorescence.

Habitat: Found occasionally locally rooted in the peaty margins of flowing channels, but may also be found floating on mats of plant matter.

Flowering: During the rainy season.

Ellery p.211.

Eriocaulaceae (piperwort family)

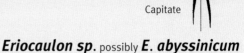
Capitate

Eriocaulon sp. possibly *E. abyssinicum*
Hochst.

Derivation: *Erion*, wool [Greek]; *kaulos*, a stem [Greek], woolly-stemmed, referring to the hairy stems of some plants of this genus; *abyssinicum*, of Abyssinia.

Identification: Minute **annual sedge-like plant**, c. 8 cm tall, growing in tufts. **Leaves** grass-like, forming a basal rosette. **Inflorescences** white and spherical, c. 3 mm dia. There is no leafy bract below the inflorescence.

Habitat: Locally rare growing in full sun in drying areas of fibrous sandy soil in the spillway and floodplain.

Flowering: During the early rains.

Cyperaceae (sedge family)

Capitate

Kyllinga buchananii C.B.Clarke

Common name: Setswana monakaladi omotonanyana.

Derivations: *Kyllinga*, for the Danish botanist Peter Kylling. *Monakaladi omotonanyana*, this is the male *monakaladi* because it does not have tubers underground.

Identification: Tufted **perennial sedge**, up to c. 50 cm tall. **Stems** triangular. **Leaves** a basal rosette, up to c. 20 × 0.3 cm, folding sharply at the mid-vein, sparse coarse **hairs** on the upper surface. **Inflorescence** a sweetly scented compact white ball of spikelets, c. 1.2 cm dia. Leafy **bracts** supporting the inflorescence up to c. 12 cm long.

Habitat: Widespread in damp open sandy areas of mixed and mopane woodland.

Flowering: During the main rains.

KS.

Cyperaceae (sedge family)

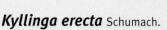
Capitate

Kyllinga erecta Schumach.

Derivation: *Kyllinga*, for the Danish botanist Peter Kylling; *erecta*, erect, upright [Latin].

Identification: Small **annual sedge**, c. 30 cm tall. **Leaves** thin and grass-like, sheathing the stems. **Inflorescence** a compact green ball of spikelets, c. 8 mm dia., with **leafy bracts** up to c. 5 cm long supporting the inflorescence.

Habitat: Common and widespread in damp sandy areas of the floodplain with some light shade.

Flowering: During the main rains.

Cyperaceae (sedge family)

Capitate

Kyllingiella microcephala (Steud.) R.W.Haines & Lye

Derivation: *Kyllingiella*, for the Danish botanist Peter Kylling, similar to the genus *Kyllinga*; *mikros*, small [Greek], *kephale*, a head [Greek], referring to the small inflorescences.

Identification: Tufted **annual sedge**, up to c. 25 cm tall. **Leaves** basal and c. half the height of the flower stalk. **Inflorescence** a compact cluster of white spikelets, c. 7 mm dia., that looks rather like a minute cauliflower, the florets being clearly delineated. Long leafy **bracts** (up to c. 6.5 cm long) support the inflorescence.

Habitat: Locally rare in an area of farmed mopane with seasonal pans.

Flowering: Early in the main rains.

Cyperaceae (sedge family)

Umbel-like clusters

Cyperus alopecuroides Rottb.

Derivation: *Cuperos* [Latin], *kypeiros* [Greek], sedge or rush; *alopecurus*, a grass like a fox's tail, *-oides*, -like [Greek], similar to a fox-tail grass.

Identification: Large vigorous **sedge** with a **rhizomous rooting** system. **Leaves** dark green, as tall as the flower stem. **Inflorescence** is umbel-like, c. 20 cm across, with large, green **spikelets** borne in dense long clusters at the ends of the branches. Basal **bracts** supporting the inflorescence are up to twice as long as the inflorescence is tall.

Habitat: Widespread growing in large stands in newly flooded channels. It appears sensitive to water depth as it soon disappears if the water becomes deeper.

Flowering: Appears to depend on the coming of the flood.

Cyperaceae (sedge family)

Umbel-like clusters

Cyperus chersinus (N.E.Br.) Kük.

Derivation: *Cuperos* [Latin], *kypeiros* [Greek], sedge or rush.

Identification: Annual **sedge** growing to a height of up to c. 75 cm. **Stem** triangular in cross-section. **Leaves** thin, c. 26 × 0.4 cm, folding at the mid-vein. **Inflorescence** is umbel-like with wet-looking tubular clusters of **spikelets**, c. 2 × 0.5 cm dia. Three long basal **bracts** support the inflorescence.

Habitat: Locally rare on the trackside in a sandy clay area of farmed mopane.

Flowering: During the main rains.

Cyperaceae (sedge family)

Umbel-like clusters

Cyperus compressus L.

Common names: Setswana motlhakana; **English** flat sedge.

Derivation: *Cuperos* [Latin], *kypeiros* [Greek], sedge or rush; *compressus*, compressed or flattened [Latin], alluding to the shape of the spikelets.

Identification: Small tufted **annual sedge**, c. 10 cm tall. **Roots** dark red. **Stem** D-shaped. **Inflorescence** is umbel-like with clusters of flattened spikelets. Spikelets red-brown with a green margin. Usually 3 broad basal **bracts**, up to c. 13 cm long, support the inflorescence.

Habitat: Widespread growing on the margins of drying areas of pans and wetlands.

Flowering: During the main rains.

Cyperaceae (sedge family)

Umbel-like clusters

Cyperus denudatus L.f.

Common names: Setswana tototwane; **English** winged sedge.

Derivation: *Cuperos* [Latin], *kypeiros* [Greek], sedge or rush; *denudatus*, stripped of leaves or hairs [Latin].

Identification: Slender erect **perennial sedge**, c. 40 cm tall but may grow to c. 1.3 m in a very wet year. **Roots** rhizomous and chestnut coloured. **Stem** sharply triangular. **Leaves** small sheaths at the base of the plant. **Inflorescence** is an umbel-like cluster of narrow chestnut-brown spikelets, c. 6 cm tall. Basal **bracts** supporting the inflorescence vestigial.

Habitat: Locally common in the sandy clay margins of the floodplain.

Flowering: Throughout the rains.

Pooley p.560.

Cyperaceae (sedge family)

Umbel-like clusters

Cyperus difformis L.

Common name: English small-flower umbrella plant.

Derivation: *Cuperos* [Latin], *kypeiros* [Greek], sedge or rush; *difformis*, of an unusual form (in relation to the rest of the genus) [Latin], possibly referring to the small size of the spikelets.

Identification: Tufted **annual sedge**, c. 35 cm tall. **Roots** dark red and white. **Leaves** clasping the stem, c. 15 × 0.6 cm. **Inflorescence** is umbel-like, c. 6 cm wide, with balls of small spikelets, c. 2 mm long. **Bracts** below the inflorescence longer than the inflorescence itself, up to c. 17 cm long.

Habitat: Found occasionally locally by seasonal pans and in damp areas of track-side.

Flowering: During the main rains.

Cyperaceae (sedge family)

Umbel-like clusters

Cyperus esculentus L.

Common names: Setswana tototwane, mohatola, tlhatlha, mizindumi, mothlathla, mokhasi; **Subiya** ngwara; **Shiyei** modwarra; **Afrikaans** hoenderuintjie, geeluintjie; **English** yellow nut sedge, nut grass, water grass, earth almond, ground almond.

Derivation: *Cuperos* [Latin], *kypeiros* [Greek], sedge or rush; *esculentus*, edible, good to eat [Latin].

Identification: Erect, hairless **perennial sedge**, up to c. 50 cm tall, arising from a rhizomous **rooting** system that terminates in fleshy tubers. **Stem** triangular. **Leaves** erect, grass-like and shiny, almost reaching to the inflorescence. **Inflorescence** umbel-like with clusters of brownish **spikelets**, supported by 3–5 short leaf-like **bracts**.

Habitat: Common and widespread cosmopolitan weed of damp areas.

Flowering: Throughout the main rains.

Uses and beliefs:
• The tubers are edible and used as *morogo* or to make an excellent curry. They have been cultivated since Egyptian times.
• Difficult to control in crops as this species regenerates from tubers.

B Van Wyk New p.274, Botweeds p.116, Hargreaves p.67. TK.

Cyperaceae (sedge family)

Umbel-like clusters

Cyperus haspan L.

Derivation: *Cuperos* [Latin], *kypeiros* [Greek], sedge or rush, *haspan* from its vernacular name in Sri Lanka.

Identification: Perennial sedge with tufts of fine, erect, triangular stems, c. 40 cm tall. **Leaves** grass-like. **Inflorescence** is umbel-like with clusters of russet-brown spikelets frequently with green tips. Leafy **bracts** below the inflorescence c. 10 cm long.

Habitat: Found occasionally in sandy clay in the spillway.

Flowering: During the main rains.

Uses: Often heavily grazed by plains game.

Cyperaceae (sedge family)

Umbel-like clusters

Cyperus iria L.

Derivation: *Cuperos* [Latin], *kypeiros* [Greek], sedge or rush.

Identification: Erect, hairless **annual sedge**, up to c. 70 cm tall. **Stem** sharply triangular. **Leaves** erect, grass-like and sheath the stems. **Inflorescence** a rather congested umbel-like cluster of long, narrow, brownish green **spikelets**, alternating up the stems, supported by 3–5 long leaf-like **bracts** of varying length.

Habitat: Locally uncommon growing on the margin of seasonal pans.

Flowering: During the main rains.

Cyperaceae (sedge family)

Umbel-like clusters

Cyperus latifolius Poir.

Common name: Setswana jarajara.

Derivation: *Cuperos* [Latin], *kypeiros* [Greek], sedge or rush; *lati-*, broad [Latin], *folius*, leaved [Latin], broad-leaved.

Identification: Large tufted **sedge** of water margins, c. 2.5 m tall. **Stem** sharply triangular. **Inflorescence** c. 26 cm tall, umbel-like with many clusters of narrow reddish-brown **spikelets**. Basal **bracts** supporting the inflorescence c. 55 cm long, triangular, stiff and borne erect.

Habitat: Locally rare in the margin of water channels.

Flowering: Early in the main rains.

Cyperaceae (sedge family)

Umbel-like clusters

Cyperus leptocladus Kunth

Derivation: *Cuperos* [Latin], *kypeiros* [Greek], sedge or rush; *lepto-*, thin or slender [Greek], *klados*, a branch [Greek], with slender stems.

Identification: Tufted **annual sedge**, c. 55 cm tall. **Leaves** grass-like, sheathing the stem, lower leaves papery. **Inflorescence** c. 5.5 cm tall, umbel-like with densely set clusters of brown-green **spikelets**. Basal **bracts** leafy, about the same length as the inflorescence.

Habitat: Widespread in sandy areas of the spillway margin in areas where there has been annual flooding for many years.

Flowering: During the main rains.

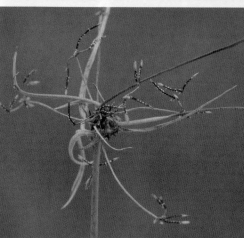

Cyperaceae (sedge family)

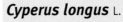
Umbel-like clusters

Cyperus longus L.

Common names: Setswana tototwane; **English** water rush, sweet cyperus.

Derivation: *Cuperos* [Latin], *kypeiros* [Greek], sedge or rush; *longus*, long [Latin], referring to the length of the spikelets.

Identification: Slender hairless **perennial sedge**, c. 80 cm tall. **Inflorescence** c. 9 cm tall, umbel-like with many clusters of long narrow reddish-brown spikelets. Inflorescence supported by basal **bracts** that are erect, quite long and slightly broad. Crushed **stems** and **roots** have the **scent** of violets.

Habitat: This cosmopolitan plant is locally rare in the margin of water channels.

Flowering: Early in the main rains.

Uses and beliefs
- *Cyperus longus* is much used in horticulture.

Cyperaceae (sedge family)

Umbel-like clusters

Cyperus papyrus L.

Common names: Setswana koma, dikoma; **Shiyeyi** koma; **English** papyrus, paper plant.

Derivation: *Cuperos* [Latin], *kypeiros* [Greek], sedge or rush; *papyrus* is the Greek name for the paper made in scrolls from the pith of this species in Ancient Egypt.

Identification: Very robust **perennial sedge** of permanent swamp and major waterways, c. 2 m tall. A stout stoloniferous **rooting** system forms huge floating mats. **Stems** rounded to triangular. **Inflorescence**, c. 35 cm dia, umbel-like with many clusters of spikelets. Basal **bracts** supporting the inflorescence are scale-like, small, brown and papery.

Habitat: Common and widespread in the permanent swamps and fringing the major rivers. It is very vigorous blocking channels and waterways.

Flowering: During the rainy season.

Uses and beliefs:
• *Cyperus papyrus* may have been introduced into the area in the early 20th Century in an effort to establish a paper-making industry.
• Wayeyi and Subiya cut the stems and split them to weave mats.
• When the stems are fresh and young in the rainy season the Wayeyi and Subiya eat them like sugar cane.
• In the Panhandle, this species is used to make rafts for transport downstream.

Ellery p.210. CM, PK.

Cyperaceae (sedge family)

Umbel-like clusters

Cyperus pseudokyllingioides Kük.

Derivation: *Cuperos* [Latin], *kypeiros* [Greek], sedge or rush; *pseudo-*, false [Greek], *kyllingioides*, similar to the genus *Kyllinga*, although they are not really alike.

Identification: Tufted slightly lax **annual sedge**, c. 65 cm tall. **Stem** sharply triangular. The bright green **inflorescence** is umbel-like with spherical clusters of spikelets. Basal **bracts** supporting the inflorescence up to c. 60 cm long with a deeply incised vein at the mid-rib.

Habitat: Locally uncommon, growing on the edge of pans, protected by other vegetation.

Flowering: Late in the main rains.

Cyperaceae (sedge family)

Umbel-like clusters

Cyperus rotundus L.

Common names: Afrikaans rooiuintjie; **English** nutsedge, nutgrass, purple nutsedge.

Derivation: *Cuperos* [Latin], *kypeiros* [Greek], sedge or rush; *rotundus*, rounded [Latin].

Identification: Perennial sedge, up to c. 1.2 m tall, with a **rhizomous tuberous rooting** system. **Stems** have 2 flat sides and 1 rounded one. **Leaves** narrow and grass-like, forming a basal rosette, about half the height of the plant. **Inflorescence** is umbel-like, c. 10 cm across. **Spikelets** an attractive pinkish brown. Basal **bracts** supporting the inflorescence almost 3 times the height of the inflorescence.

Habitat: Locally common in dry deeply sandy areas of floodplain.

Flowering: During the main rains.

Uses and beliefs:
* *Cyperus rotundus* is known as the world's worst weed! It is known as a weed in 90 countries and is difficult to eradicate because of its tuberous rhizomous rooting system, which will regenerate from small broken pieces. It can become a pest in crops.

B Van Wyk Flowers p.274, Pooley p.562.

Cyperaceae (sedge family)

Umbel-like clusters

Cyperus sphaerospermus Schrad.

Common name: Afrikaans matjiesgoed.

Derivation: *Cuperos* [Latin], *kypeiros* [Greek], sedge or rush; *sphaero*, rounded [Latin], *spermus*, seed [Latin], with rounded seeds

Identification: Tall spindly **perennial sedge** with a **rhizomous rooting** system. **Stems** sharply triangular. **Leaves** narrow, grass-like, about half the height of the plant. **Inflorescence** is an umbel-like cluster of small, dull green **spikelets**, c. 10 cm across. Basal **bracts** supporting the inflorescence up to c. 7 cm long.

Habitat: Found occasionally in damp areas of the floodplain.

Flowering: Throughout the rains.

Pooley p.504, WFNSA p.26.

Cyperaceae (sedge family)

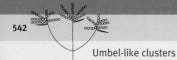
Umbel-like clusters

Cyperus squarrosus L.

Derivation: *Cuperos* [Latin], *kypeiros* [Greek], sedge or rush; *squarrosus*, with parts spreading or recurved at the ends [Latin].

Identification: Small tufted **annual sedge**, up to 15 cm tall. **Stems** triangular. Lower **leaf sheaths** and **roots** dark red. **Leaves** linear. **Inflorescence** umbel-like with egg-shaped clusters of **spikelets**. Basal **bracts** supporting the inflorescence up to c. 7 cm long.

Habitat: Found occasionally locally in damp areas near seasonal and permanent water.

Flowering: During the main rains.

Cyperaceae (sedge family)

Umbel-like clusters

Cyperus tenuiculmis Boeckeler

Derivation: *Cuperos* [Latin], *kypeiros* [Greek], sedge or rush; *tenui-*, slender, thin [Latin], *-culmus*, stem [Latin], slender-stemmed.

Identification: Perennial tufted **sedge**, up to c. 1 m tall, with slender **stems** and a rhizomous **rooting** system. **Leaves** c. one third the height of the stem. **Inflorescence** has umbel-like clusters of long narrow **spikelets**, up to c. 30 cm dia. Long basal **bracts** support the inflorescence.

Habitat: Locally rare growing in deep sand.

Flowering: During the main rains.

Uses and beliefs:
- Highly palatable and heavily grazed by game.

Cyperaceae (sedge family)

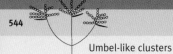
Umbel-like clusters

Cyperus zollingeri Steud.

Derivation: *Cuperos* [Latin], *kypeiros* [Greek], sedge or rush. *Zollingeri* probably for the Swiss botanist Heinrich Zollinger (1818–1859).

Identification: Tufted **annual sedge**, c. 40 cm tall. **Stem** triangular. **Leaves** grass-like, basal, may be as tall as the inflorescence. **Inflorescence** umbel-like with clusters of long pointed green **spikelets** in sparse groups on long stems. Basal **bracts** that support the inflorescence up to c. 17 cm long.

Habitat: Widespread in sandy gritty soil in the deep shade of mixed woodland.

Flowering: Early in the main rains.

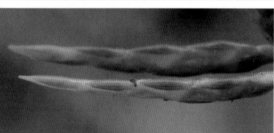

Cyperaceae (sedge family)

Umbel-like clusters

Oxycaryum cubense (Poepp. & Kunth) Lye

Derivation: *Oxys*, sharp, keen [Greek], *karyon*, a nut [Greek], referring to the shape of the fruit; *kube*, a head [Greek], head-like, probably referring to the rounded clusters of spikelets.

Identification: Vigorous **perennial sedge**, c. 1 m tall, with a **rhizomous rooting** system. **Stems** sharply triangular. **Leaves** basal, folding v-shaped at the mid-rib. **Inflorescence** umbel-like, c. 10 cm dia., composed of green balls of **spikelets**, c. 1.5 cm dia. Trailing leafy **bracts** supporting the inflorescences may be up to 60 cm long.

Habitat: Found occasionally locally on the water margin.

Flowering: During the main rains.

Ellery p.214.

Cyperaceae (sedge family)

Umbel-like clusters

Pycreus macrostachyos (Lam.) J.Raynal

Derivation: *Pycreus*, is an anagram of *Cyperus*, the inflorescences of the two genera are very similar; *macro-*, long or large [Greek], *stachys*, an ear of corn, hence a spike [Greek], with long or large spikes, alluding to the spikelets.

Identification: Rather stocky **annual sedge** of damp areas up to c. 50 cm tall. **Roots** chestnut or white. **Stem** sharply triangular. **Leaves**, sheathing the stem, up to c. 40 × 1 cm. **Inflorescence** umbel-like with many clusters of brownish-green **spikelets**. Basal **bracts** supporting the inflorescence up to c. 40 cm long and rather broad.

Habitat: Found occasionally locally in damp areas near seasonal pans and waterways.

Flowering: During the main rains.

Cyperaceae (sedge family)

Umbel-like clusters

Pycreus pelophilus (Ridl.) C.B.Clarke

Derivation: *Pycreus*, is an anagram of Cyperus, the inflorescences of the two genera are very similar; *pelophilos*, beloved, dear [Greek].

Identification: Slender tufted **annual sedge** with long fine rather lax stems, c. 35 cm tall. **Leaves** fine, grass-like, c. 25 cm × 2 mm. **Inflorescence** umbel-like with clusters of golden brown and green **spikelets**. Basal **bracts** supporting the inflorescence are about twice the height of the inflorescence.

Habitat: Widespread in sandy clay near seasonal pans.

Flowering: Throughout the main rains.

Cyperaceae (sedge family)

Umbel-like clusters

Pycreus polystachyos (Rottb.) P.Beauv. var. *polystachyos*

Derivation: *Pycreus*, is an anagram of *Cyperus,* the inflorescences of the two genera are very similar; *polys*, many *stachys*, an ear of corn [Greek], alluding to the many spikelets.

Identification: A small robust **annual sedge**, c. 30 cm tall. **Leaves** basal. An umbel-like **inflorescence** of clusters of narrow green-brown **spikelets**; supporting leafy **bracts** up to c. 10 cm long.

Habitat: Found occasionally locally in damp short grass in the spillway.

Flowering: Throughout the main rains.

Uses and beliefs:
- Heavily grazed therefore must be palatable.

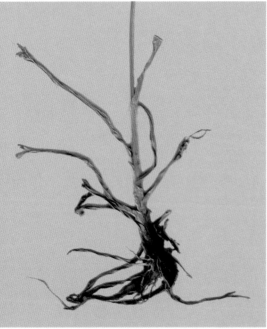

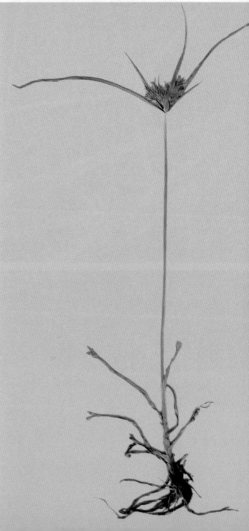

Cyperaceae (sedge family)

Umbel-like with single spikelets

Bulbostylis hispidula (Vahl) R.W.Haines subsp. *pyriformis* (Lye) R.W.Haines
(syn. *Fimbristylis hispidula* Auct.)

Common names: Setswana bojang-jwa-mmutla, bojang bomotla.

Derivation: *Bulbo*, bulb-shaped [Latin], *stylis* referring to the style, a more or less elongated part of the ovary bearing the stigma; *hispidus*, covered in coarse erect hairs or bristles [Latin].

Identification: Tufted, grass-like, **annual sedge**, usually c. 30 cm tall but up to c. 45 cm in years of heavy rainfall. **Roots** shallow and fibrous. **Stems** covered in short hairs. **Leaves** arising from around the base of the stems, c. 15 cm × 0.5 mm, folding at the mid-rib and forming a sheath around the stem. **Inflorescence** an umbel-like cluster of single-stalked dark brown **spikelets**. Very short leafy **bracts** below the inflorescence.

Habitat: Common and widespread in open sandy areas in full sun.

Flowering: Main rains through to the end of the cool dry period.

Botweeds p.118.

Cyperaceae (sedge family)

Umbel-like with single spikelets

Fimbristylis dichotoma (L.) Vahl

Derivation: *Fimbriatus*, fringed [Latin], *stylos*, a column [Greek], with a fringed style; *dichotoma*, forked in pairs, repeatedly dividing into two branches [Latin].

Identification: Small tufted **annual sedge,** c. 30 cm tall, with grass-like inflorescences. **Leaves** pubescent on both surfaces and folding at the mid-vein, forming a sheath around the stem. **Inflorescence,** c. 6 cm dia., an umbel-like cluster of single-stalked, rounded, slightly hairy, dark brown spikelets, arranged in pairs. The supporting **basal bracts** are no longer than the inflorescence.

Habitat: Widespread in light shade on the edge of the spillway.

Flowering: During the early rains.

Cyperaceae (sedge family)

Umbel-like with single spikelets

Fimbristylis longiculmis Steud.
(syn. *Fimbristylis bivalvis* (Lam.) Lye)

Derivation: *Fimbriatus*, fringed [Latin], *stylos*, a column [Greek], with a fringed style; *longus*, long [Latin], *culmus*, stem [Latin], long-stemmed.

Identification: Perennial sedge, up to c. 50 cm tall. **Leaves** very small, up to c. 15 cm tall, sheathing the base of the stem. Grass-like **inflorescence**, c. 16 cm tall, an umbel-like cluster of single, rather hairy, mahogany-brown spikelets on long stalks. Only vestigial leafy **bracts** below the inflorescence.

Habitat: Found occasionally in the peaty soil on the edge of flowing waterways.

Flowering: Early in the main rains.

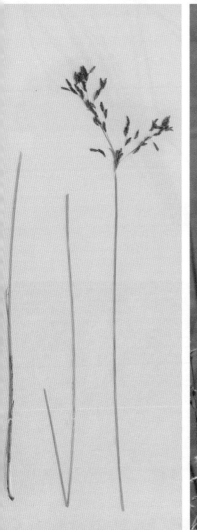

Cyperaceae (sedge family)

Unbranched pseudolateral

Schoenoplectiella senegalensis
(Steud.) Lye
(syn. *Schoenoplectus praelongatus* (Poir.) J.Raynal)

Derivation: *Schoinos*, rush or sedge [Greek], *plectus*, twisted or plaited [Greek], alluding to its use in basket-making; *senegalensis*, from Senegal.

Identification: **Annual sedge** with sharply bending stems, c. 40 cm tall. **Stems** hollow, tubular. **Inflorescence** in the form of an unbranched cluster of rounded drop-like spikelets that appears to emerge from the side of the stem, c. 10 cm up the stem (pseudolateral). The apparent continuation of the stem is in fact a bract, which may be up to 30 cm long and is tubular. **Spikelets** silvery green.

Habitat: Locally rare beside seasonal pans in *Acacia* scrub.

Flowering: Late in the main rains.

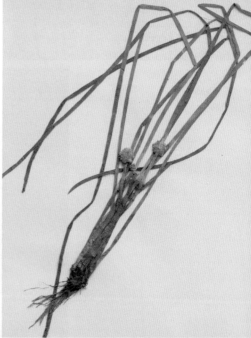

Cyperaceae (sedge family)

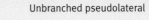
Unbranched pseudolateral

Schoenoplectus erectus (Poir.) Palla ex J.Raynal

Derivation: *Schoinos*, a rush or sedge, *plectus*, twisted or plaited [Greek], alluding to its use in basket-making; *erectus*, erect, upright [Latin].

Identification: Small **annual sedge**, c. 20 cm but may grow taller later in the season. **Inflorescence** in the form of an unbranched cluster that appears to emerge from the side of the stem below its apex (pseudolateral). The apparent continuation of the stem is in fact a bract. **Spikelets** dark brown at first, emerging green as they grow.

Habitat: Locally rare in a flat sandy open area of the spillway.

Flowering: During the early rains.

Uses and beliefs:
- Must be highly palatable as heavily grazed.

Cyperaceae (sedge family)

Branched pseudolateral

Schoenoplectus corymbosus (Roth ex Roem. & Schult.) J.Raynal

Common name: English mat sedge.

Derivation: *Schoinos*, a rush or sedge [Greek], *plectus*, twisted or plaited [Greek], alluding to its use in basket-making; *corymbosus*, provided with corymbs [Latin], clusters of flowers in which the inner flower stems are shorter than the outer ones.

Identification: Perennial sedge, up to c. 1.2 m tall with a rhizomous rooting system. **Stems** tubular, pith-filled. **Leaves** grass-like and sheath the stems. **Inflorescence** in the form of many branched clusters of pale pinky-brown spikelets with supporting scaly bracts, and one bract resembling an extension of the stem, which is in fact a bract.

Habitat: Widespread usually growing in slow-flowing water.

Flowering: Late in the main rains.

B Van Wyk New p.274, Ellery p.216.

Cyperaceae (sedge family)

Panicle

Fuirena pubescens (Poir.) Kunth

Derivation: *Fuirena*, for the Danish physician and botanist Jørgen Fuiren (1581–1628); *pubescens*, downy or hairy [Latin], alluding to the densely hairy branches and spikelets of the inflorescence.

Identification: Slender **perennial sedge** growing up to c. 1 m tall amongst other vegetation. **Roots** rhizomous. **Stem** triangular. **Leaves** grass-like, c. 21 cm tall, sheathing the stem. **Inflorescences** c. 10 cm tall panicles of large hairy spikelets. Leafy **bracts** supporting the inflorescences are usually shorter than the inflorescence.

Habitat: Common and widespread in damp peaty seasonally flooded areas of spillway and floodplain.

Flowering: Throughout the rainy period.

B Van Wyk New p.294, Ellery p.213.

Cyperaceae (sedge family)

Panicle

Fuirena umbellata Rottb.

Derivation: *Fuirena*, for the Danish physician and botanist Jørgen Fuiren (1581–1628); *umbellata*, furnished with an umbel, the flower stalks all arising from the same place.

Identification: Perennial sedge, c. 80 cm tall. **Stem** pentagonal with buttressing and a pithy core, **rooting** at nodes below the water and growing from a **bulbous rhizome** or **corm**. **Leaves** gently rolled, leaf margins have groups of bristles along them. **Inflorescence** a panicle of green balls of spikelets at the apex of the plant with leafy bracts below. Secondary inflorescences at the leaf axils.

Habitat: Found occasionally locally in peaty soil in water on the edge of flowing channels.

Flowering: Throughout the year if conditions are right.

Cyperaceae (sedge family)

Terminal elongate

Eleocharis nigrescens (Nees) Steud.

Derivation: *Helios*, marsh [Greek], *charis*, grace or beauty, beauty of the marsh [Greek]; *nigrescens*, blackish [Latin], because of its dark colour.

Identification: A small tufted **annual sedge** with a dark brown-red colouration, c. 6 cm tall. **Leaves** inconspicuous tiny papery sheaths at the base of the stems. **Inflorescence** terminal elongate, cone-like c. 3 mm long.

Habitat: Found occasionally on peaty sand in a seasonally flooded area of the spillway.

Flowering: During the early rains.

Ferns and aquatic plants

with no visible flowers

Araceae (arum family)

Lemna aequinoctialis Welw.

Common names: English duckweed.

Derivation: *Lemna*, the Greek name for some waterweeds; *aequinoctialis*, belonging to the equinoctial zone, from the equatorial regions.

Identification: A small floating mat-forming **water weed**. **Roots** fine, hair-like, floating in the water. Each **'leaflet'** asymmetrically oval, c. 2 mm long. The plants drift across still water according to the direction of the wind.

Habitat: Widespread on seasonal pans during the rains.

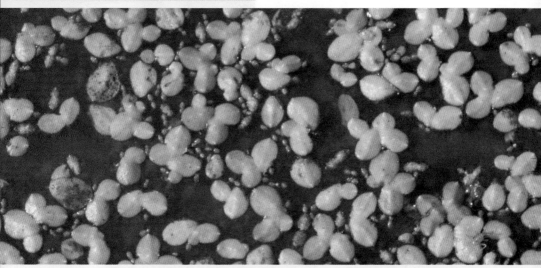

Marsileaceae

Marsilea vera Launert

Derivation: *Marsilea*, for Count Luigi Ferdinando Marsigli (1656–1730), a botanist of Bologna in Italy; *vera*, of the spring [Latin].

Identification: A mat-forming perennial **aquatic fern**. Survives briefly on land, but with much smaller leaves, as the pans dry out. **Fronds** composite, with long fine stems and 4 'leaflets' joined at a single point, borne as a basal rosette. Reproductive organs are **sporocarps** on the roots.

Habitat: Widespread in the mud of seasonal pans and in seasonally flooded marshy areas. Grows throughout the year where conditions allow.

Grows: During the main rains.

FZ Pteridophyta.

Thelypteridaceae

Cyclosorus interruptus (Willd.) H.ltô.
(syn. *Thelypteris interrupta* (Willd.) K.Iwats)

Common names: Setswana kwena, **English** bog fern.

Derivations: *kyklos*, a circle [Greek], *soros*, heap or mound [Greek], referring to the circular 'sori' (reproductive parts on the underside of the leaves); *interruptus*, interrupted, not continuous [Latin], refers to the regular lobing of the fronds. *Kwena*, crocodile, perhaps because of the scaly appearance of the leaves.

Identification: A bright green **fern** with single straight erect fronds from creeping roots, which grow through water and in marshy areas. Individual **fronds** up to c. 60 cm tall. The branches of the fronds have regular lobes. **Reproductive organs** or **sori** arranged in 2 rows on the underside of the lobes of the fronds.

Habitat: Found occasionally in this region, growing on the margin of flowing channels, often mixed with *Cyperus papyrus*.

Ellery p.176.

Contributors

During our research for this book, many people have given us the benefit of their own and their families' knowledge of plants and their experiences with them. They are our 'Contributors'. We feel that it is important to highlight the value and impact of the help given by this group of individuals and to acknowledge their contributions, many of which are found in the 'Uses and Beliefs' section of the species descriptions in this field guide.

Much of the information comes from people at Linyanti Explorations, Selinda, where we have been based. In addition, considerable help was given by the Orient Express staff at Khwai River Lodge and at Eagle Island Camp. We also had short stays at Kubu Lodge, Kasane, and with Wilderness Safaris at Vumbura Plains during which we received input from their staff.

Our thanks go to everyone for their enthusiastic help in this task, which was often given in periods of time off from their duties at camp. We hope that we have not left anyone out and that we have faithfully recorded what we have been told. We apologise in advance for any errors or misconceptions.

Gordon Matengu of Selinda Camp producing a palm frond hat.

Parks Nyame holding the background for photographing a grass.

We have allocated reference initials for each person involved for use in the field guide and we give below a brief summary of each individual listed in alphabetical order by their reference initials:-

BB	Barberton Munduu is Mbanderu (Herero) and a guide at Selinda (brother of RMu). Thanks are given to those who helped BB:- Masepa a Dipodi Selatokgotwana, a San Bushman; Monna wa tsela Dikgadikwane, Bakgalagadi; Ghasten Tuvare, Mbanderu (Herero) and Pesalema, a Motswana traditional doctor in Maun.
BN	Brookes Nkuba is Subiya and Selinda staff.
BT	Balogi Tebogo is Wayeyi and a guide at Khwai River Lodge.
CB	Caspar Bonyongo is Kalanga and a Research fellow at the University of Botswana, Harry Oppenheimer Okavango Research Centre.
CM	Charity Manyando is Subiya (from the Chobe area) and was on the staff at Selinda.
ET	Eveline Tjitunga is Mbanderu (Herero), mother of BB and RMu.
GM	Gordon Matengu is Subiya and a guide at Selinda Camp.
IM	Ishmael Mogamisi is Subiya and a manager at Selinda Camp.
JC	Johnson Chendo is Wayeyi and was a guide at Khwai River Lodge. He now has his own safari company.

Ki	Kit is Wayeyi and a boat man at Vumbura Plains.
KN	Kanawe Ntema is Subiya (from Kachicau) and a guide at Selinda.
KS	Kia Supang is Kalanga (from Chadibe, Francistown) and is manager of Motswiri Camp.
LD	Lenty Dinyando is Wayeyi (from Chobe) and Selinda staff.
LS	Leateametse Saoyene is Wayeyi (from Maun) and Selinda staff.
MA	Mompathi Aaron is Kalanga, was a guide at Selinda and is now with Conservation Corps Africa.
MI	MAP Ives is Motswana and is Conservation Manager Wilderness Safaris.
MM	Motsamai Morundu, is Hmbukushu (from Shakawe) and a guide at Selinda.
MMo	Mothupi Morutwa is Wayeyi and has retired as a guide at Khwai River Lodge.
MT	Mongedi Twiimone (Isaac) is Hmbukushu (from Shakawe) and is staff at Khwai River Lodge.
NM	Nomsa Mbere is President of the Red Cross of Botswana.
PK	Pinaepe Kelaotswe is Hmbukushu (from Shakawe) (cousin of MM). He works at Kubu Lodge.
PM	Portia Monare is Subiya and Selinda staff.
PN	Parks Nyame is Wayeyi (from Jedibe Village) and a guide at Eagle Island Camp.
RM	Robson Mashabe is San and a tracker for Selinda Walking Trails.
RMu	Russ Munduu is Mbanderu (Herero), and is a guide at Vumbura Plains (brother of BB).
RR	Rain Robson, is San and a guide at Selinda (son of RM).
SM	Soccer Molefhi is Herero (from Mahalapaye) and was Selinda staff.
SS	Summer Samakata is Wayeyi (from Sankuyo) and a trainee guide at Khwai River Lodge.
TK	Titus Kasale is Weyeyi (from Sankuyo) and a guide at Khwai River Lodge.

The above information is as things were when we collected it. Some people may have since moved on to other places or other occupations.

Bibliography

In writing this book, we started as absolute beginners and thus our references range from the elementary to the complex as our enquiries became ever more detailed. We have included them all to aid the users of our work who are following the same rewarding path.

The books are listed by reference abbreviation and then title (both in bold print), followed by Authors or Editors, publishing house, year of the edition(s) referred to in this book and, finally, ISBN number where there is one.

B Van Wyk Flowers — The Wild Flowers of the Witwatersrand and Pretoria Region
Braam van Wyk & Sasa Malan, Struik, 1988, ISBN 0-86977-814-5

B Van Wyk New — Field Guide to the Wild Flowers of the Highveld
Braam van Wyk & Sasa Malan, Struik, 1998, ISBN 1-86872-058-6

B&P van Wyk Trees — Field Guide to Trees of Southern Africa
Braam & Piet van Wyk, Struik, 1997, ISBN 1-86825-922-6

Bebbington — Describing Flowers
Anne & John Bebbington, Field Studies Council UK, 1996

Blundell — Collins Guide to the Wild Flowers of East Africa
Michael Blundell, Collins, 1987, ISBN 0-00-219812-6

Botweeds — A Guide to the Arable Weeds of Botswana
Martin Phillips, Ministry of Agriculture, Botswana, Printing and Publishing Company Botswana, 1991, ISBN 99912-1-020-2

BVW Photoguide ZA — A Photographic Guide to Wild Flowers of South Africa
Braam Van Wyk, Struik, 2000, ISBN 1-86872-390-9

Checklist — Preliminary Checklist of the Plants of Botswana
Moffat P. Setshogo, SABONET, 2005, ISBN 1-919976-18-3

Cole — Setswana — Animals and Plants
Desmond T Cole, The Botswana Society, 1995, ISBN 99912-60-24-2

CRC World Dictionary of Plant Names. Four Volumes.
Umberto Quattrocchi, CRC Press LLC, 2000, ISBNs 0-8493-2675-3, 0-8493-2676-1, 0-8493-2677-x, 0-8493-2678-8

Curtis & Mannheimer — Tree Atlas of Namibia
Barbara Curtis & Coleen Mannheimer, National Botanical Research Institute Windhoek, 2005, ISBN 99916-68-06-3

Cyperaceae of Namibia — an Illustrated Key
Nicholas Clarke & Coleen Mannheimer, National Botanical Research Institute Windheok, January 1999, ISBN 0-86976-485-3

Ellery — Plants of the Okavango Delta
Karen & William Ellery, Tsaro, 1997, ISBN 1-86840-240-1

Flowers Roodt — Common Wild Flowers of the Okavango Delta
Veronica Roodt, Shell Oil Botswana (Pty) Ltd, 1988, ISBN 99912-0-242-0

Fox & Young — Food from the Veldt Edible Plants of Southern Africa
FW Fox & ME Norwood Young, Delta Books, 1983, ISBN 0-908387-32-6

FZ — Flora Zambesiaca
This comprehensive publication covering the countries bordering the Zambezi Basin has been in preparation since 1953 and volumes are being distributed as they become available. The flora is edited and published by the Royal Botanic Gardens, Kew, UK and is available directly from them (also via www.Kewbooks.com).

Germishuizen — Illustrated Guide to the Wildflowers of Northern South Africa
Gerrit Germishuizen & Brenda Clarke, Briza Publications, 2003, ISBN1-875093-39-7

Hargreaves — Important Plants of Tsodilo
B Hargreaves & P Hargreaves, Hargreaves, 2007

Hargreaves Succulents — The Succulents of Botswana
B Hargreaves, National Museum, Monuments and Art Gallery, Botswana, 1990, ISBN 99912-1-028-8

Mabberley — The Plant Book
DJ Mabberley, Cambridge University Press, 1993, ISBN 0-521-34060-8

Müller — Grasses of South West Africa/Namibia
MAN Müller, Dir Ag & For Dept Ag & Nat Con SWA/Namibia, 1984, ISBN 0-86976-201

P Van Wyk Kr Trees — Field Guide to Trees of the Kruger National Park
Piet van Wyk, Struik, 1992, ISBN 1-86825-107-1

Palgrave — Trees of Southern Africa
Keith Coates Palgrave, Struik, 1990, ISBN 0-86977-081-0

Palgrave updated — Trees of Southern Africa (updated)
Keith & Meg Coates Palgrave, Struik, 2002, ISBN 1-86872-389-5

Plowes & Drummond — Wild Flowers of Rhodesia
DCH Plowes & RB Drummond, Longman, 1976, ISBN 0-582-64123-3

Pooley — A Field Guide to Wild Flowers of Kwazulu-Natal and the Eastern Region
Elsa Pooley, Natal Flora Publications Trust, 2005, ISBN 0-620-21500-3

Setshogo — Common Names of Some Flowering Plants of Botswana
Moffat Setshogo, Bay Publishing, 2002, ISBN 99912-970-1-4

Setshogo & Venter — Trees of Botswana: names and distribution
Moffat P Setshego & Fanie Venter, Southern African Botanical Diversity Network, 2003, ISBN 1-919795-69-3

Stearn — Botanical Latin
William T Stearn, David & Charles, 2004, ISBN 0-7153-1643-5

Stearn Dic — Stearn's Dictionary of Plant Names for Gardeners
William T Stearn, Timber Press, 2002, 0-88192-556-X

Tree Roodt — Trees & Shrubs of the Okavango Delta
Veronica Roodt, Shell Oil Botswana (Pty) Ltd, 1998, ISBN 99912-0-241-2

Tree Spotting — Sappi Tree Spotting: Low Veld Tree Identification Made Easy
Joan Van Gogh, Jacana Education Johannesburg, 1997, ISBN 1-874955-65-4

Turton — Some Flowering Plants of South Eastern Botswana
Lilian Turton, Botswana Society, 1988, ISBN 99912-60-01-3

Van Oudtshoorn 1 — Guide to Grasses of South Africa
Frits van Oudtshoorn, BRIZA, 1992, ISBN 0-620-16539-1

Van Oudtshoorn 2 — Guide to Grasses of South Africa
Frits van Oudtshoorn, BRIZA, 2002, ISBN 1-875093-176

Water Plants of Namibia — An Identification Manual
Nicholas V Clarke and Esmeralda S Klaassen, National Botanical Research Institute Windhoek, April 2001, ISBN 0-86976-520-5

W & B — Medicinal & Poisonous Plants of Southern & Eastern Africa
Watt & Breyer-Brandwijk, E&S Livingstone Ltd, 1962

WFNSA — Wild Flowers of Northern South Africa
Anita Fabian and Gerrit Germishuizen, Fernwood Press, 1997, ISBN 1-874950-29-6

Wyk — Food Plants of the World
Ben-Erik van Wyk, Timber Press, 2005, ISBN 978-0-88192-743-6

Zimbabwe Plant Checklist — A Checklist of Zimbabwean Vascular Plants
Edited by Anthony Mapaura and Johnathan Timberlake, Southern African Botanical Diversity Network Report No. 33 SABONET, 2004, ISBN 1-919976-14-0

Web Sites:-

www.kew.org
www.kew.org/epic/
Also Monocot World Checklist via www.kew.org
www.zimbabweflora.co.zw
www.ipni.org/index.html
www.plantzafrica.com
www.calflora.net/southafrica/plantnames.html
www.aluka.org
www.rogerblench.info/Ethnoscience%20data/Hausa%20plant%names.pdf

Glossary of terms

We have written this book in terms understandable by us and, we hope, other people who are not professional botanists. We also intend that people whose first language is not English will find the book easy to use. Many of the scientific words have been translated into simple English phrases, e.g. 'leaf stalk' is used for 'petiole' and 'flower stalk' for 'pedicel'.

However some technical terms are used in order to avoid lengthy essays, thus we have produced this glossary to provide the meanings of any terms that may not be familiar to the reader.

Alternate	Leaves arranged up the stem, one per node, alternately on diametrically opposite sides or spiralling up the stem.

Annual	Plants completing their life cycle in one year.
Anthela	Inflorescence of sedge with bracts and arising from a single point.
Anther	The male pollen-producing part of a flower, situated on the end of the stamen.
Apiculate	Of a leaf which ends in a short sharp point.
Aril	A fleshy covering over all or part of a seed within a fruit, often brightly coloured.
Asymmetrical	Not symmetrical.
Axil	The angle formed by the upper surface of a leaf stalk or branch stalk and the stem to which it is attached.
Axilliary	Anything which is in or grows from the axil.
Bark	The tough, sometimes woody, outer part of the stems and trunk of trees and shrubs.
Basal rosette	A circle of leaves at ground level with a stem arising from the centre.

Biennial	Plants having a life cycle of two years. Usually the plant builds up in the first year then flowers, fruits and dies in the second.
Bipinnate	A leaf with a single main stem with side branches making the whole look like a fish spine supporting the leaflets.
Bract	A leaf-like structure often at the base of a leaf stalk, flower or inflorescence.
Burr	A rough or prickly seed or seed head. Often with hooked prickles to aid seed dispersal on animals.
Calyx	The bracts or sepals surrounding the base of a flower. Usually but not always green and sometimes forming part of the attractive display of the flower.
Capsule	A fruit with more than one carpel or chamber.

Plants of Northern Botswana 567

Climax community	When a habitat contains a stable mix of plants it is said to be a climax community. Up to that stage the mix of plants will evolve year by year.
Composite (flower)	An inflorescence which appears to be a single flower but is in reality made up of more than one flower and often many, like a daisy flower.
Compound (leaf)	A leaf divided into two or more leaflets.
Cordate	Heart-shaped.
Corolla	The petals of a flower.
Corolla tube	A tube formed by the fusion or apparent fusion of the lobes of a flower.
Corymb	An inflorescence where the branches or pedicels start from different heights on the stem but reach such levels so as to form a continuous surface sometimes in the form of a flat top or a dome.
Cosmopolitan (plant)	A plant found worldwide.
Cymose	An inflorescence where the terminal bud opens first and subsequent flowers are borne on lateral stems.
Dark ages	The period of European history between 300 and 1100 AD.
Deciduous	A tree or shrub which does not keep its leaves throughout the year.
Decurrent	When a leaf margin extends down the supporting stem beyond the point where it joins, forming a 'wing'.

Decurved	Curved downwards, of a leaf.
Dehiscent or to dehisce	A splitting apart of a ripe fruit. If explosive, this forms part of the seed dispersal mechanism.
Digitate (grass)	Having parts pointing outwards like fingers from either a central point or up a stem.
Disc floret	Small florets lacking petals which form the central area of a compound flower such as a daisy.
Elliptic (leaf)	Leaf broadest in the middle.
Emarginate (leaf)	Having a notched tip.
Endemic	A plant found only in a specific geographical area.
Entire	The margin of a leaf is entire if it is smooth all the way round without teeth or indentations etc.
Epicalyx	A ring of sepal-like bracts below the real sepals.
Family	A grouping of genera sharing certain common features.
Fern	A type pf plant that reproduces by way of spores instead of seeds.
Flecked	Spotted and blotched with a different colour.
Floret	A small individual flower in a composite inflorescence.
Follicle	A fruit formed from a single carpel which opens along one side.
Forest	An area dominated by trees whose crowns meet and with a well developed understorey, giving rise to a significant reduction in light levels on the ground.

Fruit	Seed-containing structure.
Genus, genera	A grouping of like species.
Glabrous	Smooth and hairless.
Gland	An organ which secretes a substance which may be an attractant, a repellent, a sticky defence measure or simply a waste product.
Hemi-parasitic	A plant which lives attached to a host plant from which is takes water and minerals but derives some food for itself.
Herbarium	A collection of preserved plant material and associated documentation.
Imparipinnate	A pinnate leaf with leaflets arranged in pairs up the stem and with a single leaflet terminating the stem.
Inflorescence	The grouping of flowers and any supporting structures on a stem.
Insectivore	An organism which eats insects.
Internodal	The area of the stem between nodes.
Keel	As in a *Fabaceae* pea-like flower. Used to describe the angled bottom edge of the cup formed by the fusion of the two lower petals.
Latex	A clear or milky, often sticky, substance produced by plants which can be seen flowing from wounds.
Leaf margin	The edge of the leaf.
Ligule (of grasses or sedges)	The structure where the leaf blade joins the stem, often in the form of a membrane or ring of hairs.
Lobe	The division of a flower, leaf or other organ of a plant without the complete separation of the parts of the divided organ. The margin is continuous.
Morogo	A spicy cooked vegetable stew or relish eaten with mealie porridge or pap.
Mucronate (leaf tip)	A (leaf) tip with a hair growing out of it.
Mulch	A layer of organic or artificial matter placed or accumulated on the surface of the ground around a plant which cuts down the loss of moisture from the ground.
Net veining	Is when the veins of a leaf do not end abruptly at the edge or margin of the leaf but loop and join a neighbouring vein.
Node	The point on a stem where a branch or leaf is borne. On a grass stem a node is signified by a bulge and it is not always an obvious leaf or stem branching site.
Obovate	Of a leaf broadest towards the tip.
Opposite	Leaves arranged in opposing pairs on the stem. All pairs having the same orientation relative to the axis of the stem.
Opposite and decussate	Leaves arranged in opposing pairs on the stem with successive pairs up the stem being displaced by 90 degrees from the previous pair. 
Ovate	Of a leaf broadest towards the base.
Palmate	An arrangement in which leaflets radiate from a central point and look like a hand.

Palmate veining	Leaf veins or sections which radiate out from a single point to form a pattern that looks like a hand.
Pan tropical (plant)	A plant found throughout the tropics.
Panicle	Flowers arranged on stalks on the side branches of a main stem. The main stem has a growth point and is indeterminate.
Pappus	A tuft of hairs or bristles on the end of a seed. In some seeds this aids dispersal.
Parallel veining	Leaf veins which run parallel to each other along the length of the leaf.
Parasitic (plant)	A plant which lives attached to a host plant and draws its nutrition from the host.
Paripinnate	A pinnate leaf with leaflets arranged in pairs up the stem and without a single terminal leaflet.
Pedicel	The stalk of a flower joining it to the immediate supporting structure.
Perennial	Plants having a life cycle of greater than two years.
Perianth	The petals and calyx of a flower, i.e. the outer part of the flower.
Petiole	The stalk of a leaf joining it to the immediate supporting stem.
Pilose	Covered with fine hairs or down.
Pinnae	The first set of side branches of a compound leaf branching from the main axis of the leaf. These may be in the form of leaflets or branches supporting leaflets.
Pinnate	A compound leaf having leaflets growing in opposing pairs on each side of a common stem, like a feather.
Pinnatifid	Of a leaf appearing to have leaflets attached onto a central stem but not divided all the way down to that central stem.
Pioneer (plant)	A plant which is the first or amongst the first species to grow and establish itself in a newly created environment.
Poached (ground)	Wet ground turned into mud through the action of animals' feet repeatedly passing over it.
Pod	A fruit which is formed of two or more parts that split apart (dehisce) on ripening to release the seed. The splitting action may be explosive.
Prickle	Small thorns derived from the surface layer of a plant.
Protuberant	Bulging or projecting out in a rounded form beyond the surrounding surface.
Pubescent	Downy or covered with short soft hair.
Raceme	The arrangement of stalked flowers along a single unbranched stem. The youngest flower being at the tip.
Ramsar	The "Convention on Wetlands" was signed at Ramsar in Iran in 1971. Under this Convention, scheduled wetlands are entered onto what is called the "Montreux Record". The Okavango Delta is a major scheduled wetland.
Ray floret	Florets with a single large petal which grow together on the outer edge of a composite inflorescence, giving the appearance of a ring of petals.
Recurved	Bent backwards in a curve upon the previous direction.
Rhizome	A horizontal stem growing below the soil surface.
Salmon pink	A pinkish-orange colour. Plants that are this colour are in the orange, brown or red-brown section of this book.
Scabrous	Rough to the touch due to a covering of scales or short, stiff hairs or bristles.

Scorpioid	Curled like a scorpion's tail
Scrub	Area of land overgrown by bushes and shrubs.
Semi-parasitic	As hemi-parasitic. A plant which lives attached to a host plant taking water and minerals from the host but deriving some food for itself.
Sepal	A constituent part of the calyx.
Serrated	A leaf margin which is toothed.
Sessile	Of a leaf blade or a flower that emerges directly from a plant stem without a petiole or pedicel, i.e. a stalk.
Sheath (leaf)	The base of a leaf blade or stalk that encloses the supporting stem.
Spathulate	Of a leaf that is broadest above the middle.
Species	The basic level of classification of plants. Outwardly all the members of a species appear similar.
Spike	The arrangement of stalkless (sessile) flowers along a single unbranching stem. The youngest flower being at the tip.
Spikelet	The flower-bearing part of grasses and sedges.
Spine	A sharp, thin thorn formed from a modified leaf.
Stamen	Male pollen-bearing organ of a plant.
Stellate	The arrangement of hairs in a star shape, radiating outwards from a central point.
Stipule	Leaf or scale-like growth at the base of a leaf stalk, which can enclose a bud.
Stolon	A horizontal stem growing along the soil surface rooting at the nodes.
Style	A narrow stem attached to the ovary that bears the pollen-receiving stigma in the female part of the flower.
Succulent	A plant with fleshy leaves and stems which contain stored water.
Suffused	Having another colour washed softly over a base colour.
Synonym	Another name by which a plant has been or is known. Frequently a redundant official name or a name in current use in another part of the world.
Tendril	A thin finger which grows out from, for example, a plant stem or leaf, and which curls round or otherwise fixes itself to another object or plant structure thus supporting the plant.
Terminal	Situating at or forming the end or extremity of; for example, a bud at the apex of a stem.
Thorn	A general term for a hard woody projection on a plant which might be derived from a leaf, stem or branch of the plant.
Trifoliate	A compound leaf with three leaflets.
Tropics	The region of the world between the tropics of Capricorn and Cancer and the areas immediately adjacent to them.
Tuber	Swollen buried stem which acts as a food reservoir.
Understorey	Plants growing underneath the trees and bushes of woodland or forest.
Vestigial	Describing that which is left when a feature of the plant is almost lost during its evolution. The feature remains in a degenerated or an atrophied form.
Whorled	Three or more leaves arranged at the same node on the stem to form a ring.
Woodland	An area dominated by trees and grasses where the crowns of the trees do not meet and there is no distinct understorey.

Synopsis of the plants in this book

PTERIDOPHYTES

MARSILEACEAE
Marsilea vera Launert

THELYPTERIDACEAE
Cyclosorus interruptus (Willd.) H.Itô

DICOTYLEDONS

ACANTHACEAE
Asystasia gangetica (L.) T.Anderson
Barleria lugardii C.B.Clarke
Barleria mackenii Hook.f.
Blepharis integrifolia (L.f.) Schinz var. *integrifolia*
Blepharis maderaspatensis (L.) B.Heyne ex Roth
Dicliptera paniculata (Forssk.) I.Darbysh.
Duosperma crenatum (Lindau) P.G.Mey.
Hypoestes forskaolii (Vahl) R.Br.
Justicia betonica L.
Justicia bracteata (Hochst.) Zarb
Justicia divaricata (Nees) T.Anderson
Justicia exigua S.Moore
Justicia heterocarpa T.Anderson subsp. *dinteri* (S.Moore) Hedren
Ruellia patula Jacq.
Ruellia prostrata Poir.
Ruelliopsis setosa (Nees) C.B.Clarke
Ruspolia seticalyx (C.B.Clarke) Milne-Redh.
Thunbergia reticulata Nees

AIZOACEAE
Gisekia africana (Lour.) Kuntze
Sesuvium hydaspicum (Edgew.) Gonç.

AMARANTHACEAE
Achyranthes aspera L. var. *pubescens* (Moq.) Townsend
Achyranthes aspera L. var. *sicula* L.
Aerva leucura Moq.
Alternanthera pungens Kunth
Alternanthera sessilis (L.) DC.
Amaranthus graecizans L.
Amaranthus hybridus L.
Amaranthus praetermissus Brenan
Amaranthus viridis L.
Celosia trigyna L.
Cyathula orthocantha (Aschers.) Schinz
Gomphrena celosioides Mart.
Guilleminea densa (Willd.) Moq.
Hermbstaedtia linearis Schinz
Kyphocarpa angustifolia (Moq.) Lopr.
Pupalia lappacea (L.) A.Juss. var. *velutina* (Moq.) Hook.f.

ANACARDIACEAE
Rhus tenuinervis Engl.

ANNONACEAE
Friesodielsia obovata (Benth.) Verdc.

APOCYNACEAE
Duvalia polita N.E.Br.
Gomphocarpus fruticosus (L.) Aiton subsp. *rostratus* (N.E.Br.) Goyder & Nicholas
Marsdenia macrantha (Klotzsch) Schltr.
Marsdenia sylvestris (Retz.) P.I.Forst.
Orbea caudata (N.E.Br.) Bruyns subsp. *rhodesiaca* (L.C.Leach) Bruyns
Orbea huillensis (Hiern) Bruyns subsp. *huillensis*
Orbea lugardii (N.E.Br.) Bruyns
Orbea rogersii (L.Bolus) Bruyns
Orbea schweinfurthii (A.Berger) Bruyns
Orthanthera jasminiflora (Decne.) Schinz
Pergularia daemia (Forssk.) Chiov. subsp. *daemia*

ASTERACEAE
Acanthospermum hispidum DC.
Aspilia mossambicensis (Oliv.) Wild
Bidens biternata (Lour.) Merr. & Sheriff
Bidens pilosa L.
Bidens schimperi Sch.Bip. ex Walp.
Blainvillea acmella (L.) Phillipson
Calostephane divaricata Benth.
Conyza aegyptiaca (L.) Aiton
Conyza bonariensis (L.) Cronquist
Conyza stricta Willd.
Crassocephalum picridifolium (DC.) S.Moore
Dicoma schinzii O.Hoffm.
Dicoma tomentosa Cass.
Eclipta prostrata (L.) L.
Epaltes alata (Sond.) Steetz
Erlangea misera (Oliv.) & Hiern) S.Moore
Ethulia conyzoides L.f.
Flaveria bidentis (L.) Kuntze
Geigeria schinzii O.Hoffm. subsp. *rhodesiana* (S.Moore) Merxm.
Grangea anthemoides O.Hoffm.
Helichrysum argyrosphaerum DC.
Hirpicium gorterioides (Oliv.) Roessler subsp. *gorterioides*
Laggera crispata (Vahl) Hepper & J.R.I.Wood
Laggera decurrens (Vahl) Hepper & J.R.I.Wood
Melanthera scandens (Schumach. & Thonn.) Roberty subsp. *madagascariensis* (Baker) Wild
Melanthera triternata (Klatt) Wild
Nicolasia pedunculata S.Moore
Nidorella resedifolia DC.
Pechuel-loeschea leubnitziae (Kuntze) O.Hoffm.
Pseudoconyza viscosa (Mill.) D'Arcy
Pseudognaphalium luteo-album (L.) Hilliard & Burtt
Sclerocarpus africanus Jacq. ex Murray
Senecio strictifolius Hiern
Sphaeranthus peduncularis DC.
Vernonia anthelmintica (L.) Willd.
Vernonia glabrata (Steetz) Vatke var. *laxa* (Steetz) Brenan
Vernonia poskeana Vatke & Hildebr. subsp. *botswanica* G.V.Pope

BIGNONIACEAE
Kigelia africana (Lam.) Benth. subsp. *africana*

BOMBACACEAE
Adansonia digitata L.

BORAGINACEAE
Heliotropium ciliatum Kaplan
Heliotropium ovalifolium Forssk.
Heliotropium strigosum Willd.
Heliotropium supinum L.

BURSERACEAE
Commiphora edulis (Klotzsch) Engl. subsp. *edulis*
Commiphora merkeri Engl.
Commiphora mossambicensis (Oliv.) Engl.

CAMPANULACEAE
Gunillaea emirnensis (A.DC.) Thulin
Lobelia angolensis Engl. & Diels
Lobelia erinus L.

CAPPARACEAE
Boscia albitrunca (Burch.) Gilg & Gilg-Ben.
Capparis tomentosa Lam.
Cleome angustifolia Forssk. subsp. *petersiana* (Klotzsch) Kers
Cleome gynandra L.
Cleome hirta (Klotzsch) Oliv.
Cleome monophylla L.
Cleome rubella Burch.

CARYOPHYLLACEAE
Pollichia campestris Aiton
Polycarpaea eriantha Hochst. ex A.Rich. var. *efusa* (Oliv.) Turrill

CELASTRACEAE
Gymnosporia senegalensis (Lam.) Loes.
Loeseneriella africana (Willd.) N.Hallé var. *richardiana* (Cambess.) N.Hallé

COMBRETACEAE
Combretum apiculatum Sond. subsp. *apiculatum*
Combretum hereroense Schinz
Combretum imberbe Wawra
Combretum mossambicense (Klotzsch) Engl.
Pteleopsis myrtifolia (M.A.Lawson) Engl. & Diels
Terminalia sericea DC.

CONVOLVULACEAE
Astripomoea lachnosperma (Choisy) A.Meeuse
Convolvulus sagittatus Thunb.
Evolvulus alsinoides (L.) L.
Falkia oblonga C.Krauss.
Ipomoea chloroneura Hallier f.
Ipomoea coptica (L.) Roem. & Schult. var. *coptica*
Ipomoea dichroa Choisy
Ipomoea eriocarpa R.Br.
Ipomoea nil (L.) Roth
Ipomoea leucanthemum (Klotzsch) Hallier f.
Ipomoea magnusiana Schinz var. *magnusiana*
Ipomoea obscura (L.) Ker-Gawl.
Ipomoea pes-tigridis L.
Ipomoea plebeia R.Br. subsp. *africana* A.Meeuse
Ipomoea shirambensis Baker
Ipomoea sinensis (Desr.) Choisy subsp. *blepharosepala* (A.Rich.) A.Meeuse
Ipomoea sinensis (Desr.) Choisy subsp. *sinensis*
Ipomoea sp. aff. *dichroa*
Ipomoea tuberculata Ker-Gawl.
Jacquemontia tamnifolia (L.) Griseb.
Merremia pinnata (Hochst. ex Choisy) Hallier f.
Merremia verecunda Rendle
Xenostegia tridentata (L.) D.F.Austin & Staples subsp. *angustifolia* (Jacq.) Lejoly & Lisowski

CRASSULACEAE
Kalanchoe lanceolata (Forssk.) Pers.

CUCURBITACEAE
Acanthosicyos naudinianus (Sond.) C.Jeffrey
Citrullus lanatus (Thunb.) Matsum & Nakai
Coccinia adoensis (A.Rich.) Cogn.
Corallocarpus bainesii (Hook.f.) A.Meeuse
Ctenolepis cerasiformis (Stocks) Hook.f.
Cucumis africanus L.f.
Cucumis anguria L.
Cucumis metuliferus E.Mey. ex Naudin
Kedrostis abdallai A.Zimm.
Kedrostis foetidissima (Jacq.) Cogn.
Momordica balsamina L.
Mukia maderaspatana (L.) M.Roem.
Zehneria marlothii (Cogn.) R.Fern. & A.Fern.

EBENACEAE
Diospyros lycioides Desf. subsp. *sericea* (Bernh.) de Winter
Diospyros mespiliformis Hochst. ex A.DC.
Euclea divinorum Hiern

ELATINACEAE
Bergia ammannioides Roth
Bergia pentheriana Keissl.

EUPHORBIACEAE
Acalypha fimbriata Schumach. & Thonn.
Acalypha indica L.
Acalypha ornata A.Rich.
Acalypha vilicaulis Hochst. ex A.Rich.
Croton megalobotrys Müll.Arg.
Euphorbia crotonoides Boiss.
Euphorbia hirta L.
Euphorbia inaequilatera Sond.
Euphorbia polycnemoides Boiss.
Euphorbia tirucalli L.
Phyllanthus maderaspatensis L.
Phyllanthus parvulus Sond. var. *parvulus*
Phyllanthus pentandrus Schumach. & Thonn.
Pterococcus africanus (Sond.) Pax & K.Hoffm.
Ricinus communis L. var. *communis*
Schinziophyton rautanenii (Schinz) Radcl.-Sm.
Tragia okanua Pax

FABACEAE
Abrus precatorius L. subsp. *africanus* Verdc.
Acacia arenaria Schinz

Acacia erioloba E.Mey.
Acacia erubescens Welw. ex Oliv.
Acacia fleckii Schinz
Acacia hebeclada DC. subsp. *chobiensis* (O.B.Mill.) A.Schreib.
Acacia luederitzii Engl. var. *luederitzii*
Acacia mellifera (Vahl) Benth. subsp. *detinens* (Burch.) Brenan
Acacia nigrescens Oliv.
Acacia sieberiana DC. var. *woodii* (Burtt Davy) Keay & Brenan
Albizia anthelmintica Brongn.
Albizia harveyi E.Fourn.
Baphia massaiensis Taub. subsp. *obovata* (Schinz) Brummitt
Bauhinia petersiana Bolle subsp. *macrantha* (Oliv.) Brummitt & J.H.Ross
Cassia abbreviata Oliv. subsp. *beareana* (Holmes) Brenan
Chamaecrista absus (L.) H.S.Irwin & Barneby
Chamaecrista falcinella (Oliv.) Lock
Chamaecrista mimosoides (L.) Greene
Chamaecrista stricta E.Mey.
Colophospermum mopane (Benth.) J.Léonard
Crotalaria flavicarinata Baker f.
Crotalaria heidmannii Schinz
Crotalaria laburnifolia L.
Crotalaria pisicarpa Welw. ex Baker
Crotalaria platysepala Harv.
Crotalaria podocarpa DC.
Crotalaria sphaerocarpa Perr. ex DC. subsp. *sphaerocarpa*
Crotalaria steudneri Schweinf.
Dichrostachys cinerea (L.) Wight & Arn.
Dolichos junodii (Harms) Verdc.
Guibourtia coleosperma (Benth.) J.Léonard
Indigofera astragalina DC.
Indigofera charlieriana Schinz var. *charlieriana*
Indigofera charlieriana Schinz var. *scaberrima* (Schinz) J.B.Gillett
Indigofera colutea (Burm.f.) Merr.
Indigofera daleoides Benth. ex Harv. var *daleoides*
Indigofera filipes Harv.
Indigofera flavicans Baker
Indigofera trita L.f. subsp. *subulata* (Vahl ex Poir) Ali var. *subulata*
Lessertia benguellensis Baker f.
Macrotyloma daltonii (Webb) Verdc.
Neptunia oleracea Lour.
Peltophorum africanum Sond.
Philenoptera nelsii (Schinz) Schrire subsp. *nelsii*
Philenoptera violacea (Klotzsch) Schrire
Rhynchosia minima (L.) DC.
Rhynchosia totta (Thunb.) DC. var. *fenschelii* Schinz
Senna obtusifolia (L.) H.S.Irwin & Barneby
Sesbania microphylla Harms
Sesbania rostrata Bremek. & Oberm.
Stylosanthes fruticosa (Retz.) Alston
Tephrosia caerulea Baker f.
Tephrosia lupinifolia DC.
Tephrosia purpurea (L.) Pers. subsp. *leptostachya* (DC.) Brummitt var. *pubescens* Baker
Vigna luteola (Jacq.) Benth.
Vigna unguiculata (L.) Walp. subsp. *stenophylla* (Harv.) Maréchal, Mascherpa & Stainier
Vigna unguiculata (L.) Walp. subsp. *unguiculata* var. *spontanea* (Schweinf.) R.S.Pasquet
Zornia glochidiata DC.

GENTIANACEAE
Enicostemma axillare (Lam.) A.Raynal subsp. *axillare*
Sebaea grandis (E.Mey.) Steud.

GERANIACEAE
Monsonia angustifolia A.Rich.

GUTTIFERAE
Garcinia livingstonei T.Anderson

LAMIACEAE
Acrotome inflata Benth.
Clerodendrum ternatum Schinz
Clerodendrum uncinatum Schinz
Hemizygia bracteosa (Benth.) Briq.
Hoslundia opposita Vahl
Leonotis nepetifolia (L.) R.Br.
Ocimum americanum L. var. *pilosum* (Willd.) A.J.Paton
Ocimum gratissimum L. var. *gratissimum*

LENTIBULARIACEAE
Utricularia gibba L.

LORANTHACEAE
Erianthemum ngamicum (Sprague) Danser
Erianthemum virescens (N.E.Br.) Wiens & Polhill
Plicosepalus kalachariensis (Schinz) Danser

LYTHRACEAE
Ammannia baccifera L. subsp. *baccifera*
Rotala filiformis (Bellardi) Hiern

MALVACEAE
Abutilon angulatum (Guill. & Perr.) Mast. var. *angulatum*
Abutilon englerianum Ulbr.
Abutilon ramosum (Cav.) Guill. & Perr.
Gossypium herbaceum L. subsp. *africanum* (Watt) Vollesen
Hibiscus caesius Garcke
Hibiscus cannabinus L.
Hibiscus dongolensis Delile
Hibiscus lobatus (Murray) Kuntze
Hibiscus mastersianus Hiern
Hibiscus meeusei Exell
Hibiscus ovalifolius (Forssk.) Vahl
Hibiscus schinzii Gürke
Hibiscus sidiformis Baill.
Hibiscus trionum L.
Hibiscus vitifolius L. subsp. *vulgaris* Brenan & Exell
Kosteletzkya buettneri Gürke
Pavonia burchellii (DC.) R.A.Dyer
Pavonia clathrata Mast.
Pavonia senegalensis Cav.
Sida alba L.
Sida chrysantha Ulbr.
Sida cordifolia L.

MELIACEAE
Trichilia emetica Vahl
Turraea zambesica Styles & F.White

MENISPERMACEAE
Cissampelos mucronata A.Rich.
Cocculus hirsutus (L.) Diels

MENYANTHACEAE
Nymphoides forbesiana (Griseb.) Kuntze
Nymphoides indica (L.) Kuntze subsp. *occidentalis* A.Raynal

MOLLUGINACEAE
Glinus bainesii (Oliv.) Pax
Glinus lotoides L.
Glinus oppositifolius (L.) Aug.DC. var. *oppositifolius*
Limeum argute-carinatum Wawra & Peyr. var. *kwebense* (N.E.Br.) Friedr.
Limeum fenestratum (Fenzl) Heimerl
Limeum sulcatum (Klotzsch) Hutch.
Limeum viscosum (J.Gay) Fenzl var. *kraussii* Friedrich
Limeum viscosum (J.Gay) Fenzl var. *viscosum*
Mollugo cerviana (L.) Ser. ex DC.

MORACEAE
Ficus thonningii Blume

NYCTAGINACEAE
Boerhavia coccinea Mill.
Commicarpus plumbagineus (Cav.) Standl.

NYMPHAEACEAE
Nymphaea nouchali Burm.f. var. *caerulea* (Savigny) Verdc.

OLEACEAE
Jasminum fluminense Vell. subsp. *fluminense*
Jasminum stenolobum Rolfe

ONAGRACEAE
Ludwigia abyssinica A.Rich.
Ludwigia leptocarpa (Nutt.) H.Hara
Ludwigia stolonifera (Guill. & Perr.) P.H.Raven

OXALIDACEAE
Oxalis corniculata L.

PASSIFLORACEAE
Basananthe pedata (Baker f.) W.J. de Wilde

PEDALIACEAE
Ceratotheca sesamoides Engl.
Dicerocaryum eriocarpum (Decne.) Abels
Harpagophytum procumbens (Burch.) DC. ex Meisn. subsp. *procumbens*
Harpagophytum zeyheri Decne. subsp. *sublobulatum* (Engl.) Ihlenf. & Harm.
Sesamum angustifolium (Oliv.) Engl.
Sesamum triphyllum Asch.

PLUMBAGINACEAE
Plumbago zeylanica L.

POLYGALACEAE
Polygala erioptera DC.

POLYGONACEAE
Oxygonum alatum Burch.
Oxygonum sinuatum (Meisn.) Dammer
Persicaria glomerata (Dammer) S.Ortiz & Paiva
Persicaria limbata (Meisn.) H.Hara
Polygonum decipiens (R.Br.) K.L.Wilson
Polygonum plebeium R.Br.

PORTULACACEAE
Portulaca hereroensis Schinz
Portulaca kermesina N.E.Br.
Portulaca oleracea L. subsp. *oleracea*
Portulaca quadrifida L.
Talinum caffrum (Thunb.) Eckl. & Zeyh.
Talinum crispatulatum Dinter

RHAMNACEAE
Berchemia discolor (Klotzsch) Hemsl.
Ziziphus mucronata Willd. subsp. *mucronata*

RUBIACEAE
Gardenia volkensii K.Schum. subsp. *spatulifolia* (Stapf & Hutch.) Verdc.
Kohautia caespitosa Schnizl. subsp. *brachyloba* (Sond.) D.Mantell
Kohautia subverticillata (K.Schum.) D.Mantell subsp. *subverticillata*
Kohautia virgata (Willd.) Bremek.
Oldenlandia capensis L.f. var. *capensis*
Oldenlandia corymbosa L. var. *caespitosa* (Benth.) Verdc.
Pavetta cataractarum S.Moore
Pentodon pentandrus (Schumach. & Thonn.) Vatke
Spermacoce senensis (Klotzsch) Hiern

SAPINDACEAE
Cardiospermum halicacabum L.

SCROPHULARIACEAE
Alectra orobanchoides Benth.
Alectra picta (Hiern) Hemsl.
Aptosimum decumbens Schinz
Craterostigma plantagineum Hochst.
Cycnium tubulosum (L.f.) Engl.
Jamesbrittenia elegantissima (Schinz) Hilliard
Sopubia mannii Skan var. *tenuifolia* (Engl. & Gilg) Hepper
Striga asiatica (L.) Kuntze
Striga gesnerioides (Willd.) Vatke

SOLANACEAE
Solanum panduriforme E.Mey. ex Dunal
Solanum tarderemotum Bitter
Solanum tettense Klotzsch var. *renchii* (Vatke) R.E.Gonçalves
Withania somnifera (L.) Dunal

STERCULIACEAE
Hermannia eenii Baker f.
Hermannia glanduligera K.Schum.
Hermannia kirkii Mast.
Hermannia modesta (Ehrenb.) Planch.
Hermannia quartiniana A.Rich.
Melhania forbesii Mast.
Waltheria indica L.

TILIACEAE
Corchorus tridens L.
Grewia bicolor Juss.
Grewia flavescens Juss.
Triumfetta pentandra A.Rich.

TURNERACEAE
Streptopetalum serratum Hochst.
Tricliceras glanduliferum (Klotzsch) R.Fern.
Tricliceras lobatum (Urb.) R.Fern.

VAHLIACEAE
Vahlia capensis (L.f.) Thunb. subsp. *vulgaris* Bridson var. *vulgaris*

VERBENACEAE
Lantana angolensis Moldenke
Phyla nodiflora (L.) Greene

VITACEAE
Ampelocissus africana (Lour.) Merr.
Cyphostemma congestum (Baker) Wild & R.B.Drumm.

ZYGOPHYLLACEAE
Tribulus terrestris L.

MONOCOTYLEDONS

ALISMATACEAE
Caldesia parnassifolia (L.) Parl.

ALLIACEAE
Scadoxus multiflorus (Martyn) Raf.

AMARYLLIDACEAE
Crinum harmsii Baker
Chlorophytum sphacelatum (Baker) Kativu subsp. *milanjianum* (Rendle) Kativu
Chlorophytum sphacelatum (Baker) Kativu subsp. *sphacelatum*
Pancratium tenuifolium Hochst. ex A.Rich.

ARACEAE
Lemna aequinoctialis Welw.

ARECACEAE
Hyphaene petersiana Klotzsch ex Mart.
Phoenix reclinata Jacq.

ASPARAGACEAE
Asparagus africanus Lam.
Asparagus nelsii Schinz

Sansevieria aethiopica Thunb.
Sansevieria pearsonii N.E.Br.

COLCHICACEAE
Gloriosa superba L.

COMMELINACEAE
Commelina africana L. var. *barberae* (C.B.Clarke) C.B.Clarke
Commelina diffusa Burm.f.
Commelina forsskalii Vahl
Commelina macrospatha Gilg & Ledermann ex Mildbr.
Commelina petersii Hassk.
Commelina subulata Roth
Cyanotis foecunda DC. ex Hassk.
Floscopa glomerata (Schult. & Schult. f.) Hassk.

CYPERACEAE
Bulbostylis hispidula (Vahl) R.W.Haines subsp. *pyriformis* (Lye) R.W.Haines
Cyperus alopecuroides Rottb.
Cyperus chersinus (N.E.Br.) Kük
Cyperus compressus L.
Cyperus denudatus L.f.
Cyperus difformis L.
Cyperus dubius Rottb.
Cyperus esculentus L.
Cyperus haspan L.
Cyperus iria L.
Cyperus latifolius Poir.
Cyperus leptocladus Kunth
Cyperus longus L.
Cyperus margaritaceus Vahl
Cyperus papyrus L.
Cyperus pectinatus Vahl
Cyperus pseudokyllingioides Kük
Cyperus rotundus L.
Cyperus sphaerospermus Schrad.
Cyperus squarrosus L.
Cyperus tenuiculmis Boeckeler
Cyperus zollingeri Steud.
Eleocharis nigrescens (Nees) Steud.
Fimbristylis dichotoma (L.) Vahl
Fimbristylis longiculmis Steud.
Fuirena pubescens (Poir.) Kunth
Fuirena umbellata Rottb.
Kyllinga buchananii C.B.Clarke
Kyllinga erecta Schumach.
Kyllingiella microcephala (Steud.) R.W.Haines & Lye
Oxycaryum cubense (Poepp. & Kunth) Lye
Pycreus macrostachyos (Lam.) A.Raynal
Pycreus pelophilus (Ridl.) C.B.Clarke
Pycreus polystachyos (Rottb.) P.Beauv. var. *polystachyos*
Schoenoplectus corymbosus (Roem. & Schult.) J.Raynal
Schoenoplectus erectus (Poir.) J.Raynal
Schoenoplectus senegalensis (Steud.) J.Raynal

ERIOCAULACEAE
Eriocaulon ?abyssinicum Hochst.

ERIOSPERMACEAE
Eriospermum bakerianum Schinz subsp. *bakerianum*

HYACINTHACEAE
Albuca abyssinica Jacq.
Dipcadi glaucum (Burch. in Ker-Gawl) Baker
Dipcadi longifolium (Lindl.) Baker
Dipcadi marlothii Engl.
Ledebouria revoluta (L.f.) Jessop

HYDROCHARITACEAE
Ottelia ulvifolia (Planch.) Walp.

IRIDACEAE
Ferraria glutinosa (Baker) Rendle
Lapeirousia odoratissima Baker
Lapeirousia schimperi (Asch. & Klatt) Milne-Redh.

ORCHIDACEAE
Ansellia africana Lindl. subsp. *africana*

POACEAE
Andropogon eucomus Nees
Andropogon gayanus Kunth
Aristida adscensionis L.
Aristida hordeacea Kunth
Aristida junciformis Trin. & Rupr.
Aristida meridionalis (Stapf) Henrard
Aristida scabrivalvis Hack.
Aristida stipitata Hack.
Aristida stipoides Lam.
Brachiaria deflexa (Schumach.) Robyns
Brachiaria dura Stapf
Brachiaria grossa Stapf
Brachiaria humidicola (Rendle) Schweick.
Brachiaria nigropedata (Ficalho & Hiern) Stapf
Cenchrus biflorus Roxb.
Cenchrus ciliaris L.
Chloris virgata Sw.
Chrysopogon nigritanus (Benth.) Veldkamp
Cymbopogon caesius (Hook & Arn.) Stapf
Cynodon dactylon (L.) Pers.
Dactyloctenium aegyptium (L.) Willd.
Dactyloctenium giganteum B.S.Fisher & Schweick.
Digitaria debilis (Desf.) Willd.
Digitaria milanjiana (Rendle) Stapf
Digitaria sanguinalis (L.) Scop.
Digitaria velutina (Forssk.) P.Beauv.
Echinochloa colona (L.) Link
Echinochloa jubata Stapf
Echinochloa stagnina (Retz.) P.Beauv.
Eleusine indica (L.) Gaertn.
Enneapogon cenchroides (Roem. & Schult.) C.E.Hubb.
Enteropogon macrostachyus (Hochst. ex A.Rich.) Munro ex Benth.
Eragrostis aspera (Jacq.) Nees
Eragrostis cilianensis (All.) Vignolo ex Janch.
Eragrostis cylindriflora Hochst.
Eragrostis echinochloidea Stapf
Eragrostis inamoena K.Schum.
Eragrostis lappula Nees
Eragrostis pallens Hack.
Eragrostis pilosa (L.) P.Beauv.
Eragrostis porosa Nees
Eragrostis pusilla Hack.
Eragrostis superba Peyr.
Eragrostis viscosa (Retz.) Trin.
Eulalia aurea (Bory) Kunth
Heteropogon contortus (L.) Roem. & Schult.
Hyperthelia dissoluta (Steud.) Clayton
Leersia hexandra Sw.
Leptochloa fusca (L.) Kunth
Melinis kallimorpha (Clayton) Zizka
Melinis repens (Willd.) Zizka subsp. *grandiflora* (Hochst.) Zizka
Microchloa caffra Nees
Oplismenus burmannii (Retz.) P.Beauv.
Panicum coloratum L. var. *coloratum*
Panicum fluviicola Steud.
Panicum hirtum Lam.
Panicum kalaharense Mez
Panicum maximum Jacq.
Panicum subalbidum Kunth
Paspalidium obtusifolium (Delile) N.D.Simpson
Paspalum scrobiculatum L.
Pennisetum macrourum Trin.
Perotis patens Gand.
Phragmites australis (Cav.) Trin. ex Steud.
Pogonarthria fleckii (Hack.) Hack.
Pogonarthria squarrosa (Roem. & Schult.) Pilg.
Schizachyrium jeffreysii (Hack.) Stapf
Schmidtia pappophoroides Stued. ex J.A.Schmidt
Setaria pumila (Poir.) Roem. & Schult.
Setaria sagittifolia (A.Rich.) Walp.
Setaria sphacelata (Schumach.) M.B.Moss
Setaria verticillata (L.) P.Beauv.
Sorghastrum nudipes Nash
Sporobolus cordofanus (Steud.) Coss.
Sporobolus fimbriatus (Trin.) Nees
Sporobolus iocladus (Trin.) Nees
Sporobolus macranthelus Chiov.
Sporobolus panicoides A.Rich.
Sporobolus pyramidalis P.Beauv.
Stipagrostis uniplumis (Roem. & Schult.) de Winter
Trachypogon spicatus (L.f.) Kuntze
Tragus berteronianus Schult.
Tragus racemosus (L.) All.
Tricholaena monachme (Trin.) Stapf & C.E.Hubb.
Trichoneura grandiglumis (Nees) Ekman
Urochloa brachyura (Hack.) Stapf
Urochloa trichopus (Hochst.) Stapf
Vossia cuspidata (Roxb.) Griff.

POTAMOGETONACEAE
Potamogeton nodosus Poir.

VELLOZIACEAE
Xerophyta humilis (Baker) T.Durand & Schinz

XANTHORRHOEACEAE
Trachyandra arvensis (Schinz) Oberm.

Common names index

!gou, 381
!ola, 380

//ah, 257
//ha, 381
//kowa, 381
//wa, 381
/ana, 257
/kane, 378

aambeibossie, 334
aambeiwortel, 348
acacia, 11,12
acacia mistletoe, 177
acrotome, 392
aerva, 334
African cucumber, 244
African ebony, 84
African foxtail, 508
African herbage, 358
African mangosteen, 228
African rosewood, 254
Afrika-geelmelkhout, 228
Afrikaans witbiesie, 521
almond, 532
  earth, 532
  ground, 532
alsbos, 176
amaranthus weed, 335
anaka, 348
angelbossie, 176
annual bristle grass, 502
annual three-awn, 502
appelblaar, 53
apple, 52, 53, 59, 237, 244, 246
  apple-leaf lance pod, 52
  apple-leaf tree, 53
  balsam apple, 244
  bitter apple, 59, 237
  grey bitter apple, 59
  Kalahari apple tree, 52
  Kalahari star apple, 246
  monkey apple, 237
  poison apple, 59
  red star apple, 246
  Sodom apple, 59
arge-flowered sebaea, 272
arrow grass, 499
ash vetch, 174
assegaaigras, 460
asystasia, 324
awnless barnyard grass, 435

badingwana, 54
balloon vine, 422
balsam apple, 244
balsam pear, 244
balsam tree, 253
balsamina, 244
bamotshai, 113
baobab, 352
bark cloth fig, 106
bastard brandy bush, 315

bastard dwaba-berry, 72
bastard jute, 282
bastard mustard, 358
bastard umbrella thorn, 380
bastard yellow wood, 81
baster-haak-en-steek, 380
bastermopanie, 254
baswabile, 165,260,402
batchelor's button, 337
bean, 127, 165, 175, 375
  black-eyed bean, 175
  castor bean, 127
  love bean, 165
  lucky bean climber, 165
  wild coffee bean, 375
beesdubbeltjie, 180,181
beestedoorn, 181
beggar sticks, 205
Benghal commelina, 43
Bermuda grass, 448, 455
berry, 59, 72, 78, 84, 89, 110, 240, 316
  bastard dwaba-berry, 72
  donkey berry, 316
  fever-berry, 89
  gooseberry cucumber, 240
  inkberry, 59
  jackal-berry, 84
  northern dwaba-berry, 72
  poisonous gooseberry, 110
  river fever-berry, 89
  waxberry, 78
besemgras, 466
bird plum, 107
bird's brandy, 197
bird's grass, 435
bistort, 188
bitter apple, 59, 237
bitter wild cucumber, 239, 241
bitterappel, 59
bittermelon, 237
bitterossie, 34
blaasklimop, 422
black ironwood, 253
black jack, 204, 206
black thorn, 381
black winter cherry, 422
black-eyed bean, 175
black-eyed susan, 289
black-footed brachiaria, 434
black-footed grass, 434
blackjack, 205
bladder hibiscus, 289
bladderweed, 289
bladderwort, 276
bladdoring, 378
blade thorn, 378
blanket-stabbers, 205
bleekblaarboom, 384
blink(h)aarboesmansgras, 489
blinkblaar-wag-'n-bietjie, 108
bloublom, 195
bloubos, 246

bloubuffelsgras, 508
blouhaak, 377
blousaad, 482
blousaadgras, 490
blousaadsoetgras, 482
blouselblommetjie, 43
blouwaterlelie, 55
blue buffalo grass, 508
blue bush (hairy), 246
blue grass, 447,458
blue haze, 49
blue lotus, 55
blue powder puff, 48
blue thorn, 377
blue water lily, 55
blue weed, 31
blue-seed tricholaena, 490
bobbejaan-se-dood, 349
boes-manstee, 251
bog fern, 560
bogoma, 4, 69, 70, 409, 413, 513, 514
bojang bomotla, 549
bojang-jwa-dipitse, 498
bojang-jwa-mahupu, 506
bojang-jwa-mmutla, 549
bojang-jwa-phiri, 510
bojang-jwa-tau, 465
bokbaardgras, 463
bokunogu, 294
bomama, 70
boot protector plant, 180
bosveld crotalaria, 261
bosveldkatjiepiering, 416
bottle brush grass, 510
bowstring hemp, 348
branching abutilon, 279
breëblaarterpentyngras, 459
bristle grass, 504
bristly foxtail, 513
broad-leaved turpentine grass, 459
broom love grass, 472
brown ivory, 107
brown top buffel-grass, 482
bruin-ivoor, 107
buffalo grass, 508,512
buffalo thorn, 108
buffelsgras, 482
bur bristle grass, 513
bur cucumber, 240
bur grass, 513
bur marigold, 205, 206
burgrass, 507,514
Burkis fig, 106
burnut, 322
burweed, 64, 150
bush, 34, 77, 87, 92, 127, 212, 246, 229, 259, 315, 329, 334, 339, 346, 357, 420
  bastard brandy bush, 315
  blue bush (hairy), 246
  caper bush, 77
  castor-oil bush, 127

578 Plants of Northern Botswana

catbush, 346
dye bush, 92
false brandy bush, 315
foam bush, 334
haemorrhoid bush, 334
hare's tail bush, 339
kudu-bush, 229
poison bush, 87
sickle bush, 259
smelter's bush, 212
stinkbush, 34
sweat bush, 34
white ribbon bush, 329
woolly caper bush, 357
Zambezi brides-bush, 420
bush buffalo grass, 482
bushman grass, 489
bushveld crotalaria, 261
bushveld dropseed, 500
bushveld gardenia, 416
bushwillow, 81, 229, 230, 360, 361
knobbly creeping bushwillow, 360
large fruited bushwillow, 81
myrtle bushwillow, 361
red bush-willow (hairy), 229
russet bush-willow, 230
stink bushwillow (two-winged), 361
butterfly tree, 253
buttonweed, 421

calthrop (land), 322
camel thorn, 257
camel's foot, 375
candle pod acacia, 379
candle thorn, 379
caper bush, 77
carrot weed, 338
carrot-seed grass (common), 514
Carter's curse, 31
castor bean, 127
castor oil plant, 127
castor-oil bush, 127
cat's claw, 136,393
cat's tail, 151,513
cat's tail dropseed, 501
cat's tail grass, 501, 510
cat's whiskers, 358
catbush, 346
catjang, 175
cau, 71
cawaq, 53
cha, 236
chaff flower, 63, 150
chaodabi, 124
cherry, 83, 100, 422
black winter cherry, 422
cherry vine, 83
winter cherry, 110
chibiringa, 202
chico, 165
chimiwane, 376
chinyevana, 41
chiromiundi, 360
choba-choba, 347
Chobe candle acacia, 379

chotho, 389
chovachova, 346
Christmas tree grass, 488
chunan, 243
citron, 237
citron melon, 237
climbing lily, 120
coast button grass, 449
cobbler's pegs, 205
cock's comb, 66
coffee neat's foot, 375
coloku, 82
combretum, 81, 230, 360
knobbly combretum, 360
large fruited combretum, 81
mouse-eared combretum, 230
Mozambique combretum, 360
shaving brush combretum, 360
commelina, 43, 231
Benghal commelina, 43
yellow commelina, 231
commiphora, 71, 224, 225, 226
commiphora rhus, 71
pepper-leaved commiphora, 226
rough-leaved commiphora, 224
zebra-bark commiphora, 225
common bristle grass, 512
common buffalo grass, 482
common crowfoot, 449
common dropseed, 500
common false-thorn, 384
common gardenia, 416
common mealie-witchweed, 194
common pigweed, 66
common reed, 464
common sopubia, 193
common turpentine grass, 459
common wild fig, 106
confetti spikethorn, 79
confetti tree, 79
cong, 81
copalwood (Rhodesian), 254
coppery three awn, 465
coral bead plant, 165
corkwood, 224, 225, 226
pepper-leaved corkwood, 226
rough-leaved corkwood, 224
zebra corkwood, 225
cornflower vernonia, 37
couch grass, 448
cowpea, 175
crab finger grass, 453
crab grass, 453, 455
crab's eyes, 165
crane's bill, 176
creeping false paspalum, 433
creeping lady's sorrel, 302
creeping milkweed, 91
creeping oxalis, 302
creeping paspalum, 441
creeping signal grass, 433
creeping sorrel, 302
crop grass, 453
cross grass, 443
crotalaria, 264
crowfoot, 449

crowfoot grass, 455
cucumber, 236, 239, 240, 241
bitter wild cucumber, 239, 241
bur cucumber, 240
gemsbok cucumber, 236
herero cucumber, 236
horned cucumber, 239, 240, 241
cud weed, 221
cudwa, 375
cyathula, 69

dagga, 137
daisy, 207, 213, 218, 219, 424
daisy lawn, 424
South American daisy, 219
wing stem daisy, 207, 213
yellow water-daisy, 218
damaqoq, 277
dayflower, 231
delele, 314
delele-kwakwa, 289
devil's claw, 181
devil's horsewhip, 150
devil's thorn, 322
dhobi, 184
diamond leaf, 85
diamond-leaved euclea, 85
diangamoti, 103
dikgose, 375
dikoma, 538
dikori, 34
dikurubede, 277
dimhwa, 55
dinawa, 175
dingofo, 449
dinyangombe, 314
ditch grass, 441
dithyana, 80
diviya, 55
divuyu, 352
dixombombo, 34,217
diyanambo, 347
diyanga, 337
doboma, 85
dog's tail, 455,512
doll's protea, 152
dongola hibiscus, 283
donkey berry, 316
donkey flower, 374
dove milk, 126,164
dropseed, 486, 500, 501
bushveld dropseed, 500, 501
cat's tail dropseed, 501
common dropseed, 500
dropseed grass, 500
narrow-plumed dropseed, 501
pan dropseed, 486
drui, 360
dubbeltjie, 322,415
duck grass, 449
duckweed, 559
duiwelhaak, 181
duiwelsdis, 180
dukurukane, 277
dwarf date palm, 202
dwarf sage, 217

Plants of Northern Botswana **579**

dye bush, 92
dzau, 362

earth almond, 532
ebony, 84
    African ebony, 84
    ebony diospyros, 84
    Mozambique ebony, 84
    Rhodesian ebony, 84
    Transvaal ebony, 84
eclipta, 351
edible seed melon, 237
eenjarige denneboomgras, 442
eenjarige steekgras, 502
Egyptian finger grass, 449
elandsdoring, 180
elbow buffalo grass, 483
elephant trunk, 81
elephant's ear, 277
elmboogbuffelsgras, 483
emm, 352
empunga, 497
endumba, 392
ennyennyane, 180
epalamela, 177
eshoshong, 322
everlasting weed, 215

fairy grass, 478
false brandy bush, 315
false mopane (large), 254
false plumbago, 409
false signal grass, 491
false spurry, 419
famine grass, 488
feather finger grass, 447
feather palm, 202
feather top chloris, 447
feathertop grass, 447
featherweight tree, 248
fertility plant, 69
fever-berry, 89
fig, 106
    bark cloth fig, 106
    Burkis fig, 106
    common wild fig, 106
    Peters fig, 106
    strangler fig, 106
finger grass, 447, 449, 451, 453, 454
    crab finger grass, 453
    Egyptian finger grass, 449
    feather finger grass, 447
    flaccid finger grass, 454
    long plumed finger grass, 454
    makarikari finger grass, 452
    milanje finger grass, 452
fireball lily, 113
fish-bone cassia, 251
flaccid finger grass, 454
flame lily, 120
flannel weed, 294
flat sedge, 529
flat-topped thorn, 382
flax-leaf fleabane, 209
floating heart, 401
flower-of-an-hour, 289

fluitjiesriet, 464
foam bush, 334
fog grass, 424
forget-me-not, 354
foxglove, 179
    false, 179
    wild, 179
foxtail buffalo grass, 508
fynsaadgras, 500

g!u, 380
G//are, 377
gagara, 22
gare, 378
gcan, 71
geel-cleome, 227
geelappel, 59
geelboslelie, 120
geelhout, 362
geeluintjie, 532
gemsbok cucumber, 236
gemsbok horn, 349
gemsbok komkommer, 236
gentian, 272, 297, 401
    primrose gentian, 272
    water gentian, 401
    yellow water gentian, 297
gewone wildevy, 106
gewone-mielierooiblom, 194
gherkin, 240
ghushika, 228
ghushosho, 362
giant crowfoot, 450
giant orchid, 301
giant spear grass, 463
giant three awn, 465
gifappel, 59
ginger grass, 459
giraffe thorn, 257
gisekia, 148
gkaa, 241
globe amaranth, 337
go, 257, 380
gobe-jwatlhoa, 78
golden bristle grass, 512
golden setaria, 512
golden timothy grass, 512
golden tumbling grass, 512
gombossie, 142
gondovoro, 120
gongoni three-awn, 504
gonya grass, 445
goose grass, 455
gooseberry cucumber, 240
goshwe, 258
grapple plant, 181, 182
grapple thorn, 181
green mistletoe, 101
grey bitter apple, 59
grey love grass, 469
grey sour grass, 506
grey tussock grass, 463
grey-leaf heliotrope, 354
grootklits, 150
grootkoorsbessie, 89
grootsaad-sporobolus, 488

ground almond, 532
gubagu, 223
guexwe, 205
guinea grass, 482
gumane, 253
gwee, 63
gwi, 349
gxoxe, 316
gxwi, 254

haemorrhoid bush, 334
hairy dicoma, 152
hairy stemmed acalypha, 124
hand-palm ipomoea, 367
hardekool, 81
hare's tail bush, 339
hay grass, 447
heart pea, 422
heart seed, 422
heartleaf sida, 294
hell's curse, 66
herero cucumber, 236
herringbone grass, 442, 443
hibiscus, 283, 289
    bladder hibiscus, 289
    dongola hibiscus, 283
hippo grass, 430, 437
hoary basil, 54
hoenderuintjie, 532
hook thorn, 381
horned cucumber, 239, 240, 241
horseweed, 209
hqang'arci, 59
hyena grass, 510

ibangu, 180
ibozu, 352
idarere, 313,314
igogo-chitukunu, 322
ilala palm, 345
imboke, 70
Indian acalypha, 87
Indian ginseng, 110
Indian love grass, 473
Indian millet, 455
ingongo, 248
inkberry, 59
insekwasekwa, 405
insika, 228
isika, 228
isiko, 55
isotho, 297,401
isuma, 84
isunde, 385
ituhatuha, 374
ivangogu lye ingombe, 181
ivory tree, 81
ivozu, 352
iziye, 107
izungwe, 154

jackal-berry, 84
jakkalsbessie, 84
jarajara, 535
jasmine pea, 385
jelly melon, 239, 240, 241

Jersey cudweed, 221
Jimson weed, 302
jujube, 108
jungle rice, 435

ka, 236
ka jisa, 236
kabunga, 416
kafblom, 150
kaguhe, 334
kakiedubbeltjie, 64
kakoma, 345
Kalahari (sand) acacia, 380
Kalahari-appelblaar, 52
Kalahari apple-leaf, 52
Kalahari Christmas tree, 259
Kalahari currant, 71
Kalahari panicum, 481
Kalahari sand kweek, 498
Kalahari sand quick grass, 498
Kalahari sand thorn, 380
Kalahari star apple, 246
kalake, 232
kalanchoe, 122
kameeldoring, 257
kamongamba, 344
kangarangana, 380
kangungwe, 450
kaniyangngombe, 237
kanlyanzovu, 237
kanniedood, 304
karkoer, 237
karo, 243
karu, 243
karuarua, 82
kasamu komoliro, 177
kashe, 414
katamanwusea, 240
katcama, 237
katoenbossie, 280
katstert-mannagras, 512
katstertgras, 510
katstertjie, 151
kavangu, 416
kavimba, 81
kawa, 236
kaXhee, 409
kazungula, 154, 236
keme, 237
kenaf, 282
kgaba, 75
kgane, 367
kgato, 66
kgatwe, 239
kgelegetla, 305, 306
kgengwe, 237
kgo-tuduwa, 293
kgobe-tsa-badisana, 197
kgopo, 44
kgopokgolo, 247
kgotodua, 179, 184, 220
kgwakazen-guro, 244
khaki weed, 64
khezane, 498
khonkhorose, 76
kierieklapper, 230

kleinrolgras, 444
kleinskraalhans, 209
klettgrass, 513
kleurbossie, 92
klitsgras, 513
klosaarbossie, 405, 406, 407
knapsekerel, 205
knob-thorn, 258
knobbly combretum, 360
knobbly creeping bushwillow, 360
knoppiesdoring, 258
knoppiesklimop, 360
knotweed, 187
kobo, 99, 120
koedoebos, 229
koekbossie, 294
koffiebeesklou, 375
koma, 538
kothuinzovo, 277
kraalnaboom, 90
kremetartboom, 352
krulblaartaaibos, 71
kubutona, 187
kudu-bush, 229
kukuruthwe, 315
kweek, 448
kwena, 560

lady flower, 327
laeveldse geelmelkhout, 228
laloentjie, 244
lance tree, 53
land grass, 453, 455, 512
large carrot-seed grass, 515
large devil's thorn, 180
large fountain bristle grass, 467
large fruited bushwillow, 81
large fruited combretum, 81
large plume eragrostis, 468
leadwood, 81
leafy bract justicia, 330
leatla, 392
lebelebele, 509
lechachalanoga, 242
ledelele, 314
ledutla?, 273
leetsane, 319
lefswe, 75, 200
legabala, 239
legapu, 237
legatapitse, 180, 181
legatapitsi, 182
legau, 237
legonnyane?, 393
leilane, 75
lekanangwane, 180
lekatane, 237
lekatse, 180
lekawa, 236
leketa, 358
leloba, 280
lematla, 180
lemon grass, 459
lenapa, 478
lengakapitse, 181
lengakare, 182

leopard orchid, 301
lepheto, 443
lephutse, 113
leralagori, 178
lesotho, 297
letakana, 120
letetemetso, 187
letlhajwa, 246
letlhaka, 464
letseta, 280, 283
letswai-la-khudu, 414
lily, 55, 113, 120, 390, 396
    blue water lily, 55
    climbing lily, 120
    fireball lily, 113
    flame lily, 120
    pincushion lily, 113
    royal shaving brush lily, 113
    spider lily, 390
    superb lily, 120
    vlei lily, 396
loetsane, 91
lofse?, 266
lola palm, 345
lole, 107
long plumed finger grass, 454
long tail cassia, 249
long-awned grass, 505
long-awned water grass, 437
lovangolo nzovu, 180
lovangu, 322
lovanzovu, 95
love bean, 165
love grass, 468, 469, 471, 472, 473, 476
    broom love grass, 472
    grey love grass, 469
    Indian love grass, 473
    rough love grass, 468
    sticky love grass, 476
    stink love grass, 469
lowveld mangosteen, 228
lucky bean climber, 165

m- -fetola?, 171
mabele, 512
mabophe, 165
madikolo, 81
mafavuke, 43
magaga, 236
magic guarri, 85
mago, 217
magogodi-a-noka, 325
mahango, 509
mahogany (Rhodesian), 254
mahwaa, 55
makakarana, 182
makakare, 181
makarikari finger grass, 452
makarikari vingergras, 452
makgolela, 77
makgolela-a-dinaka-tsa-pudi, 348
makgonatshe, 124
makgonatsotlhe, 124
makhudugwane, 392
makoholi, 471

Plants of Northern Botswana   **581**

makungara, 55
malalamakatse, 180
malisaka, 238
malisako, 244
malomaagorothwe, 264
malomaagwerothwe, 40, 264
malomaarothe, 42
malomaarotho, 40
malomagorotwe, 264
mambalane, 315
mamnyati, 174
mampi-pinyane, 242
mangobombo, 217
mangosteen, 228
    African mangosteen, 228
    lowveld mangosteen, 228
mangqore, 316
manketti tree, 248
mankettiboom, 248
mantshegi, 244
manXatura, 344
manXore, 316
mapanda, 53
maramata, 180
mareko, 77
marethe, 292
marsh grass, 435
masepaabanyana, 277
masigomabe, 142, 413
masika, 228
masiku maave, 142
masitis, 178
masogomabe, 142
masupegane, 49, 215
mat sedge, 554
matabele, 194
matebelwe, 194, 309
matinose, 291
matjiesgoed, 541
matlebilo, 421
matopie, 77
matsogomabe, 143
maXatora, 244
mayanga-ombwa, 120
mayungu anyambi, 343
mbabagulo, 95
mbabashulo, 95
mbgweti, 81
mbiriri, 84
mbuyu, 352
mealie crotalaria, 264
medlar, 84
meerjarige denneboomgras, 443
meidebossie, 313
melkbos, 73
melon, 236, 237, 239, 240, 241
    bittermelon, 237
    citron melon, 237
    edible seed melon, 237
    jelly melon, 239, 240, 241
    tsama melon, 237
    wild melon, 236
    wild watermelon, 237
merjarige beesgras, 445
merremia, 235
mfafa, 482

mfafu, 378
mfhafha, 482
mfuthe, 127
mhaha?, 482,497
mhahu, 378
mhalatsamaru, 346,347
mhata, 52,53
mhatla, 53
mhudiri, 229
mielie-crotalaria, 264
mierbossie, 337
milanje finger grass, 452
milanje grass, 452
milanje vingergras, 452
milk weed, 73
mimosa, 257
miniature morning glory, 235
misbredie, 66
mistletoe, 100, 101, 177
    acacia mistletoe, 177
    green mistletoe, 101
    Ngami mistletoe, 100
mitwa-ya-ntse, 79
mizindumi, 532
mmabashete, 285
mmabasi, 285
mmadikokwana, 178
mmampimpinyane, 244
mmankgar-wane, 285
mmapupu, 244
mmapuupuu, 244
mmola, 384
mmonyana, 204
mmumo, 106
mnondo, 206
moana, 352
moanzabalo, 361
moarasope, 110
mobonona, 362
mobuyu, 352
mochancha, 375
mochenje, 84
mochinga, 72
mochingachinga, 72
mochope, 375
modan-danyane, 371
modi, 500
modidmo, 55
modikangwetsi, 508
modikaseope, 110
modiyangwe, 378
modupaphiri, 71
modwarra, 532
modyangwe, 357
moelethaga, 282
mofala, 490
mofofo, 230
mofungi, 361
mogabala, 239, 241
mogaga, 340
mogalori-kodumela, 124
mogamapodi, 514
mogapu, 236
mogat-apeba, 296
mogata, 52
mogatapeba, 512

mogatawapeba, 296
mogatla-wa-tau, 467
mogatla-wapeba, 512
mogatla-watau, 489
mogatlawammutla, 339
mogato, 66
mogatololo, 54
mogoka, 379
mogokatau, 376
mogoliri, 229
mogonono, 362
mogose, 375
mogotlho, 257
mogotse, 348
mogotswe, 375
mogwana, 315
mohaha, 482
mohahu, 378
moharatshwene, 88
mohata, 52,53
mohatla, 52
mohatola, 532
moherasope, 110
mohosi, 375
mohubuhubu?, 320
mohudiri, 229
mohulapitse, 338
mohulere, 229
mohuthi, 375
mohwidiri, 229
moithimodiso, 122
mojakubu, 430,510
mojaphuti, 306
mojatangombe, 66
mok-gomatha, 316
mok-wankusha, 249
moka, 380
mokabe, 230
mokabi, 230
mokaikai, 237,296
mokala, 257,258,380
mokalu, 108
mokamakama, 459
mokanja, 496
mokankele, 316
mokanonga, 228
mokapa, 241
mokapana, 4, 236, 241
mokapane, 236
mokata, 230
mokate, 237
mokau, 103
mokazan-molotho, 40
mokeketi, 108
mokera o shoro, 339
mokerete, 107
moketekete, 108
mokgabi, 229
mokgalo, 108
mokgompatha, 316
mokgotshe, 348
mokgotshe ômontonanyana, 349
mokgôtshe-ô-monamagadi, 348
mokgotshi, 349
mokgwelekgwele, 380
mokha, 380

mokhasi, 532
mokhesa, 136
mokhure, 127
mokhutsomu, 84
mokidi, 316
mokoba, 258
mokochong, 84
mokode, 34
mokodi, 34, 217
mokojane, 34
mokoka, 378
mokoko, 378
mokokwane, 378
mokolane, 345
mokololo, 53
mokolwane, 345
mokomoto, 224
mokon-dekonde, 72
mokon-kono, 228
mokongwa, 248
mokonkolwane, 345
mokononka, 228
mokororo, 53
mokoshi, 375
moktshumo, 84
moku, 282, 380
mokuba, 362
mokuku, 120
mokulane, 345
mokut-shumo, 84
mokutemu-tembuze, 79
mokwelenyane, 204
molalakgaka, 384
molalaphage, 498
molekangwetsi, 506, 508
molelwana, 194
molemo wa segogwane, 187
molemowanonyane, 49
mollo-wa-badimo, 109, 307
molora, 63, 150
moloto, 377
mompondo, 316
monabo, 315
monakaladi, 521
monakaladi omotonanyana, 524
monamagadi, 217
monawana, 175
monepenepe, 249
mongalangala, 292
mongana, 381
monganga, 108
mongave, 230
monghongho, 248
mongoma, 184
mongongo, 248
mongongo nut, 248
mongwa, 414
monka, 381
monkey apple, 237
monkey guava, 84
monkey pepper, 243
mono, 127
monoga, 383
monomane, 77
monontshane, 145
monthe, 385

monwana, 230
monxidi, 430
monyaku, 240
monyana, 205, 206
monyapula, 416
monyondo, 81
mooinooiientjie, 42
mooka, 380
mooku, 380
moono, 127
moonyana, 204
moonyane, 205
mopanda, 52, 53
mopane grass, 428
mophane, 253
mophethe, 165
mophithi, 165
mophuratshukudu, 376
mopipi, 77
moponda, 52
mopondopondo, 375
mopororo, 53
moporota, 154
moraanoga, 82
morala, 416
moralana, 228
morararupe, 110
morarasope, 110
moravi, 416
more-o-mabele, 258, 381
more-o-mosetlha, 382
more-wanoga, 82
moremotala, 90
moretho-thobi, 221
moroka, 224, 226
morola?, 59
morolwana, 59, 60
morolwane, 59
moromoswane, 209
morongo, 89
morumosetlha, 382
morupaphiri, 71
morupe, 266
morwanyeru, 79
mosamo womosana, 344
mosaoka, 380
mosasawana, 71
moseka, 472
moseka?, 452
mosekangwetsi, 508
mosela-watshwene, 508
moselesele, 259
mosetlho, 322
moshanje, 500
mosharashagana, 293
moshawa, 286, 287, 292
moshetondo, 85
moshika, 228
moshikiri, 102
moshitondo, 85
moshosho, 362
moshumo, 239
mosiama, 151
mosiki, 102
mosikili, 102
mosimama, 206, 223

mosit-wasitwane, 346
mositanoka, 337
mositanokana, 73, 268
mositi, 178
mositwane, 346
mosokalatshebe, 348
mosokela-tsebeng, 348
mosokelateng, 376
mosokelatsebing, 348
mosokotsala, 232
mosonowamuthantshê, 339
mosu, 257
mosukujane, 54
mosungula, 154
mosupogane, 49
mot-shentshe, 375
motaloga, 375
motangtanyane, 232
motantanyane, 235, 296, 311
motatija, 385
motawana, 357
motawaphesa, 512
motetene, 434
mothakeja, 246
mother-in-law's tongue, 348
mothlathla, 532
mothlwa-o-jewa, 375
motindi, 361
motlalemetse, 221
motlalemetsi, 34
motlhabana, 148
motlhahola, 85
motlhaje, 246
motlhajwa, 246
motlhakana, 529
motlhakola, 85, 345
motlhakua, 85
motlhatswapelo, 354
motlho, 448
motlhomaganyane, 184
motlhono, 79
motlhwa, 448
motlhwakeja, 385
motlôpi, 77
motokwane, 137
motombolo, 360
motono, 315
motsaodi, 228, 254
motsaudi, 228, 254
motsebe, 89
motsentsela, 107, 345
motsetse, 90
motshe-wa-badimo, 192
motsheketsane, 360
motshikiri, 472
motshwarakgano, 63, 150
motsiara, 230
motsididi, 232, 369
motsikiri, 472
motsintsila, 107
motsitsi, 90
motsope, 375
motsore, 81
motsot-sojane, 316
motswalak-goro, 313
motswalakgoro, 285, 294

motswe, 414
motsweketsane, 103, 360, 410
motswere, 81
motswiri, 81
motulu, 295
motuu, 315
motwakidja, 375
motweng, 63
moumo, 106
mouse-eared combretum, 230
movimba, 81
movunguvungu, 154
mowana, 352
moXhane, 316
moXhwewe, 165
moXhXo, 152
moxinxa-mokulane, 202
Mozambique combretum, 360
Mozambique ebony, 84
mozinzila, 107
mozungula, 154
mpangale, 259
mpani, 253
mphaga, 482
mpitipiti, 165
mpumutwi, 85
mpuzu, 316
mubesuba, 360
muchenje, 84
muchima, 312
muchira ukaza, 151
mudhangwe, 410
mughandutji, 258
mughombe, 382
mugorokoko, 71
mugutswe, 375
muhane, 253
muhorono, 362
muhoto, 257
muhotsi, 375
mukololo, 52,53
mukona, 378,381
mumpaumpa, 229
munde, 385
mungave, 230
mungcinda, 379
munge, 385
mungongo, 248
muparapara, 376
mupidi, 170
mupondo, 375
murengambo, 377
musihantabwe, 246
musikili, 102
mutemo, 299
mutsintsila, 107
muvichi, 84
muwanduwehi, 258
muwane, 352
muzauli, 254
mwanduchi, 258
mweye, 259
myrtle bushwillow, 361

N!ã, 377
n//a, 379

n//ah, 379
n//an, 237
n/uah, 236
n?a, 237
n?o'nu, 239
nachwa, 120
nama, 514
namele, 452
narrow-plumed dropseed, 501
Natal mahogany, 102
Natal red top, 478
natalse rooipluim, 478
nawa-yanaga, 175, 267
nchecheni, 108
ndcwa, 375
ndongo, 312
nettle spurge, 95
ngaganyama, 249
Ngami mistletoe, 100
nganje, 448
ngarakashe, 99
ngawa, 258
ngharara, 450
ngocha, 90
ngondovuro, 120
ngongo, 248
ngotza, 90
ngwara, 532
ngweti, 81
ngxaio, 448
nidorella, 220
nine-awned grass (common), 506
nium, 352
nkogo, 258
nkogwana, 381
nkoshwana, 381
nlala, 202
nnala, 416
nnyo-yammutla, 339
nondothinde, 452
nonura, 59
noordelike dwababessie, 72
nordelike lalapalm, 345
northern dwaba-berry, 72
northern ilala palm, 345
northern lala palm, 345
nqodi, 385
nqoli, 385
nsekesa, 375
nshangule, 85
nsibi, 254
nsuru-ntukunu, 362
nsusu, 362
nswazwi, 230
ntangabe, 240
ntatatjiba, 294
ntewa, 315
nthare, 230
nthoka-mosare, 205
ntjetjeni, 108
ntopi, 77
ntotoba, 59
ntshin-gitsha, 229
ntshingidtza, 229
ntshingitshi, 229
ntuntulwa, 59

ntutuba, 236
nunukwa, 59
nunura, 59
nut grass, 532
nutgrass, 540
nutsedge, 540
nyevi, 358
nyololo, 238
n≠ahli, 378
n≠eng, 378

ogumane, 253
oi?, 137
ojeete, 259
okae, 89
okapi, 339
okaXhee, 409
old man's beard, 457
old-land's grass, 447,512
omotonanyana, 34
ompopusa, 348
omuhengehenge, 382
omum-bonde, 257
omumbuti, 229
omungondo, 380
omungongomwi, 377
omunxumuhari, 268
omupapaku, 205
omusaona, 381
omuseasetu, 362
omutaurammmbuku, 378
omutjaitjai, 40
onduri, 183
ongawu, 345
onshabwe, 148
opane, 253
opblaa-boontjie, 422
orarakanga, 384
orchid, 301
  giant orchid, 301
  leopard orchid, 301
  tree orchid, 301
orogu?, 82
oruejo, 445, 446
oruzenga, 344
oshoma, 84
otjihangatene, 181, 182
otjiraura, 177
otjitjandoko, 277
ovambenderu omutjivi, 89
ovumo, 106
oxalis, 302
oxhone, 202
oyondo, 81

paddle-pod, 80
palamêla, 100, 101, 103, 177, 301
palm, 202, 301, 345, 367
  dwarf date palm, 202
  feather palm, 202
  hand-palm ipomoea, 367
  ilala palm, 345
  lola palm, 345
  nordelike lalapalm, 345
  Northern ilala palm, 345
  Northern lala palm, 345

real fan palm, 345
   vegetable-ivory palm, 345
   wild date palm, 202
pan dropseed, 486
panfynsaadgras, 486
panicled amaranth, 66
paniculated spot flower, 350
panvingergras, 452
paper plant, 538
paper plume, 330
paper thorn, 64
paperbark acacia, 382
paperbark thorn, 382
papierbasdoring, 382
papyrus, 538
para cress, 350
Paraguay cress, 350
pea, 165, 175, 385, 422
   cowpea, 175
   heart pea, 422
   jasmine pea, 385
   rosary pea, 165
pearl millet, 508
pekolola, 374
pelo botlhoko, 389
pelobotlhoko, 152
peperblaarkanniedood, 226
pepper-leaved commiphora, 226
pepper-leaved corkwood, 226
persimmon, 84
Peters fig, 106
peters-se-vy, 106
peultjiesbos, 227
phalatsi, 113
pharaspikiri, 376
phefo, 37
phesana tsa bathwana, 421
phesana-yangwana, 421
phiho, 37
philo, 37
phoka, 445,446
phuduhudu, 25
phuka, 445
phusana, 176
pig's senna, 128
pigweed, 303
pincushion grass, 429
pincushion lily, 113
pingping-tshegatshega, 82, 244
poison apple, 59
poison bush, 87
poison onion, 96
poisonous gooseberry, 110
polished star, 117
poo-khunung, 313
poprosie, 215
porselein, 304
pretty lady, 40,42
primrose gentian, 272
prince's feather, 66
prostrate globe amaranth, 337
pteleopsis, 361
puka, 441,482,497
puncture vine, 322
purple nutsedge, 540
purple pan weed, 153

purple spike grass, 510
purple three-awn, 466
purple witchweed, 195
purple-throated ipomoea, 371
purslain, 303
purslane, 304
purslane (common), 303
pusky, 303
pusley, 303, 304
python climber, 103
python vine, 103

quick grass, 448

ragwort, 223
rain gauge, 184
rain tree, 53
raisin, 315, 316
   rough-leaved raisin, 316
   sandpaper raisin, 316
   white-leaved raisin, 315
ramarungwana, 176
rankklits, 23
ranksuring, 302
rapolodwane, 344
rasp grass, 496
rattle pod, 261
real fan palm, 345
red amaranth, 66
red bush-willow (hairy), 229
red spike thorn, 79
red star apple, 246
red top grass, 478
rekuXhwa, 79
rethajwa, 246
rewawan-jovu, 95
Rhodesian blue grass, 458
Rhodesian ebony, 84
rice grass, 496
riri-satau, 465
river fever-berry, 89
riverbank pennisetum, 509
riverbed grass, 509
roerkuid, 221
rolling grass, 444
rooi-agurkie, 240,241
rooi-komkommer, 241
rooiaarbossie, 151
rooiblom, 194
rooiblommetjie, 195
rooibos(wilg), 229
rooiboslelie, 120
rooiessenhout, 102
rooikomkommer, 240
rooipendoring, 79
rooiuintjie, 540
ropuiti, 236
rosary pea, 165
rotho, 358
rothwe, 358
rough bryony, 245
rough love grass, 468
rough-leaved commiphora, 224
rough-leaved corkwood, 224
rough-leaved raisin, 316
royal shaving brush lily, 113

rubber hedge euphorbia, 90
rurithi, 66
rusperbossie, 41
russet bush willow, 230
rutunguza, 59
rwedthi, 65
rwithe, 314

sage, 34, 217
   dwarf sage, 217
   silky sage, 217
   wild sage, 34
samurai, 180
sand acacia, 376
sand camwood, 385
sand thorn, 376
sanddoring, 376
sandkamhout, 385
sandmelktou, 344
sandpaper raisin, 316
sandveld asparagus, 347
sanyane, 478
sausage tree, 154
savanna gardenia, 416
sebabatsane, 95
sebabatswane, 95
sebabet-sane, 95
sebeditona, 54
seboka, 116
sebrabaskanniedood, 225
sedge, 529, 530, 532, 540, 554
   flat sedge, 529
   mat sedge, 554
   nutsedge, 540
   purple nutsedge, 540
   winged sedge, 530
   yellow nut sedge, 532
sedupapula, 393
segobe, 470
segokwe, 99
segowe, 514
sehokgwe, 120
sekelbos, 259
sekelgras, 443
sekhi, 379
sekhumba, 55
sekoba, 313
selaole, 184
selekangwetsi, 455
seloka, 443, 460, 466, 467, 502, 505
semonamone, 137
semonye, 122
sengaparile, 181, 182
senyane, 478
senyaparele, 181
sephalane, 76
sepodise, 64
serepe, 190, 303
seretlwana, 313
seriri, 465
seriri-sa-tau, 465
seromo, 392
sesetlho, 280
seshangane, 91
seshungwa, 358

sesunkwane, 54
sethare-se-tala, 90
setlepetlepe, 66
setlhabakolobe, 76, 204, 205, 206
setlhabi, 178
setou, 464
setsee, 379
setshe, 379
setshi, 379
sexhodo, 113
shaving brush combretum, 360
shepherd's tree, 77
shi, 254
shiny-leaf-wait-a-bit, 108
shôi?, 37
shosho, 322
shrub jasmine, 411
sickle bush, 259
sickle grass, 443
sickle-fruit, 63
sickle-leaved albizia, 384
sickle-leaved false-thorn, 384
sifonkola, 249
signal grass, 446
silky autumn grass, 462
silky burweed, 339
silky bushman grass, 489
silky sage, 217
silver cluster-leaf, 362
silver terminalia, 362
silver thread grass, 457
sina, 205
single-leaved cleome, 41
sinshungwa, 358
sjam-bokpeul, 249
sjambok pod, 249
skurweblaarkanniedood, 224
skurwerosyntjie, 316
slangwortel, 188
slender meadow grass, 473
small buffalo grass, 497
small bushman grass, 489
small carrot-seed grass, 514
small-flower umbrella plant, 531
smelter's bush, 212
smelterbossie, 212
smooth creeping spurge, 163
snake root, 188
snotterbelletjie, 358
snowflake grass, 457
Sodom apple, 59
South American daisy, 219
Spanish needles, 205
sparrowgrass, 346
spear grass (common), 460
spider flower, 41
spider lily, 390
spider wisp, 358
spiky mother-in-law's tongue, 349
spindlepod, 41
spiny sida, 400
spot flower, 350
spotted signal grass, 434
staggers grass, 498
star jasmine, 344

starbur, 76
  bristly, 76
  upright, 76
steenboksuring, 302
stekeltaaiman, 400
stick grass, 460
sticky bristle grass, 513
sticky love grass, 476
stiff bowstring, 349
stinging nettle, 95
stink bushwillow (two-winged), 361
stink eragrostis, 469
stink grass, 469
stink love grass, 469
stinkboswilg, 361
stinkbush, 34
strangler fig, 106
studthorn, 180
suikerteebossie, 78
sulu, 416
suma, 84
superb lily, 120
swarthaak, 381
sweat bush, 34
sweet cyperus, 537
sweethearts, 70,205

tadwa, 450
tall bushman grass, 489
tallapoa, 180
tanglehead, 460
tattoo plant, 409
tchuvongololo, 357
teak (Rhodesian), 254
tedutsabaana, 70
teebossie, 176
teesuikerkaroo, 78
terblanbossie, 289
thatching grass, 472
thepe, 65
thiba-di-molekane, 136
thikerva, 202
thitha, 34, 221
thobega, 241
thola, 59
tholwana, 59
tholwanakgomo, 59
tholwane, 59
thontholwana, 59
thorn, 64, 79, 108, 180, 181, 257,
  258, 322, 376, 377, 378, 379,
  380, 381, 382, 383, 384
  bastard umbrella thorn, 380
  black thorn, 381
  blade thorn, 378
  blue thorn, 377
  buffalo thorn, 108
  camel thorn, 257
  candle thorn, 379
  common false-thorn, 384
  confetti spikethorn, 79
  devil's thorn, 322
  flat-topped thorn, 382
  giraffe thorn, 257
  grapple thorn, 181

  hook thorn, 381
  Kalahari sand thorn, 380
  knob-thorn, 258
  large devil's thorn, 180
  paperbark thorn, 382
  paper thorn, 64
  red spike thorn, 79
  sand thorn, 376
  sickle-leaved false-thorn, 384
  studthorn, 180
  thorn tree, 257
  wait-a-bit thorn, 381
  worm-bark false-thorn, 383
  worm-cure false thorn, 383
thotamadi, 305
thotamadi yomotunanyana, 306
thulwathulwane, 59
tick grass, 493
tiger-foot ipomoea, 160
tite grass, 494
tlatlana, 476
tlăba, 253
tlhale, 280
tlhatlha, 532
tlhogotshweu, 334
tnotsantsa, 375
tobacco witchweed, 195
tobi, 145
toothbrush tree, 85, 246
tototwane, 530, 532, 537
towerghwarrie, 85
Transvaal ebony, 84
Transvaal gardenia, 416
trassiedoring, 379
tree, 53, 77, 79, 81, 85, 154, 246,
  248, 253, 257, 259, 295, 301,
  352, 488
  apple-leaf tree, 53
  balsam tree, 253
  black ironwood, 253
  butterfly tree, 253
  confetti tree, 79
  Christmas tree grass, 488
  featherweight tree, 248
  ivory tree, 81
  Kalahari Christmas tree, 259
  lance tree, 53
  manketti tree, 248
  rain tree, 53
  sausage tree, 154
  shepherd's tree, 77
  thorn tree, 257
  toothbrush tree, 85, 246
  tree orchid, 301
  upside-down tree, 352
  Zambezi honeysuckle tree, 295
tsakokhee, 69
tsama, 237
tsamai?, 337
tsama melon, 237
tsamma, 237
tsaodi, 254
tsaro, 202
tsatsalopane, 176
tsaudi, 254

tsebe-yatlou, 277
tsebeatoje, 267
tsetwane, 152
tshero, 182
tshetlho, 322
tshetlho-e-tonanyana, 180
tshetlho-ya-dinku, 180
tshika-dithate, 170
tshikadithata, 170, 277
tshikitshane, 489
tshitladingwetsi, 447
tshobatshobane, 346
tshoo-la-khudu, 231
tshuga, 341
tshwang, 498
tsizyna, 107
tsube, 498
tswa, 452
tswaitswai?, 302
tswigho, 95
tswii, 55, 297
tszini, 246
tuinranksuring, 302
tumbleweed, 392, 444
two-winged pteleopsis, 361

ugandu, 258
umbolo, 404
umbwiti, 236
unganda, 381
ungqo, 53
ungulu, 161
upanda, 53
upside-down tree, 352
ushi, 254
ushuu, 230, 257
utata, 106
utunda, 84
uurblom, 289
uvunguvungu, 154
uwara, 53
uxhoro, 154
uXhouwa, 362

vaalboom, 362
vaalsuurgras, 506
varkkos, 303
vato-ohoni, 165
veergras, 457
vegetable-ivory palm, 345
veld paspalum, 441
vetiver grass, 492
vine, 83, 103, 322, 422
    balloon vine, 422
    cherry vine, 83
    puncture vine, 322
    python vine, 103

visgraat-cassia, 251
vlei ink-flower, 192
vlei lily, 396
voëlent, 177

wait-a-bit thorn, 381
wandering jew, 43, 231
wanoka, 71
wasiwa, 40
water gentian, 401
water grass, 437, 532
water lettuce, 274
water rush, 537
water snowflake, 401
waterleaf, 68
waxberry, 78
white bauhinia, 375
white buffalo grass, 497
white head, 334
white herring-bone grass, 501
white plumbago, 413
white ribbon bush, 329
white-leaved grewia, 315
white-leaved raisin, 315
white-tipped hemizygia, 394
wild asparagus, 346, 347
wild basil, 54
wild bindweed, 364
wild coffee bean, 375
wild cotton, 73, 280
wild cucumber, 240
wild dagga, 137
wild date palm, 202
wild everlasting, 215
wild evolvulus, 49
wild jasminum, 410
wild lucerne, 263, 264, 270
wild melon, 236
wild onion, 96
wild petunia, 232
wild plum, 228
wild plumbago, 413
wild purslane, 303, 304
wild sage, 34
wild sesame, 184
wild spinach, 238
wild stockrose, 282, 285, 286
wild watermelon, 237
wilde komkommer, 240
wilde-bos-ganna, 42
wilde-komkommer, 241
wilde-stokroos, 282
wildedadelboom, 202
wildepatat, 232
wilderysgras, 496
wildkatoen, 280
willow herb, 300

window seed, 404
wing stem daisy, 207, 213
winged sedge, 530
winter cherry, 110
wire grass, 504
witblaarrosyntjie, 315
witchweed, 194, 195
    common mealie-witchweed, 194
    purple witchweed, 195
    tobacco witchweed, 195
witgat, 77
witpluimgras, 447
womuzuka, 161
wood sorrel, 302
woodland gardenia, 416
woolflower, 151
woolly caper bush, 357
worm-bark false-thorn, 383
worm-cure albizia, 383
worm-cure false thorn, 383
worsboom, 73
wortelsaadgras, 514
wurmbasvalsdoring, 383

xaa, 36
xao, 151
xaoqoo, 152
xarexo, 205
xgargum, 79
xhaa, 65
xonequm, 165
xuamudsa, 63
xuase, 63
xummu, 151
xumo, 248
xwate, 181

yellow commelina, 231
yellow cosmos, 206
yellow ipomoea, 232
yellow mouse whiskers, 227
yellow nut sedge, 532
yellow pan weed, 260
yellow sorrel, 302
yellow thatching grass, 461
yellow water gentian, 297
yellow water-daisy, 218
yo monamagadi, 305

Zambezi brides-bush, 420
Zambezi honeysuckle tree, 295
Zambezi-bruidsbos, 420
zang, 77
zebra corkwood, 225
zebra-bark commiphora, 225
zhinca, 258

Scientific names index

Accepted species names are listed in italics, synonyms in plain text. Major entires for each species have their page numbers in plain text, the page numbers for minor entries are in italics

Abrus precatorius L., 260, *402*
   subsp. *africanus* Verdc., 165
Abutilon angulatum (Guill. & Perr.) Mast. var. *angulatum*, 277
Abutilon englerianum Ulbr., 278
Abutilon ramosum (Cav.) Guill. & Perr., 279
Acacia arenaria Schinz, 376
Acacia erioloba E.Mey., *11*, *18*, 257
Acacia erubescens Welw. ex Oliv., 377, *378*
Acacia fleckii Schinz, 377, *378*
Acacia hebeclada DC., *11*, 257
   subsp. *hebeclada*, 379
   subsp. *chobiensis* (O.B.Mill.) A.Schreib., 379
Acacia luederitzii Engl. var. *luederitzii*, *11*, 380
Acacia mellifera (Vahl) Benth., 258
   subsp. *detinens* (Burch.) Brenan, 381
Acacia nigrescens, *11*, *15*, *17*, *18*, *19*, 301
Acacia sieberiana DC. var. *woodii* (Burtt Davy) Keay & Brenan, 382
Acalypha fimbriata Schumach. & Thonn., 86
Acalypha indica L., *86*, 87
Acalypha ornata A.Rich., 88
Acalypha villicaulis Hochst. ex A.Rich., 124
Acanthosicyos naudinianus (Sond.) C.Jeffrey, *4*, 236
Acanthospermum hispidum DC., 76
Achyranthes aspera L., 73
   var. *sicula* L., 63
   var. *pubescens* (Moq.) Townsend, *63*, 150
Acrotome inflata Benth., 392
Adansonia digitata L., 352
Aerva leucura, *152*, *389*
Albizia anthelmintica Brongn., 383
Albizia harveyi E.Fourn., *136*, 384
Albuca abyssinica Jacq., 273
Alectra orobanchoides Benth., 308
Alectra picta (Hiern) Hemsl., 309
Aloe sp., 345
Alternanthera sessilis (L.) DC., 335
Amaranthus graecizans L., 65
Amaranthus hybridus L., 66
Amaranthus praetermissus Brenan, 67
Amaranthus viridis L., 68
Ammannia baccifera L. subsp. *baccifera*, 138
Ampelocissus africana (Lour.) Merr., 146
Andropogon eucomus Nees, 457
Andropogon gayanus Kunth, 458
Ansellia africana Lindl. subsp. *africana*, 301
Anthericum milanjianum Rendle, 342
Anthericum whytei Baker, 341
Aptosimum decumbens Schinz, 57
Aristida adscensionis L., 502
Aristida hordeacea Kunth, 503
Aristida junciformis Trin. & Rupr., 504
Aristida meridionalis (Stapf) Henrard, 465, *467*
Aristida scabrivalvis Hack., 466
Aristida stipitata Hack., 505
Aristida stipoides Lam., 467
Asparagus africanus Lam., *63*, *346*, 347

Asparagus nelsii Schinz, 347
Aspilia mossambicensis (Oliv.) Wild, 203
Astripomoea lachnosperma (Choisy) A.Meeuse, 363
Asystasia gangetica (L.) T.Anderson, 324

Baphia massaiensis Taub. subsp. *obovata* (Schinz) Brummitt, 385
Barleria mackenii Hook.f., *20*, 22
Basananthe pedata (Baker f.) W.J.de Wilde, 412
Bauhinia petersiana Bolle subsp. *macrantha* (Oliv.) Brummitt & J.H.Ross, 375
Berchemia discolor (Klotzsch) Hemsl., *13*, *107*, *345*
Bergia ammannioides Roth, 123
Bergia pentheriana Keissl., 162
Bidens biternata (Lour.) Merr. & Sheriff, 204, *205*
Bidens pilosa L., *204*, 205
Bidens schimperi Sch. Bip. ex Walp., 206
Blainvillea acmella (L.) Philipson, *20*, 350
Blepharis integrifolia E.May. & Drège, 20
   var. *integrifolia*, 23
Blepharis maderaspatensis (L.) B.Heyne ex Roth, 326
Boerhavia coccinea Mill., 178
Boerhavia diffusa, 178
Boscia albitrunca (Burch.) Gilg & Gilg-Ben., *10*, 77
Brachiaria deflexa (Schumach.) Robyns, 491
Brachiaria dura Stapf, 431
Brachiaria grossa Stapf, 432
Brachiaria humidicola (Rendle) Schweick., 433
Brachiaria nigropedata (Ficalho & Hiern) Stapf, 434
Bulbostylis hispidula (Vahl) R.W.Haines subsp. *pyriformis* (Lye) R.W.Haines, 549

Caldesia parnassifolia (L.) Parl., 333
Caldesia reniformis (D.Don) Makino, 333
Calostephane divaricata Benth., 207
Capparis tomentosa Lam., 357
Caralluma lugardii N.E.Br., 115
Caralluma schweinfurthii A.Berger, 116
Cardiospermum halicacabum L., 422
Cassia abbreviata Oliv. subsp. *beareana* (Holmes) Brenan, 249
Cassia mimosoides L., 251
Cassia obtusifolia L., 256
Celosia trigyna L., 336
Cenchrus biflorus Roxb., 507
Cenchrus ciliaris L., 508
Ceratotheca sesamoides Endl., 179
Chamaecrista absus (L.) H.S.Irwin & Barneby, 128
Chamaecrista falcinella (Oliv.) Lock, 250, *251*, *252*
Chamaecrista mimosoides (L.) Greene, *250*, 251, *252*
Chamaecrista stricta E.Mey., *250*, *251*, 252
Chloris virgata Sw., *19*, 447
Chlorophytum sphacelatum (Baker) Kativu, 341, *342*
   subsp. *sphacelatum*, 341, *342*
   subsp. *milanjianum* (Rendle) Kativu, 342
Chrysopogon nigritanus (Benth.) Veldkamp, 492
Cissampelos mucronata A.Rich., 296
Citrullus lanatus (Thunb.) Matsum. & Nakai, 237

Cleome angustifolia Forssk. subsp. *petersiana* (Klotzsch) Kers, 227
Cleome gynandra L., 358
Cleome hirta (Klotzsch) Oliv., 40, *42*
Cleome monophylla L., 41
Cleome rubella Burch., 42
Clerodendrum ternatum Schinz, 393
Clerodendrum uncinatum Schinz, 16, 136, *384*
Coccinia adoensis (A.Rich.) Cogn., 238
Cocculus hirsutus (L.) Diels, 103
Colophospermum mopane (Benth.) J.Léonard, 10, 11, 12, 14, 15, 17, 20, 253
Combretum apiculatum Sond. subsp. *apiculatum*, 229, *241*
Combretum hereroense, 12, 16, 19
Combretum imberbe Wawra, 16, 17, 81
Combretum mossambicense (Klotzsch) Engl., 15, 19, 259, 360
Commelina africana L. var. *barberae* (C.B.Clark) C.B.Clark, 231
Commelina benghalensis L., 43
Commelina diffusa Burm.f., 44
Commelina forsskalii Vahl, 45
Commelina macrospatha Gilg & Ledermann ex Mildbr., 46
Commelina petersii Hassk., 47
Commelina subulata Roth, 14, 121
Commicarpus plumbagineus (Cav.) Standl., 409
Commiphora merkeri Engl., 11, 225
Commiphora mossambicensis (Oliv.) Engl., 226
Convolvulus sagittatus Thunb., 364
Conyza aegyptiaca (L.) Aiton, 208
Conyza stricta, 17
Conyza transvaalensis Bremek., 208
Corallocarpus bainesii (Hook.f.) A.Meeuse, 82
Corchorus tridens L., 314
Crassocephalum picridifolium (DC.) S.Moore, 211
Craterostigma plantagineum Hochst., 58
Crinum harmsii Baker, 340
Crotalaria flavicarinata Baker f., 386
Crotalaria heidmannii Schinz, 387
Crotalaria laburnifolia L., 261
Crotalaria pisicarpa Welw. ex Baker, 262
Crotalaria platysepala Harv., *5*, 263
Crotalaria podocarpa DC., 129
Crotalaria steudneri Schweinf., 265
Croton megalobotrys Müll. Arg., *12*, *13*, *15*, *19*, 89
Ctenolepis cerasiformis (Stocks) Hook.f., 83
Cucumis africanus L.f., 239
Cucumis anguria L., 240
Cucumis metuliferus E.Mey. ex Naudin, *4*, *229*, 241
Cyanotis foecunda DC. ex Hassk., 48
Cyathula orthocantha (Aschers.) Schinz, 69, *346*
Cyclosorus interruptus (Willd.) H.Itô., *18*, 561
Cycnium tubulosum (L.f.) Engl., 192
Cymbopogon caesius (Hook. & Arn.) Stapf, 459
Cymbopogon excavatus (Hochst.) Burtt Davy, 459
Cynodon dactylon (L.) Pers., 448
Cyperus alopecuroides Rottb., 527
Cyperus chersinus (N.E.Br.) Kük., 528
Cyperus compressus L., 529
Cyperus denudatus L.f., 530
Cyperus difformis L., 531

Cyperus dubius Rottb., 520
Cyperus esculentus L., 532
Cyperus haspan L., 533
Cyperus iria L., 534
Cyperus latifolius Poir., 535
Cyperus leptocladus Kunth, 536
Cyperus longus L., 537
Cyperus margaritaceus Vahl, 521
Cyperus nudicaulis Poir., 522
Cyperus papyrus L., *13*, *156*, *186*, *298*, 538
Cyperus pectinatus Vahl, 522
Cyperus pseudokyllingioides Kük., 539
Cyperus rotundus L., 540
Cyperus sphaerospermus Schrad., 541
Cyperus squarrosus L., 542
Cyperus tenuiculmis Boeckeler, 543
Cyperus zollingeri Steud., 544
Cyphostemma congestum (Baker) Wild & R.B.Drumm., 320

Dactyloctenium aegyptium (L.) Willd., 449
Dactyloctenium giganteum B.S.Fisher & Schweick., 450
Dicerocaryum eriocarpum (Decne.) Abels, 180
Dichrostachys cinerea, 360
Dicliptera paniculata (Forssk.) I.Darbysh., 327
Dicoma schinzii O.Hoffm., 118
Dicoma tomentosa Cass., 152
Digitaria debilis (Desf.) Willd., 451
Digitaria milanjiana (Rendle) Stapf, 452
Digitaria sanguinalis (L.) Scop., 453
Diospyros lycioides Desf. subsp. *sericea* (Bernh.) de Winter, *13*, *15*, *244*, 246
Diospyros mespiliformis Hochst. ex A.DC., *15*, 84
Dipcadi glaucum (Burch. in Ker-Gawl.) Baker, 96
Dipcadi longifolium (Lindl.) Baker, 97
Dipcadi marlothii Engl., 98
Dolichos junodii (Harms) Verdc., 166
Duosperma crenatum (Lindau) P.G.Mey., *20*, 328
Duvalia polita N.E.Br., 117

Echinochloa colona (L.) Link, 435
Echinochloa jubata Stapf, 436
Echinochloa stagnina (Retz.) P.Beauv., 437
Eclipta prostrata (L.) L., 351
Eleocharis nigrescens (Nees) Steud., 557
Eleusine indica (L.) Gaertn., 455
Enicostema axillare (Lam.) A.Raynal subsp. *axillare*, 389
Enneapogon cenchroides (Roem & Schult.) C.E.Hubb., 506
Enteropogon macrostachyus (Hochst. ex A.Rich.) Munro ex Benth., 428
Epaltes alata (Sond.) Steetz, 29
Eragrostis aspera (Jacq.) Nees, 468
Eragrostis cilianensis (All.) Vignolo ex Janch., 469
Eragrostis cylindriflora Hochst., 470
Eragrostis echinochloidea Stapf, 493
Eragrostis inamoena K.Schum., 494
Eragrostis lappula Nees, 471
Eragrostis pallens Hack., 472
Eragrostis pilosa (L.) P.Beauv., 473
Eragrostis porosa Nees, 474
Eragrostis pusilla Hack., 475
Eragrostis superba Peyr., 495

Eragrostis tef, 473
Eragrostis viscosa (Retz.) Trin., 476
Erianthemum ngamicum (Sprague) Danser, 100
Erianthemum virescens (N.E.Br.) Wiens & Polhill, 101
Eriocaulon sp., 523
Ericaulon abyssinicum Hochst., 523
Erlangea misera (Oliv. & Hiern) S.Moore, 30
Ethulia conyzoides L.f., 31
Euclea divinorum Hiern, 85, *136*, *345*, *384*
Eulalia aurea (Bory) Kunth, 456
Euphorbia crotonoides Boiss., 125
Euphorbia hirta L., 126
Euphorbia inaequilatera Sond., 163
Euphorbia polycnemoides Boiss., 164
Evolvulus alsinoides (L.) L., 49

Falkia oblonga C.Krauss., 365
Ferraria glutinosa (Baker) Rendle, 135
Ficus burkei (Miq.) Miq., 106
Ficus thonningii Blume, 106
Fimbristylis bivalvis (Lam.) Lye, 551
Fimbristylis dichotoma (L.) Vahl, 550
Fimbristylis hispidula Auct., 549
Fimbristylis longiculmis Steud., 551
Floscopa glomerata (Schult. & Schult.f.) Hassk., 156
Friesodielsia obovata (Benth.) Verdc., 72
Fuirena pubescens (Poir.) Kunth, 555

Garcinia livingstonei T.Anderson, *10*, *15*, 228
Gardenia volkensii K.Schum. subsp. *spatulifolia* (Stapf & Hutch.) Verdc., 416
Geigeria schinzii O.Hoffm. subsp. *rhodesiana* (S.Moore) Merxm., 213
Gisekia africana (Lour.) Kuntze, 148
Gisekia pharnacioides, 148
Glinus bainesii (Oliv.) Pax, *260*, 402
Glinus lotoides L., 104
Glinus oppositifolius (L.) Aug. DC. var. *oppositifolius*, 104, 403
Gloriosa superba L., *34*, 120
Gomphocarpus fruticosus (L.) Aiton, *63*
  subsp. *rostratus* (N.E.Br.) Goyder & Nicholas, *63*, 73
Gomphrena celosioides Mart., 337
Gossypium herbaceum L. subsp. *africanum* (Watt) Vollesen, 280
Grangea anthemoides O.Hoffm., 214
Grewia flavescens Juss., *11*, 316
Guibourtia coleosperma (Benth.) J.Léonard, 254
Guilleminea densa (Willd.) Moq., 338
Gunillaea emirnensis (A.DC.) Thulin, 119
Gymnema sylvestre (Retz.) Schult., 200
Gymnosporia senegalensis (Lam.) Loes., 79
Gynandropsis gynandra (L.) Briq.), 358

Haemanthus multiflorus Martyn, 113
Harpagophytum procumbens (Burch.) DC. ex Meisn., *182*
  subsp. *procumbens*, 181
Harpagophytum zeyheri Decne., *181*
  subsp. *subloblatum* (Engl.) Ihlenf. & Hartm., 182
Helichrysum argyrosphaerum DC., 215
Heliotropium ciliatum Kaplan, 353
Heliotropium ovalifolium Forssk., *19*, 354
Heliotropium strigosum Willd., 355

Heliotropium supinum L., *13*, 356
Hemizygia bracteosa (Benth.) Briq., 394
Hermannia angolensis K.Schum., 141
Hermannia eenii Baker f., 141
Hermannia glanduligera, *143*
Hermannia kirkii Mast., *142*, 143
Hermannia modesta (Ehrenb.) Planch., 196
Hermannia quartiniana A.Rich., 311
Hermbstaedtia linearis Schinz, 151
Heteropogon contortus (L.) Roem. & Schult., 460
Hibiscus caesius Garke, 281
Hibiscus calyphyllus, 286
Hibiscus cannabinus L., 282
Hibiscus dongolensis Delile, 283
Hibiscus lobatus (Murray) Kuntze, 397
Hibiscus mastersianus Hiern, 284
Hibiscus meeusei Exell, 285
Hibiscus ovalifolius (Forssk.) Vahl, 286
Hibiscus schinzii Gürke, 287
Hibiscus sidiformis Baill., 5, 288
Hibiscus trionum L., 289
Hibiscus vitifolius L. subsp. *vulgaris* Brenan & Exel, 290
Hippocratea africana (Willd.) Loes. var. *richardiana* (Cambess.) N.Robson), 80
Hirpicium gorterioides (Oliv. & Hiern) Roessler subsp. *gorterioides*, 216
Hoslundia opposita Vahl, 395
Hyperthelia dissoluta (Steud.) Clayton, 461
Hyphaene petersiana Klotzsch ex Mart., *13*, *202*, *301*, *345*, *395*
Hypoestes forskaolii (Vahl) R.Br., 329

Indigofera astragalina DC., 130
Indigofera charlieriana Schinz var. *scaberrima* (Schinz) J.B.Gillett, 168
Indigofera charlieriana Schinz var. *charlieriana*, 167
Indigofera colutea (Burm.f.) Merr., 131
Indigofera daleoides Benth. ex Harv. var. *daleoides*, 169
Indigofera filipes Harv., 132, *195*
Indigofera flavicans Baker, 170, *195*
Indigofera trita L.f. subsp. *subulata* (Vahl ex Poir) Ali var. *subulata*, 133
Ipomoea chloroneura Hallier f., 366
Ipomoea coptica (L.) Roem. & Schult. var. *coptica*, 367
Ipomoea dichroa Choisy, 157, *158*
Ipomoea eriocarpa R. Br., 159
Ipomoea leucanthemum (Klotzsch) Hallier f., 368
Ipomoea magnusiana Schinz var. *magnusiana*, 369
Ipomoea nil (L.) Roth, 50
Ipomoea obscura (L.) Ker-Gawl., 232
Ipomoea pes-tigridis L., 160
Ipomoea shirambensis Baker, 161
Ipomoea sinensis (Desr.) Choisy subsp. *blepharosepala* (A.Rich.) A.Meeuse, 371
Ipomoea sinensis (Desr.) Choisy subsp. *sinensis*, 372
Ipomoea tuberculata Ker-Gawl., 233
Ipomoea sp., 158

Jacquemontia tamnifolia (L.) Griseb., 51
Jamesbrittenia elegantissima (Schinz) Hilliard, 310
Jasminum fluminense Vell. subsp. *fluminense*, 410
Jasminum stenolobum Rolfe, 411
Justicia bracteata (Hochst.) Zarb, 24

Justicia divaricata (Nees) T.Anderson, 25
Justicia exigua S.Moore, 331
Justicia forskaolii Vahl, 329
Justicia heterocarpa T.Anderson subsp. *dinteri* (S.Moore) Hedren, 26

Kalanchoe lanceolata (Forssk.) Pers., *14*, 122
Kedrostis abdallai A.Zimm., 242, 244
Kedrostis foetidissima (Jacq.) Cogn., 243
Kigelia africana, 10
Kohautia caespitosa Schnizl. subsp. *brachyloba* (Sond.) D.Mantell, 307
Kohautia subverticillata (K.Schum.) D.Mantell subsp. *subverticillata*, 109, 307
Kohautia virgata (Willd.) Bremek., 417
Kosteletzkya buettneri Gürke, 398
Kyllinga buchananii C.B.Clarke, 524
Kyllinga erecta Schumach., 525
Kyllingiella microcephala (Steud.) R.W.Haines & Lye, 526
Kyphocarpa angustifolia (Moq.) Lopr., 339

Laggera crispata (Vahl) Hepper & J.R.I.Wood, 32
Laggera decurrens (Vahl) Hepper & J.R.I.Wood, *12*, *17*, 217
Lantana angolensis Moldenke, 197
Lapeirousia odoratissima Baker, 390
Lapeirousia schimperi (Asch. & Klatt) Milne-Redh., 391
Ledebouria revoluta (L.f.) Jessop, 99
Leersia hexandra Sw., 496
Lemna aequinoctialis Welw., 559
Leonotis nepetifolia (L.) R.Br., 137
Lessertia benguellensis Baker f., 171
Limeum argute-carinatum Wawra & Peyr. var. *kwebense* (N.E.Br.) Friedr., 105
Limeum fenestratum (Fenzl) Heimerl, 404
Limeum sulcatum (Klotzsch) Hutch., 405
Limeum viscosum (J.Gay) Fenzl var. *kraussii* Friedrich, 406
Limeum viscosum (J.Gay) Fenzl var. *viscosum*, 407
Lobelia angolensis Engl. & Diels, 155
Lobelia erinus L., 39
Loeseneriella africana, 15
Loeseneriella africana (Willd.) N.Hallé var. *richardiana* (Cambess.) N.Hallé, 80
Lonchocarpus capassa Rolfe, 53
Lonchocarpus nelsii (Schinz) Schinz), 52
Ludwigia abyssinica A.Rich., 298, *299*, *300*
Ludwigia leptocarpa (Nutt.) H.Hara, *298*, *299*, *300*
Ludwigia stolonifera (Guill. & Perr.) P.H.Raven., *18*, *19*, *298*, *299*, *300*

Macrotyloma daltonii (Webb) Verdc., 388
Mariscus dubius (Rottb.) Kük. ex C.E.C.Fisch., 520
Marsdenia macrantha (Klotzsch) Schltr., 343
Marsdenia sylvestris (Retz.) P.I.Forst., 200
Marsilea vera Launert, *13*, 560
Maytenus senegalensis (Lam.) Exell), 79
Melanthera marlothiana, 219
Melanthera scandens (Schumach. & Thonn.) Roberty subsp. *madagascariensis* (Baker) Wild, 218
Melanthera triternata (Klatt.) Wild, 219
Melhania forbesii Mast., 312

Melinis kallimorpha (Clayton) Zizka, 477
Melinis repens (Willd.) Zizka subsp. *grandiflora* (Hochst.) Zizka, 478
Merremia pinnata (Hochst. ex Choisy) Hallier f., 234
Merremia tridentata subsp. *angustifolia* (Jacq.) Ooststr., 235
Merremia verecunda Rendle, 373
Microchloa caffra Nees, 429
Mollugo cerviana (L.) Ser. ex DC., 408
Momordica balsamina L., *242*, *244*, *246*
Monechma debile (Forssk.) Nees), 24
Monechma divaricatum (Nees) C.B.Clarke), 25
Monsonia angustifolia A.Rich., 176
Mukia maderaspatana (L.) M.Roem., 245

Neptunia oleracea Lour., *13*, 260, *402*
Nicolasia pedunculata S.Moore, 33
Nidorella resedifolia DC., 220
Nymphaea capensis Thunb., 55
Nymphaea lotus, *16*, *17*, *19*
Nymphaea nouchali Burm.f. var. *caerulea* (Savigny) Verdc., 55
Nymphoides forbesiana (Griseb.) Kuntze, *16*, 297
Nymphoides indica (L.) Kuntze subsp. *occidentalis* A.Raynal, 401

Ocimum americanum L. var. *pilosum* (Willd.) A.J.Paton, 54
Ocimum canum Sims, 54
Ocimum gratissimum L. var. *gratissimum*, 275
Oldenlandia capensis L.f. var. *capensis*, 418
Oldenlandia corymbosa L. var. *caespitosa* (Benth.) Verdc., 419
Oplismenus burmannii (Retz.) P.Beauv., 439
Orbea caudata (N.E.Br.) Bruyns subsp. *rhodesiaca* (L.C.Leach) Bruyns, 74
Orbea huillensis (Hiern) Bruyns subsp. *huillensis*, 114
Orbea lugardii (N.E.Br.) Bruyns, 115
Orbea rogersii (L.Bolus) Bruyns, 201
Orbea schweinfurthii (A.Berger) Bruyns, 116
Orbeopsis caudata subsp. *rhodesiaca* (L.C.Leach) L.C.Leach, 74
Orthanthera jasminiflora (Decne.) Schinz, 344
Ottelia ulvifolia (Planch.) Walp., 274
Oxalis corniculata L., 302
Oxycaryum cubense (Poepp. & Kunth) Lye, 545
Oxygonum alatum Burch., 414
Oxygonum sinuatum (Meisn.) Dammer, 415

Pancratium tenuifolium Hochst. ex A.Rich., 396
Panicum, 491
Panicum coloratum L. var. *coloratum*, 497
Panicum fluviicola Steud., 479
Panicum hirtum Lam., 480
Panicum kalaharense Mez, 481
Panicum maximum Jacq., 482
Panicum subalbidum Kunth, 483
Paspalidium obtusifolium (Delile) N.D.Simpson, 440
Paspalum scrobiculatum L., 441
Pavetta cataractarum S.Moore, 420
Pavonia burchellii (DC.) R.A.Dyer, 291
Pavonia clathrata Mast., 399

Pavonia senegalensis Cav., 292
Pechuel-loeschea leubnitziae (Kuntze) O.Hoffm., *12*, 16, *17*, 34, *73*
Peltophorum africanum Sond., 255
Pennisetum glaucocladum Stapf & C.E.Hubb., 509
Pennisetum macrourum Trin., *14*, 509
Pentodon pentandrus (Schumach. & Thonn.) Vatke, 56
Pergularia daemia (Forssk.) Chiov. subsp. *daemia*, 75
Peristrophe paniculata (Forssk.) Brummitt, 327
Perotis patens Gand., 510
Persicaria glomerata (Dammer) S.Ortiz & Paiva, 186
Persicaria limbata (Meisn.) H.Hara, 187
Philenoptera nelsii (Schinz) Schrire, *10*, *16*, *53* subsp. *nelsii*, 52
Philenoptera violacea (Klotzsch) Schrire, *10*, *12*, *13*, *16*, *18*, *19*, *52*, 53
Phoenix reclinata, 345
Phragmites australis (Cav.) Trin. ex Steud., *14*, *422*, 464
Phyla nodiflora (L.) Greene, 424
Phyllanthus maderaspatensis L., 91
Phyllanthus parvulus Sond. var. *parvulus*, 92
Phyllanthus pentandrus Schumach. & Thonn., 93
Plicosepalus kalachariensis (Schinz) Danser, 177
Plukenetia africana Sond., 94
Plumbago zeylanica L., 413
Pogonarthria fleckii (Hack.) Hack., 442
Pogonarthria squarrosa (Roem. & Schult.) Pilg., 443
Pollichia campestris Aiton, 78
Polycarpaea corymbosa (L.) Lam. var. effusa Oliv., 359
Polycarpaea eriantha Hochst. ex A.Rich. var. *effusa* (Oliv.) Turrill, 359
Polygala erioptera DC., 185
Polygonum decipiens (R.Br.) K.L.Wilson, 188
Polygonum plebeium R.Br., 189
Portulaca hereroensis Schinz, 190
Portulaca kermesina N.E.Br., 140
Portulaca oleracea L. subsp. *oleracea*, 303
Portulaca quadrifida L., 304
Potamogeton nodosus Poir., 191
Potamogeton thunbergii, *16*, *17*, *19*
Protasparagus africanus (Lam.) Oberm., 346
Pseudoconyza viscosa (Mill.) D'Arcy, 35
Pseudognaphalium luteo-album (L.) Hilliard & Burtt, 221
Pteleopsis anisoptera, 361
Pteleopsis myrtifolia (M.A.Lawson) Engl. & Diels, *11*, 361
Pterococcus africanus (Sond.) Pax & K.Hoffm., 94
Pupalia lappacea (L.) A.Juss var. *velutina* (Moq.) Hook.f., 70
Pycreus macrostachyos (Lam.) J.Raynal, 546
Pycreus pelophilus (Ridl.) C.B.Clarke, 547
Pycreus polystachyos (Rottb.) P.Beauv. var. *polystachyos*, 548

Rhus tenuinervis Engl., 71
Rhynchelytrum grandiflorum Hochst., 478
Rhynchelytrum kallimorphon Clayton, 477
Rhynchosia minima (L.) DC., 266
Rhynchosia totta (Thunb.) DC. var. *fenchelii* Schinz, *195*, 267
Ricinodendron rautanenii Schinz, 248
Ricinus communis L. var. *communis*, 127
Rotala filiformis (Bellardi) Hiern, 139

Rotala heterophylla A.Fern & Diniz, 139
Rotheca uncinata (Schinz) Herman & Retief, 136
Ruellia prostrata Poir., 27
Ruelliopsis setosa (Nees) C.B.Clarke, 28
Ruspolia seticalyx (C.B.Clarke) Milne-Redh., 112

Sansevieria deserti, 258
Sansevieria aethiopica Thunb., *345*, 348
Sansevieria pearsonii N.E.Br., 349
Scadoxus multiflorus (Martyn) Raf., 113
Schinziophyton rautanenii (Schinz) Radcl.-Sm., 248
Schizachyrium jeffreysii (Hack.) Stapf, 462
Schmidtia pappophoroides Steud. ex J.A.Schmidt, 498
Schmidtia bulbosa Stapf, 498
Schoenoplectiella senegalensis (Steud.) Lye, 552
Schoenoplectus corymbosus (Roth ex Roem. & Schult.) J.Raynal, 554
Schoenoplectus erectus (Poir.) Palla ex J.Raynal, 553
Schoenoplectus praelongatus (Poir.) J.Raynal, 552
Sclerocarpus africanus Jacq. ex Murray, 222
Sebaea grandis (E.Mey.) Steud., 272
Senecio strictifolius Hiern, 223
Senna obtusifolia (L.) H.S.Irwin & Barneby, 256
Sesamum angustifolium (Oliv.) Engl., *183*, *184*
Sesamum triphyllum Asch., *183*, 184
Sesbania microphylla Harms, 268
Sesbania rostrata Bremek & Oberm., *13*, 269
Sesuvium hydaspicum (Edgew.) M.L.Gonç., 149
Setaria pumila (Poir.) Roem. & Schult., 511
Setaria sagittifolia (A.Rich.) Walp., 499
Setaria sphacelata, *14*
Setaria ustilata de Wit, 511
Setaria verticillata (L.) P. Beauv., 513
Sida alba L., 400
Sida chrysantha Ulbr., 293
Sida cordifolia L., 294
Solanum panduriforme Drège ex Dun., 59
Solanum tarderemotum Bitter, 423
Solanum tettense Klotzsch var. *renschii* (Vatke) R.E.Gonçalves, 60
Sopubia mannii Skan var. *tenuifolia* (Engl. & Gilg) Hepper, 193
Sorghastrum nudipes Nash, 484
Spermacoce senensis (Klotzsch) Hiern, 421
Sphaeranthus peduncularis DC., *13*, 153
Sporobolus cordofanus (Steud.) Coss., 485
Sporobolus fimbriatus (Trin.) Nees, 500
Sporobolus ioclados (Trin.) Nees, 486
Sporobolus macranthelus Chiov., 487
Sporobolus panicoides A.Rich., 488
Sporobolus pyramidalis P.Beauv., 501
Stipagrostis uniplumis (Roem. & Schult.) de Winter, 489
Streptopetalum serratum Hochst., 144
Striga asiatica (L.) Kuntze, 194
Striga gesnerioides (Willd.) Vatke, 195
Stylosanthes fruticosa (Retz.) Alston, 270

Talinum caffrum (Thunb.) Eckl. & Zeyh., 305
Talinum crispatulatum Dinter, 306
Tephrosia caerulea Baker f., 172
Tephrosia lupinifolia DC., 173
Tephrosia purpurea (L.) Pers. subsp. *leptostachya* (DC.) Brummitt var. *pubescens* Baker, *174*, *195*

Terminalia sericea DC., *16, 19,* 362
Thelypteris interrupta (Willd.) K.Iwats), 561
Thunbergia reticulata Nees, 199
Trachyandra arvensis (Schinz) Oberm., 321
Trachypogon capensis (Thunb.) Trin., 463
Trachypogon spicatus (L.f.) Kuntze, 463
Tragia okanyua Pax, 95
Tragus berteronianus Schult., 514
Tragus racemosus (L.) All., 515
Tribulus terrestris L., 322
Trichilia emetica Vahl, 102
Tricholaena monachne (Trin.) Stapf & C.E.Hubb., 490
Trichoneura grandiglumis (Nees) Ekman, 444
Tricliceras glanduliferum (Klotzsch) R.Fern., 145
Tricliceras lobatum (Urb.) R.Fern., 318
Triumfetta pentandra A.Rich., 317
Turraea zambesica Styles & F.White, 295

Urochloa brachyura (Hack.) Stapf, 445, 446
Urochloa mosambicensis, 446
Urochloa oligotricha, 445
Urochloa trichopus (Hochst.) Stapf, *19,* 446
Utricularia gibba L., 276

Vahlia capensis (L.f.) Thunb. subsp. *vulgaris* Bridson var. *vulgaris,* 319

Vellozia humilis Baker, 61
Vernonia anthelmintica (L.) Willd., 36
Vernonia glabra (Steetz) Vatke var. *laxa* (Steetz) Brenan, *12,* 37
Vernonia poskeana Vatke & Hildebr. subsp. *botswanica* G.V.Pope, 38
Vigna luteola (Jacq.) Benth., *13, 18,* 271
Vigna unguiculata (L.) Walp. subsp. *stenophylla* (Harv.) Maréchal, Mascherpa & Stainier, 175
Vigna unguiculata (L.) Walp. subsp. *unguiculata* var. *spontanea* (Schweinf.) R.S.Pasquet, 175
Vossia cuspidata (Roxb.) Griff., 430

Waltheria indica L., 313
Withania somnifera (L.) Dunal, 110

Xenostegia tridentata (L.) D.F.Austin & Staples subsp. *angustifolia* (Jacq.) Lejoly & Lisowski, 235
Xerophyta humilis (Baker) & Schinz, 61
Ximenia caffra, 259, 360

Zehneria marlothii (Cogn.) R.Fern. & A.Fern., 374
Ziziphus mucronata Willd. subsp. *mucronata, 13, 34, 73,* 108